Soil Health Series: Volume 5 Soil Health and Sustainability in Spain and Portugal

SERIES EDITOR
Douglas L. Karlen

EDITORS
Iñigo Virto and Rodrigo Antón

CONTRIBUTORS
Maria Almagro
Institute of Agricultural Research and Training, Fishing, Food and Organic Farming of Andalusia (IFAPA), Camino de Purchil s/n, Granada, Spain
Rodrigo Antón
Área Edafología y Química Agrícola, Departamento de Ciencias, IS- FOOD, Universidad Pública de Navarra (UPNA), 31006, Pamplona, Spain
Ander Arias-González
NEIKER - Basque Institute for Agricultural Research and Development, Department of Forest Sciences, 48160, Derio, Spain
Carolina Boix-Fayos
Soil and Water Conservation Research Group, Centro de Edafología y Biología Aplicada del Segura, CEBAS-CSIC, Campus Universitario de Espinardo Murcia, Spain
Jessica Cuartero
Swiss Federal Research Institute Wsl, Birmensdorf, Switzerland
Isabel S. de Soto
Área Edafología y Química Agrícola, Departamento de Ciencias, IS- FOOD, Universidad Pública de Navarra (UPNA), 31006, Pamplona, Spain
Delphine Derrien
UMR Sol, Agro et hydrosystème, Spatialisation (SAS), INRAE- Rennes, 35042, Rennes, France
Elvira Díaz-Pereira
Soil and Water Conservation Department, CEBAS-CSIC (Spanish Research Council), Campus de Espinardo, Murcia, Spain
Alberto Enrique
Área Edafología y Química Agrícola, Departamento de Ciencias, IS- FOOD, Universidad Pública de Navarra (UPNA), 31006, Pamplona, Spain
Rafael Espejo
Department of Agrarian Production, Universidad Politécnica de Madrid. Avda. Puerta de Hierro 2, 28040, Madrid, Spain
António Fabião
Forestry Research Centre, Instituto Superior de Agronomia, Lisboa, Portugal
Juan A. Fernández
Department of Agricultural Engineering, Universidad Politécnica de Cartagena, Cartagena, Spain
Instituto de Biotecnología Vegetal (IBV), Campus Muralla del Mar, Edificio I+D+I, Universidad Politécnica de Cartagena, Cartagena, Spain
Nahia Gartzia-Bengoetxea
NEIKER - Basque Institute for Agricultural Research and Development, Department of Forest Sciences, 48160, Derio, Spain
Noelia Garcia-Franco
Chair of Soil Science, TUM School of Life Sciences Weihenstephan, Technical University of Munich, Freising, Germany
Chiquinquirá Hontoria
Department of Agrarian Production, Universidad Politécnica de Madrid. Avda. Puerta de Hierro 2, 28040, Madrid, Spain
Centro de Estudios e Investigación para la Gestión de Riesgos Agrarios y Medioambientales, CEIGRAM, Universidad Politécnica de Madrid, Senda del Rey, 13, 28040, Madrid, Spain
Miguel Itarte
Área Edafología y Química Agrícola, Departamento de Ciencias, IS- FOOD, Universidad Pública de Navarra (UPNA), 31006, Pamplona, Spain
Manuel Madeira
Forestry Research Centre, Instituto Superior de Agronomia, Lisboa, Portugal
Ignacio Mariscal-Sancho
Department of Agrarian Production, Universidad Politécnica de Madrid. Avda. Puerta de Hierro 2, 28040, Madrid, Spain
Centro de Estudios e Investigación para la Gestión de Riesgos Agrarios y Medioambientales, CEIGRAM, Universidad Politécnica de Madrid, Senda del Rey, 13, 28040, Madrid, Spain
Inazio Martínez de Arano
European Forest Institute, Bioeconomy Program, Bioregions Facility, Joensuu, North-Karelia, Finland
María Martínez-Mena
Soil and Water Conservation Department, CEBAS-CSIC (Spanish Research Council), Campus de Espinardo, Murcia, Spain
Afonso Martins
Universidade de Trás- os- Montes e Alto Douro (UTAD), Apartado 1013, Vila Real, Portugal
Luis Orcaray
Instituto Navarro de Infraestructura y Tecnologías Agroalimentarias (INTIA), Villava, Spain
Onurcan Özbolat
Department of Agricultural Engineering, Universidad Politécnica de Cartagena, Cartagena, Spain
Instituto de Biotecnología Vegetal (IBV), Campus Muralla del Mar, Edificio I+D+I, Universidad Politécnica de Cartagena, Cartagena, Spain
José Antonio Pascual
Soil and Water Conservation Department, CEBAS-CSIC (Spanish Research Council), Campus de Espinardo, Murcia, Spain

Remigio Paradelo Núñez
CRETUS, Departamento de Edafoloxía e Química Agrícola, Universidade de Santiago de Compostela, Facultade de Farmacia, Santiago de Compostela, Spain
Fernando Peregrina
Department of Agrarian Production, Universidad Politécnica de Madrid. Avda. Puerta de Hierro 2, 28040, Madrid, Spain
Centro de Estudios e Investigación para la Gestión de Riesgos Agrarios y Medioambientales, CEIGRAM, Universidad Politécnica de Madrid, Senda del Rey, 13, 28040, Madrid, Spain
Fernando Raimundo
Universidade de Trás- os- Montes e Alto Douro (UTAD), Apartado 1013, Vila Real, Portugal
Ana Raquel Rodriguesinigo.virto
Adega Cooperativa de Dois Portos, D.R.L., Dois Portos- Torres Vedras, Portugal
Margarita Ros
Soil and Water Conservation Department, CEBAS-CSIC (Spanish Research Council), Campus de Espinardo, Murcia, Spain
Virginia Sánchez-Navarro
Department of Agricultural Engineering, Universidad Politécnica de Cartagena, Cartagena, Spain
Ana Simoes-Mota
Área Edafología y Química Agrícola, Departamento de Ciencias, IS- FOOD, Universidad Pública de Navarra (UPNA), 31006, Pamplona, Spain
Paula Soares
Forestry Research Centre, Instituto Superior de Agronomia, Lisboa, Portugal
Henar Urmeneta
Departamento Estadística, Informática y Matemáticas, Universidad Pública de Navarra, 31006, Pamplona, Navarra, Spain
Iñigo Virto
Área Edafología y Química Agrícola, Departamento de Ciencias, IS- FOOD, Universidad Pública de Navarra (UPNA), 31006, Pamplona, Spain
Armelle Zaragüeta
Área Edafología y Química Agrícola, Departamento de Ciencias, IS- FOOD, Universidad Pública de Navarra (UPNA), 31006, Pamplona, Spain
Raúl Zornoza
Department of Agricultural Engineering, Universidad Politécnica de Cartagena, Cartagena, Spain
Instituto de Biotecnología Vegetal (IBV), Campus Muralla del Mar, Edificio I+D+I, Universidad Politécnica de Cartagena, Cartagena, Spain

EDITORIAL CORRESPONDENCE
Soil Science Society of America
5585 Guilford Road, Madison, WI 53711-58011, USA

Soil Health Series: Volume 5 Soil Health and Sustainability in Spain and Portugal

Edited by

Iñigo Virto and Rodrigo Antón

SSSA
Soil Science Society of America

WILEY

Soil Health Insights from Spain and Portugal

Douglas L. Karlen, Soil Health Series Editor

Soil Health and Sustainability in Spain and Portugal is the fifth volume in the Soil Health Series being published by Wiley and the Soil Science Society of America (SSSA). Volumes one and two provide background and general methods for assessing soil biological, chemical, and physical properties and processes affecting soil health, while volumes three and four explore those factors, indicators, and management practices affecting soil health in Brazil and India, respectively. This volume, edited by Professor Iñigo Virto and Professor Rodrigo Antón draws upon their research experience in Spain, France and the U.S. on soil health, the soil C cycle, and its response to management to further our global evaluations of the concept. Collaborating with scientists from Spain, Portugal, and other countries, this volume focuses on soil health and sustainability throughout the Iberian Peninsula.

For those not familiar with this geographic area, the Iberian Peninsula, encompasses 583,254 km^2 and is the second-largest European peninsula by area after the Scandinavian Peninsula. It is the westernmost of the three major southern European peninsulas and of the larger Eurasian landmass and has a population of roughly 53 million. There are three dominant climate types (Mediterranean, oceanic, and highland alpine) with a topography that is very contrasting and has uneven relief.[1]

Chapter 1 identifies the most relevant soil health issues and provides a general overview of the natural conditions of the Iberian Peninsula. The diversity and distribution of major soil types, climates, and land uses are examined to evaluate inter-relationships between agricultural management and soil health. Chapter 2 examines different sustainable agricultural management practices for soil carbon

[1] The Iberian Peninsula. https://en.wikipedia.org/wiki/Iberian_Peninsula

sequestration and stabilization, as an indicator of soil health for permanent crops under semiarid conditions. Green manuring or retaining pruning residues are shown to be effective for coping with climate change, but there are no a tailored-made solutions for these semiarid regions. Chapter 3 examines crop diversity in Southeastern Spain with three cash crops (rainfed almonds, irrigated mandarins, and vegetables) by quantifying the variation in different soil physical, chemical and biological indicators of soil health. The results show that crop diversity enhances soil health through the incorporation of plant biomass which creates wider diversity in soil microorganisms with multiple functions.

Chapters 4 and 8 examine soil health in an urban environment and after 25 years of sewage sludge application in Northern Spain. The urban study was done in Santiago de Compostela (Spain) and shows that maintaining and enhancing soil health in those environments can be quite challenging because many soils have been disturbed by anthropic activities and contamination. A study established in 1992 close to Pamplona (Spain) was used to evaluate the impact of sewage sludge recycling, one of the circular economy and European Green Deal goals. The study quantifies several soil physical, chemical and biological indicators, as well as their relation to crop yield and quality in a four-year rotation.

Chapter 5 and 6 examine soil health of highly weathered degraded acid soil and the potential to use irrigation to enhance soil health, respectively. Acid soils are primarily located in the western third of the Iberian Peninsula, on the siliceous domain of the Hesperian massif. They are poor soils that can function to support arboreal plants or arbustive vegetation, which consists of woody plants that are smaller than trees but larger than herbaceous plants but are very vulnerable when cultivated. Chapter 5 examines the long-term effects of integrated farming treatments on a selection of chemical, physical and biological soil properties, measured at different depths, and their relationships with crop development and greenhouse emissions. Chapter 6 is important, since due to climate uncertainty in many areas, including the Mediterranean basin, up to 80% of agricultural production comes from irrigated land. However, the use of irrigation can also increase the environmental footprint of agriculture, which opens questions on the sustainability of irrigation in some areas. This chapter examines soil health during a transition from dryland agriculture to long-term irrigation for different soil and climate conditions within the Iberian Peninsula.

Finally, Chapters 7 and 9 focus on planted and managed forests in Northern Spain and forest and agroforestry systems in Portugal, respectively. In N Spain and Portugal, *Pinus radiata* and *Pinus pinaster* plantations occupy more than 290,000 ha and 2 million ha, respectively, with rotation lengths between 35 and 40 years. Conventional site preparation consists of the partial removal of logging residues followed by down-slope ripping or blading which pushes logging residues and the humus layer away from the site. This depletes nutrients, increases soil loss, and degrades soil physical properties. Visual assessment is used to evaluate those problems from a soil health perspective.

Overall, Iñigo, Anton and their colleagues have compiled an excellent volume that addresses soil health and sustainability throughout the Iberian Peninsula. Thank you for your excellent contribution to our Soil Health Series.

Preface

This volume in the Soil Health series was written at the invitation of Dr. Douglas L. Karlen and the Soil Science Society of America, and is the fifth in the series, after the first two edited by Dr. Karlen and his collaborators on general perspectives and methodologies.

The objective of this series is to provide a comprehensive overview of experience-based knowledge on soil health assessment under different soil, climate and management conditions globally. Under the umbrella of the first two volumes, the aim is to provide an approach focused on studies developed in different areas of the world, where systematized soil health studies have been conducted. In this case, a selection of case studies in the Iberian Peninsula is offered, which try to be representative of the diversity of natural conditions of our territory, and of the majority uses of the soil that occur in it. Thus, the book includes works of soil health assessment in agricultural (rainfed and irrigated) land, forest and urban soils, in a gradient of climates that corresponds to the most representative conditions of the Peninsula.

Corresponding to this heterogeneity, the approaches to the study of soil health in the different chapters are also diverse. Thus, there are chapters focused on the study of a few selected indicators, and others that use wider ranges, and even a chapter focused on the visual assessment of soil condition in forest management.

This diversity of approaches corresponds to that described in the first book of the series regarding the evolution of soil quality assessment, and even of the concept itself. As described in the first chapter, the definition of Soil Health, and of the methods for its assessment, has changed in recent decades from an approach focused mainly on the productivity of agricultural and forest soils, to an approach that considers soil functions within the (agro)ecosystem, as well as its role in the provision of ecosystem services that contribute to the maintenance of the environment and human well-being. An idea that actually develops in contemporary times, and attempts to systematize, the concern for soil quality that has accompanied humankind since its origin, and in particular since the establishment of sedentary communities whose survival depends on the conservation of natural resources.

This book has also been written in a context of growing awareness on the part of society as a whole, and of the Administration, of the importance of the soil resource for our well-being. In particular, the European Union has developed several initiatives in recent years with the aim of increasing knowledge and regulating the health of Europe's soils (with initiatives such as the European Soil Protection Strategy, the EJP-Soil Joint European Program, the Soil Mission, the C Agriculture Regulation, and above all, the proposed Soil Monitoring and Resilience Directive). In addition, both environmental and agricultural policies include in the Union different mechanisms that try to protect more or less indirectly the health of Europe's soils.

Undoubtedly, all these initiatives have been and are the subject of debate among policymakers, stakeholders, and even within the scientific community itself, so we can speak of a moment of special intensity in our continent in the debate on strategies and tools for soil health assessment, not only among professionals, but even in the regulations themselves. This book is thus more a "state of the art" than a closed work, which attempts to provide quantitative and science-based information on the particular case of some soils of continental Spain and Portugal. For this reason, priority has been given to the search for long-term studies and/or comprehensive assessments of the agricultural or forestry systems considered.

The long trajectory of many agricultural and forestry research and consultancy institutions in our territory would undoubtedly allow the development of several more volumes on this subject. Following the example of the first volume, in this book we have decided to start somewhere, and open to the public some representative cases, instead of embarking on the impossible task of trying to cover all possible situations and approaches.

Iñigo Virto & Rodrigo Antón

1

Soil Health in the Iberian Peninsula

Iñigo Virto, Rodrigo Antón, Miguel Itarte, and Alberto Enrique

Chapter Overview

This chapter summarizes the most relevant geographical and physical characteristics of the Iberian Peninsula including its soils and land uses. It also introduces the general context in which the case studies described in the following chapters have been developed, and their relevance to soil health.

The chapter also includes an overview of soil health studies in the Peninsula in recent years, and the changes occurred in the topics of interest within these studies.

Finally, the chapter offers an overview of the information explained in the book chapters, grouped in relation to the land use they describe (agriculture, forest and urban soils).

Introduction

Soil health has become a unifying concept, tool, and goal for all kinds of soil-related users, stakeholders, and decision-makers in the past several decades (Karlen et al., 2021a, 2021b). In this book, we aimed to explore and present long-term experiences dealing with different angles of soil health assessment and monitoring in the Iberian Peninsula, home to mainland Spain and Portugal. In this chapter, we introduce first the geophysical and land use context in this corner of Europe, revise the evolution of scientific production related to the topic, and summarize the contributions developed in the book's other eight chapters.

Soil Health Series: Volume 5 Soil Health and Sustainability in Spain and Portugal, First Edition.
Edited by Iñigo Virto and Rodrigo Antón.

Published 2025 by Soil Science Society of America, Inc. and John Wiley & Sons, Inc.

This work has been carried out in a fast-changing context, both in terms of changes affecting soils, and in terms of the increase in general awareness and regulatory proposals about the protection of soils in the European Union. Indeed, soil health assessment is becoming paramount for understanding the response of soils to the so-called contemporary "global change" due to biophysical planetary-scale changes (e.g., atmospheric circulation and climate, biogeochemical cycles, the hydrological cycle, biodiversity) and changes in the human population, economy, use of resources, land use and cover, urbanization, and other human-related activities. Owing to the key role of soils in terrestrial ecosystems and as the basis of agriculture, and thus the changes in use associated with primary production, soils appear at the center of global change, both in relation to its causes and its effects. This position implies that these changes directly affect the functioning of soils, and require new evaluation systems to assess soils as a resource, determine their responses to changes, and identify possible mitigation and adaptation strategies.

The increase in soil heath studies responds to this need and to the evidence that the role of soils in the face of global change must be approached from a global and holistic perspective. So far, the most widespread trend has been to evaluate the effect of specific soil management measures in isolation. This reductionist approach is common in scientific research and useful in local studies, but it sometimes has limited practical relevance to the wider farming and soil-using community. For instance, the limitation of these approaches becomes clear in the case of agrarian systems, which are based on an interrelated and interdependent complex of options and measures that can only be slightly modified or not modified at all without affecting each other (Faber et al., 2022).

In fact, the basis for the development of concepts linked to soil quality or soil health is that the relationship between soil functioning and the services that soils provide is not univocal (different functions can provide the same service), and vice versa (the same service can depend on several functions), so establishing simple relationships is not easy or, in many cases, adequate. Thus, the definition of the concept of "soil quality" has varied from its first formal conceptions in the 1990s, based specifically on the condition of agricultural soils for production, towards a broader conception, which emphasizes sustainable management of the land and provides a holistic approach (Karlen et al., 2003). Indeed, throughout the history of soil science, concepts such as "soil productivity" (already cited in 1910), "soil conservation," "soil security," or "soil resilience" have been developed and have contributed to increase our awareness of the need to understand and protect the complexity of soil functioning beyond the study of a few or several isolated properties. These new concepts have emerged within the soil science community through multiple actions, such as reflection on the social, cultural, and/or political needs that exist outside this community, or by being progressive and relevant to new emerging interests such as global health or security (Mizuta et al., 2021).

In a recent attempt to compile the different approaches defined in the last years, the Intergovernmental Technical Panel on Soils (ITPS) of the Global Soil Alliance, led by the Food and Agriculture Organization of the United Nations, released a summary definition of soil health: "the capacity of the soil to sustain the productivity, diversity and environmental services of terrestrial ecosystems" (ITPS, 2021). This definition fits the core concept defined by Janzen et al. (2021), who concluded that the multiple definitions of this concept have at least three things in common. First, soil health has at its core the notion of functionality, that is, the ability of a soil to perform its vital functions normally. Second, the soil is considered as a living and dynamic system, with a torrent of connected and intertwined processes. Third, an essential element of the definition of soil health is sustainability or resilience, which acknowledges that a healthy soil maintains the functioning of the ecosystem over time.

However, the term soil health continues to appear to some as a theoretical and complex idea, essentially a "popular metaphor" (Janzen et al., 2021). A translation from metaphors into concrete facts and data, possibly applicable to quantitative assessment of soil health relevant for policy-making and to regulations to rule this assessment, is needed to move from the emotional sphere to a political one (Panagos et al., 2022). In this context, the European Union is taking steps towards a concrete formulation for the protection of European soils. The European Commission has thus been preparing a Soil Health Law since 2020. This legal framework will contribute to (a) achieving the objectives of the European Soil Strategy 2030, with the aim of granting soils the same level of protection as water and air, and (b) radically improving the condition of soils to better provide the services that we depend on.

In compliance with this strategy, the European Union Soil Mission proposes that by 2030 at least 75% of the soils of each EU member state will be healthy or will show significant improvement towards compliance of the accepted thresholds of the indicators selected to support ecosystem services. This objective corresponds to an increase in healthy soils with respect to a reference baseline established by each member state. The determination of this reference must be made based on specific indicators. The indicators proposed by the EU Mission for Soil Health (*A Soil Deal for Europe*, European Commission, 2023) cover measurable physical, chemical, and biological soil properties and landscape parameters. At the same time, measurements will have to follow agreed protocols and thresholds that consider the variability of soil types, land use, and climate, with reference points defined by the member states themselves.

In this framework, consensus exists that the assessment of soil health needs to be done in the context of specific geographic, biophysical, and management conditions

(Costantini et al., 2020). The Iberian Peninsula, because of its geographical location, physical characteristics, and long history of land use and land use changes, is home to a great diversity of landscapes, including many based on different uses of soils. A recent compilation of land of soil degradation pathways in Europe (Prăvălie et al., 2024) has shown that the peninsula is also one of the most threatened territories in Europe in this respect, with many areas included as those affected by three or more soil degradation problems. In particular, soil loss by water and wind erosion, soil compaction, nutrient imbalances, or acidification can be identified in different areas. Groundwater decay and aridity at critical levels in most of the southern half of the peninsula are additional relevant threats to soil functioning.

This book therefore presents some of the most representative situations in the peninsula in this respect and highlights the approaches used to assess soil health in each.

General Geophysical Characteristics and Land Uses

The Iberian Peninsula, considered in this chapter as the joint surface of continental Spain and Portugal, is located at the southwestern limit of the European continent, and extends over 582,400 km^2. The peninsula displays extensive landscape and biological diversity owing to its geographical position and its geological, geomorphological, and bioclimatic characteristics. These characteristics also result in a great soil diversity.

From the geological and geomorphological point of view, as described by Maestro et al. (2013), the complex geological history of Iberia reflects its location between the African and European plates and is recorded by rocks ranging in age from Precambrian to Quaternary within a complex and diverse geological setting (Figure 1.1). The peninsula hosts several mountain ranges exceeding 2,000–2,500 m (Cantabrian Mountains, northwestern ranges, Central Range, and Iberian Range) and two ranges with peaks greater than 3,000 m (Pyrenees and Sierra Nevada in the Betic Range).

Despite this complexity, the peninsula can be divided into three main geologic domains based on lithological and hydrologic characteristics (Martín-Serrano et al., 2005; Figure 1.1). These domains include (a) the so-called Iberian or Hesperian Massif (western sector, including most of Portugal), (b) the mountain ranges of the Alpine orogeny (including the Pyrenees and other major chains of the same age) and other Mesozoic formations, and (c) the Cenozoic basins (including the sedimentation basins of the four major rivers: Tajo/Tejo, Duero/Douro, Ebro, and Guadalquivir). In some small areas, relatively recent volcanic complexes are also present, although they represent a minimal surface in the central, northeast, and northwest areas of continental Spain.

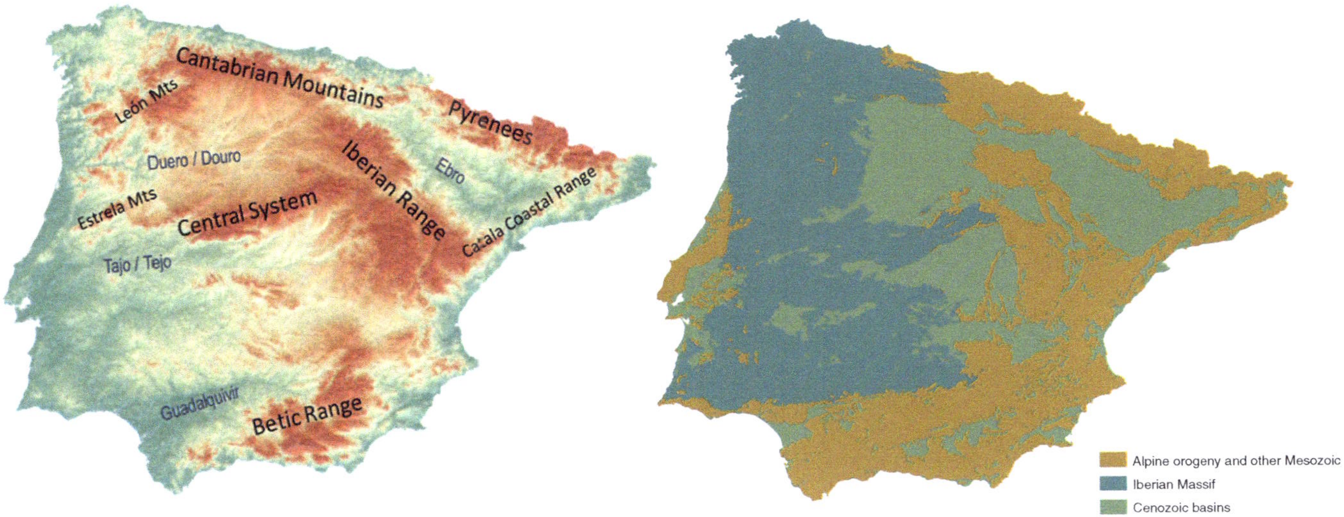

Figure 1.1 Major systems and river basins (left) and main geologic domains (right) in the Iberian Peninsula. Own elaboration from Martín-Serrano et al. (2005) and IGME (2015).

The lithology of these domains can be also summarized in major lithological units (Gallardo, 2016; Figure 1.2). As such, siliceous metamorphic rocks (mainly slates, schists, and quartzites) and intrusive rocks (granites and similar) are the predominant materials in the Iberian Massif (Calvo de Anta et al., 2020). The Cenozoic basins are formed mostly of calcareous materials (limestones, dolostones, marls), and sedimentary rocks, which in most cases also include calcareous materials from their source areas. Major mountain chains display diverse lithologies related to their tectonic and orogeny processes (Vera et al., 2004). In the north, in the Pyrenees and Cantabrian Mountains (Figure 1.1), limestone and other calcareous materials alternate with older, metamorphized materials, especially in the axial zones of the Pyrenees, and pre-Cenozoic materials. Other major ranges in the central and eastern areas (Iberian Mountains, Catalan Coastal Range, Figure 1.1) also contain significant calcareous massifs, and at some points, older metamorphic outcrops. In the south, the Betic Mountains are part of the Alpine ranges surrounding the Mediterranean (Mediterranean Alpine Orogen), with Triasic sedimentary materials in the North, and some metamorphic materials outcropping in the south.

The location and geography of the Iberian Peninsula also results in a multiplicity of climates. Because most wet fronts arrive from the Atlantic Ocean and flow from west to east, all northern and northwestern areas are, in general, rainier than the south and, especially than the east and southeast (Gallardo, 2016). Mountain chains in the north and in northern Portugal (Figure 1.1) also contribute to hinder the entry of these fronts to the inner part of the peninsula. As a result, there is a

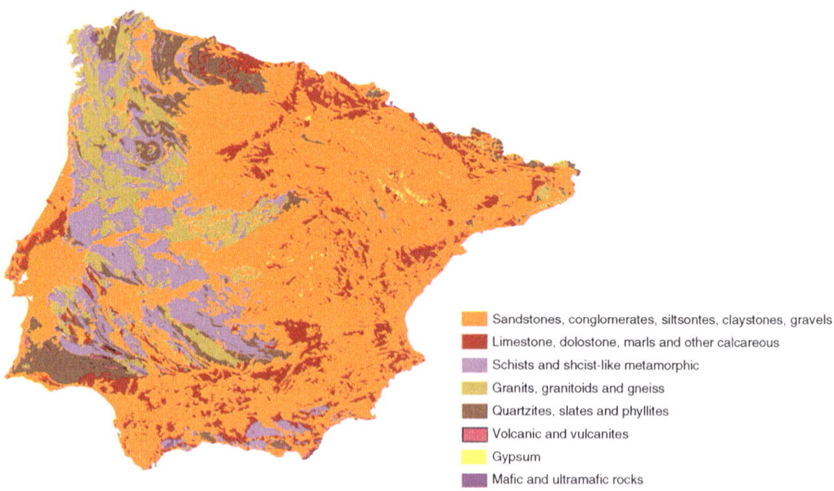

Figure 1.2 Lithologic domains in the Iberian Peninsula. Own elaboration from IGME (2015).

gradient in average annual rainfall from the northwest to the southeast (Figure 1.3). In terms of temperature, latitude, continentality, and altitude all play a role, resulting in a generally warmer southern half (with the highest temperatures recorded in the southeast and the Guadalquivir Basin; Figure 1.1), and milder temperatures in the northern half and mountain areas (Figure 1.3). In addition, the proximity to the ocean or sea makes coastal temperatures less variable throughout the year, whereas a larger contrast between the cold and warm seasons exists in the central area. Altitude also results in colder temperatures corresponding with the major ranges (Figures 1.1 and 1.3). This results in clearly different areas with contrasting annual water balance (moist or dry) and average temperatures (Figure 1.3).

These areas correspond to a large extent to those defined by the global Köppen climate classification (Figure 1.3). Most of the peninsula displays temperate (C-type) climates, with differences in the amounts and distribution of precipitation. While most of the north is under moist C climates (Cf), the central and southern parts and a large area around the east coast display some type of summer drought (Cs). The temperature in the summer is lower in the northern half (Csb) than in the south and southeast (Csa). Csa climate corresponds to the classical Mediterranean climate (mild winters and dry, warm summers, and most rain falling in late winter or in spring or fall). In contrast, relevant areas in the eastern half (mostly corresponding to the Ebro depression and areas in the southeast) are arid enough to be defined as dry climates by Köppen (B-type), in particular steppe-like (Bs), differing owing to whether the annual temperature is milder (Bsk, $<18\,°C$) or warmer (Bsh, $>18\,°C$).

These differences in climate have consequences in land uses, types of natural vegetation, and especially in agriculture. While moist and fresh areas in the north and northwest have been traditionally devoted to forest and grasslands, plains and gentle-slope areas in the central and eastern parts comprise most of the agricultural land. In many cases, irrigation is used on this land as a tool to increase yields and enable the productivity of crops.

In addition, the combination of these diverse soil forming factors results in a high diversity of soil types in Iberia. As depicted in Figure 1.4, major soil types comprise Cambisols of different types and Regosols and Leptosols, also with different qualifiers. Luvisols and Podzols are also present, mostly in the western half of the peninsula (Figure 1.5). Cambisols are relatively young soils, with little or no profile differentiation (either because of limited age, slow pedogenesis, or rejuvenation of the soil material), with at least an incipient subsurface soil formation and evidence of horizon differentiation based on changes in structure, color, clay, or carbonate content (IUSS Working Group WRB, 2022). These soils occur in a wide variety of environments around Europe and under many kinds of vegetation (Soil Atlas of Europe, 2005). The USDA

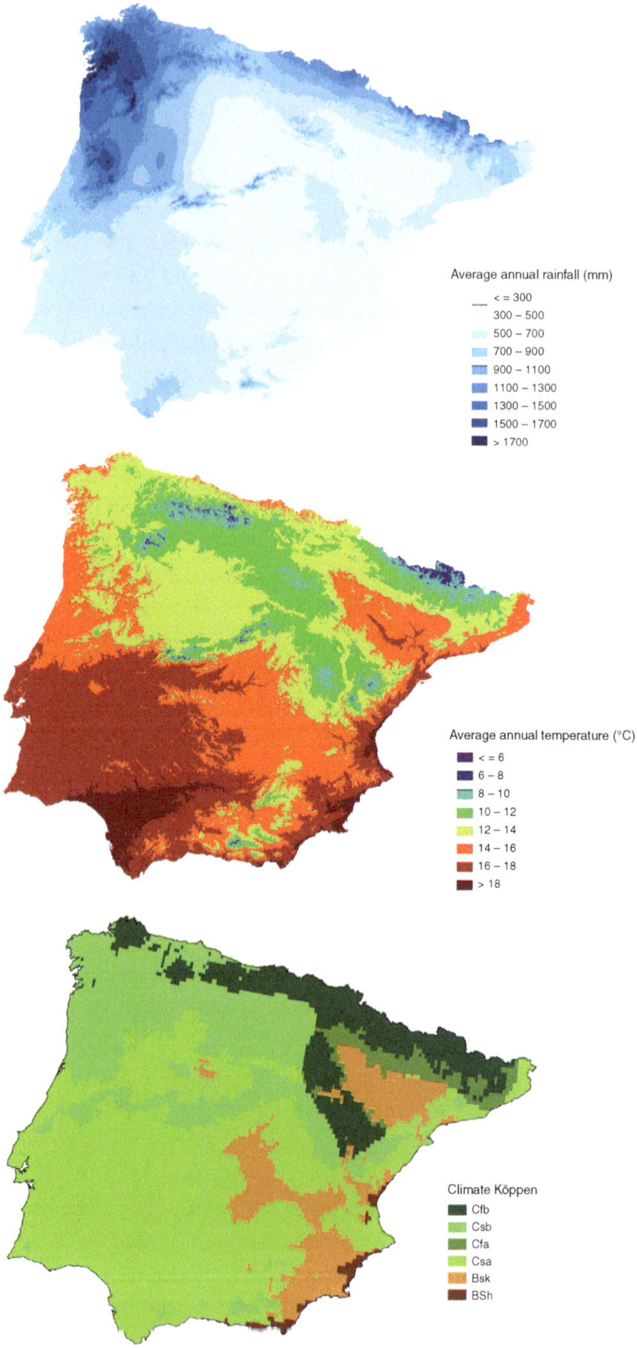

Figure 1.3 Distribution of average annual rainfall (top), and temperature (center), and Köppen climate areas (bottom) in the Iberian Peninsula. Own elaboration from E-OBS (Cornes et al., 2018), CRU-TS 4.06 (Fick & Hijmans, 2017), and World Map of Köppen-Geiger Climate Classification (https://koeppen-geiger.vu-wien.ac.at/).

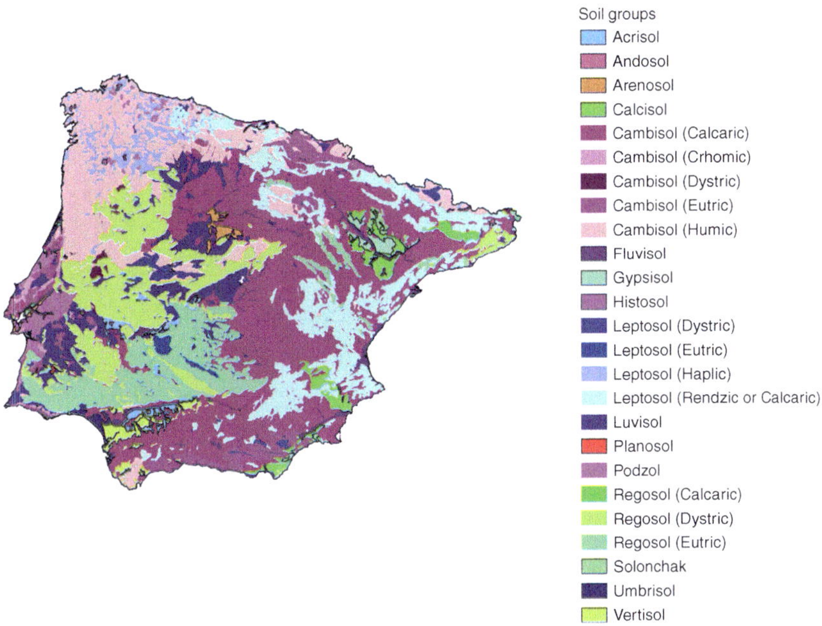

Figure 1.4 Distribution of soil groups of the World Reference Base in the Iberian Peninsula (IUSS Working Group, 2015). Own elaboration from ESDAC, ESDB v2.0, 2004.

Soil Taxonomy classifies most Cambisols as Inceptisols. Depending on their parent material, climate, and landscape positions, Cambisols can be differentiated in soils that have contrasting characteristics and behavior in terms of their response to management or potential use. In the Iberian Peninsula, two types are dominant according to the European Soil Data Center (ESDAC, ESDB v2.0, 2004, Figure 1.4). In the eastern part, corresponding to the areas with calcareous and sedimentary dominating lithologies, Calcaric Cambisols dominate (Figure 1.5). These are Cambisols having calcaric material between 20 and 50 cm from the surface, or between 20 cm and the bedrock, or a cemented layer (IUSS Working Group WRB, 2015). Other soils with calcaric or calcium-rich features (such as Calcaric Leptosols and Regosols, and of course Calcisols) are also widespread in this part of the peninsula (Figure 1.5). In fact, in the Soil Atlas of Europe, these areas appear as occupied mostly by Calcisols (with substantial secondary accumulation of carbonates). These areas correspond to soils with a Mediterranean type of climate, including Calcixerepts and Haploxerepts according to the USDA Soil Taxonomy, which together cover more than 140,000 km^2 in Spain (Gómez-Miguel and Badía-Vilas, 2016). These findings imply that the presence of carbonates in different forms in the soil

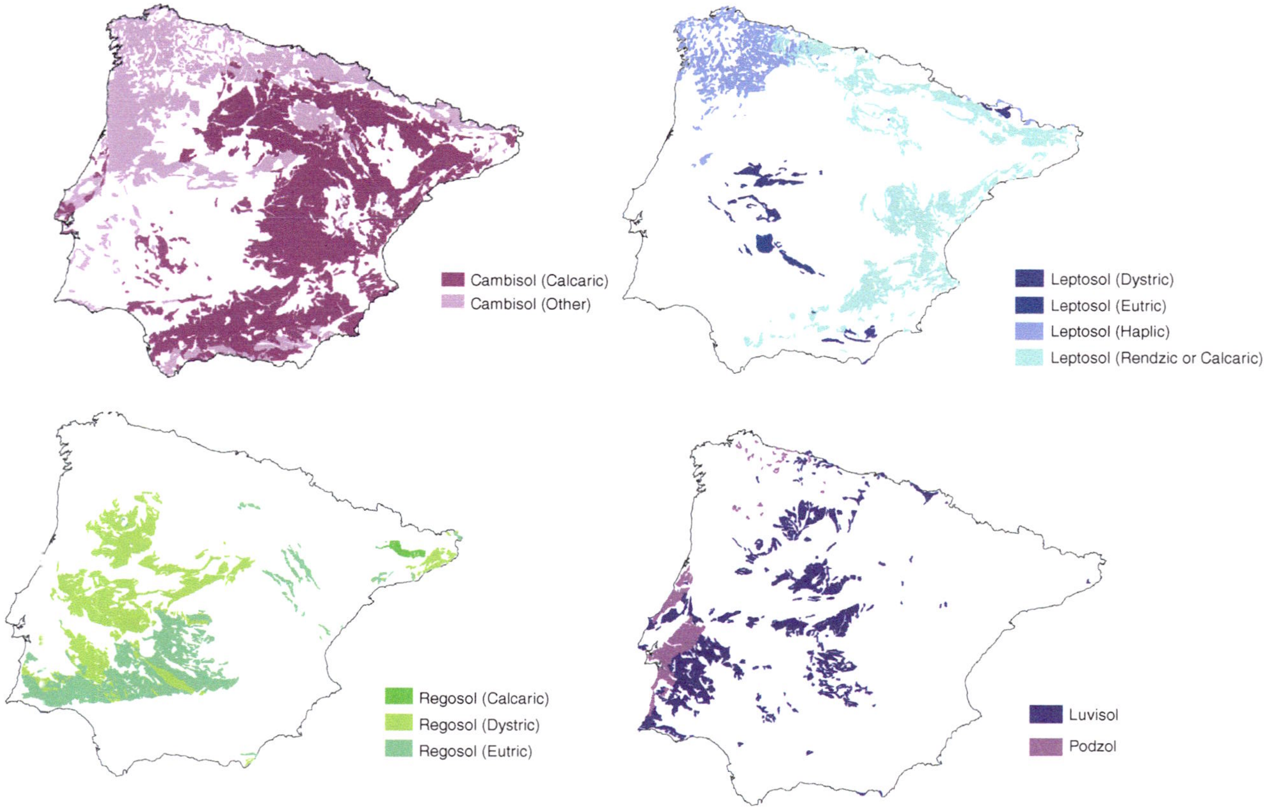

Figure 1.5 Distribution of Cambisols, Leptosols, Regosols, Luvisols, and Podzols in the Iberian Peninsula (IUSS Working Group, 2015). Own elaboration from ESDAC, ESDB v2.0, 2004.

profile is a common trait of soils in this area of the peninsula, as many are developed on calcareous and limestone-derived parent materials, with climates that hinder complete leaching of carbonates.

In contrast, Cambisols in the northwest and areas with certain altitude appear as Dystric and/or Humic containing organic-matter rich upper horizons (Figure 1.5), in many cases displaying a well-structured, dark-colored surface horizon with varying base saturation. The Soil Atlas of Europe (2005) depicts this area as mostly covered by Umbrisols with a dark, acid surface horizon rich in organic matter, which corresponds to climate in these regions, where precipitation considerably exceeds evapotranspiration, especially in the northern areas.

In a wide portion of the west and the southwest, Dystric and Eutric Regosols are also abundant (Figure 1.5). These are very weakly developed mineral soils occurring on unconsolidated materials in many cases and with limited development explained by climate limitations such as prolonged dryness, the parent material characteristics, or erosion. Regosols are extensive in eroding lands, in particular, in arid and semi-arid areas and in mountainous regions (Soil Atlas of Europe, 2005). They correspond to Entisols in the USDA Soil Taxonomy and appear as Xerorthents in the description on the soils of Spain by Gómez-Miguel and Badía-Vilas (2016).

Finally, soils with leaching processes leading to the accumulation of clay (Luvisols) and/or organic matter (Podzols) appear mostly in the western half of the peninsula (Figure 1.5), the latter being especially abundant in central/western Portugal. Luvisols present a marked textural difference in the profile, with a subsurface horizon of high activity clay accumulation and high base saturation, corresponding to Alfisols in the USDA Soil Taxonomy. A wide range of parent materials and environmental conditions lead to great diversity in this soil group (Soil Atlas of Europe, 2005), which includes several subgroups (from Calcic and Chromic to Ferric and Gleyic) in the Iberian Peninsula. Podzols are bleached acid soils with a B horizon that accumulates aluminum, iron, and/or organic compounds. They mostly develop in humid, well-drained areas and are common in Europe under vegetation with acidic litter (Soil Atlas of Europe, 2005). Other soil types can be dominant locally, such as Vertisols in the Guadalquivir area (Figure 1.4).

Finally, with respect to land cover and land uses, forests and shrublands (including some agroforestal areas such as the so-called *dehesas* or *montados*) represent the most widespread land cover (252,550 km^2, 42% of the geographic area), followed by farmlands used for different purposes (189,400 km^2, 32%) and grasslands and badlands (102,506 km^2, 32%). Less than 8% of the surface is associated with other types of uses (urban land, infrastructure, or water bodies; Figure 1.6). Agricultural and forest uses of soil are therefore the most widespread, and the ones with the greatest interactions with soil and soil health in this territory.

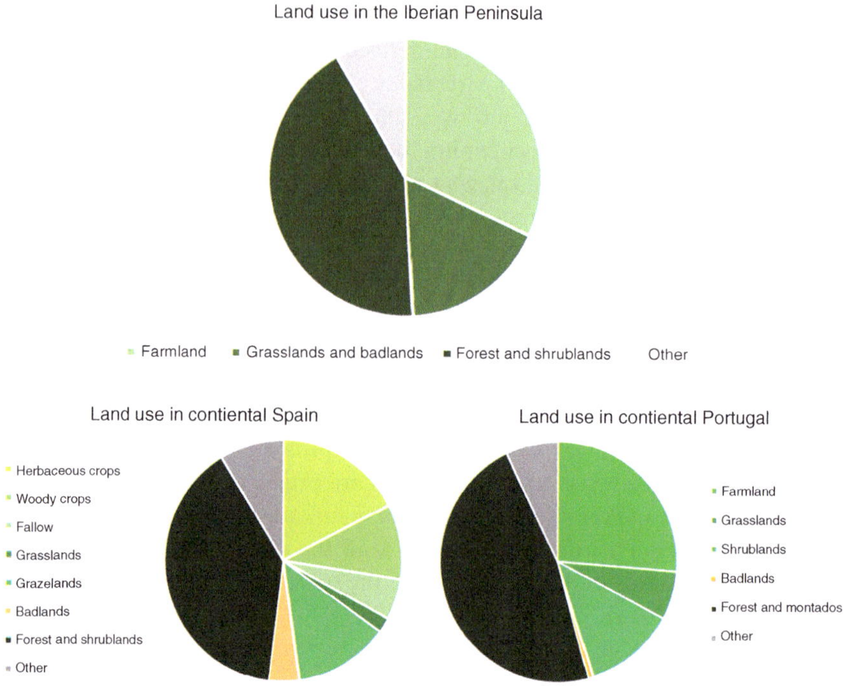

Figure 1.6 Land use in the Iberian Peninsula (top), and distribution of land use typologies in continental Spain (bottom left) and Portugal (bottom right) in 2021 and 2018, respectively. *Sources*: Ministerio de Agricultura, Pesca y Alimentación (Spain) and Instituto Nacional de Estatística (Portugal) https://www.mapa.gob.es/es/estadistica/temas/estadisticas-agrarias/agricultura/distribucion-general-suelo/ https://www.ine.pt/xportal/xmain?xpid=INE&xpgid=ine_destaques&DESTAQUESdest_boui=435668469&DESTAQUESmodo=2&xlang=pt.

In relation to agriculture, available statistics differ between Spain and Portugal. In Spain, data from the Ministry of Agriculture (Ministerio de Agricultura, Alimentación y Medio Ambiente, 2022) indicate that herbaceous crops represent 53% of the total agricultural land, and woody permanent crops, 30.5%. The other 16.7% includes fallow or abandoned agricultural land. Almost one-fourth (23.2%) of this land is irrigated. Grasslands and grazelands (included grazed badlands) account for 10.6% and 8.4%, respectively, of this land use, with less than 1.5% being irrigated (Figure 1.6). Forest and shrublands represent 39.5% of the continental land cover of Spain (~200,000 km^2).

In Portugal, 132,562 ha of land was irrigated in 2013 (Governo do Portugal, 2014), which represents 5.7% of the total agricultural land, while forests and shrublands (including *montados*) represent 59.5% of the total surface. Farmland as a whole account for 26.2%.

All these data support the idea that major threats to soil health in the Iberian Peninsula arise from the agricultural and forestry use of the soils.

Historical Assessment

To describe the evolution of scientific publications developed in the Iberian Peninsula concerning the concepts related to soil health, a bibliometric analysis was conducted. We collected data on the evolution of publications that refer not only to soil health but also to those related to soil quality, in order to observe the evolution of these concepts. Specific publications on both topics indexed in the Scopus database (Elsevier, Germany) until May 2024 were searched, avoiding publications in which only the concept of soil quality or soil health was mentioned without deeper exploration. The first search criterion was TITLE ("soil health" OR "soil quality"). A second search criterion TITLE-ABS-KEY (Spain OR Portugal OR iberi* OR mediterr*) was applied to focus on studies conducted in the Iberian Peninsula. Preliminary filtering was necessary to discard publications outside the desired geographic scale, especially among those collected with the "mediterr*" criterion. The results were filtered by type of document, focusing on articles, reviews, and book chapters. In addition to categorizing the studies by topic, the R package *bibliometrix* (Aria & Cuccurullo, 2017) was used to create a simple scientific map that provided information concerning the evolution of annual scientific production and the most frequently employed keywords and terms.

Although many references in the so-called gray literature were likely missing, and even some works referred in *Scopus* may have also escaped filtering, the database generated is representative of the trends observed in Portugal and Spain in the past few decades. The database contained 115 papers published from 1999 to 2024, with 107 related to soil quality and eight to soil health. Two of the papers proposed a mixed evaluation of both concepts and were included in the soil health analysis. The documents comprised 113 scientific articles and two book chapters. Seven conference papers were discarded. The number of publications showed a positive and growing trend in research related to soil quality since the late 1990s, with a later emergence of publications related to soil health starting in the 2010s and becoming more prominent from 2017 onwards (Figure 1.7).

Among the publications on soil quality, 78 (73%) focused on agricultural soils, eight (7%) on forest soils, and seven (7%) on mountain soils (Figure 1.8). Of the mountain soil publications, five specifically examined mid-mountain agricultural systems. Six publications (6%) addressed soil quality in degraded soils, including studies in industrial and mining areas. Finally, 10 publications investigated mixed land uses, evaluating soils under various uses or targeting different ecosystems and topics. Examples include studies on soil quality in fluvial systems

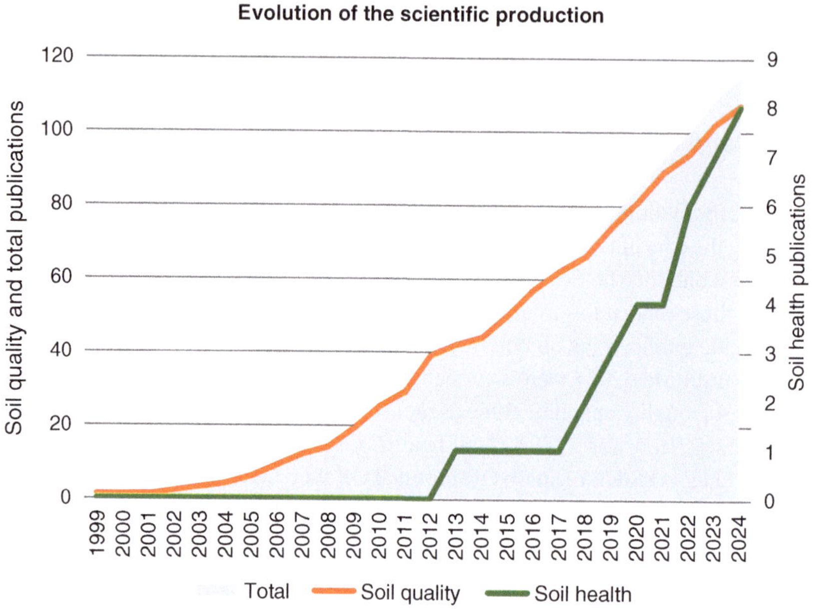

Figure 1.7 Evolution of scientific production on soil quality (red line, primary axis), soil health (green line, secondary axis), and total publications (gray shade, primary axis).

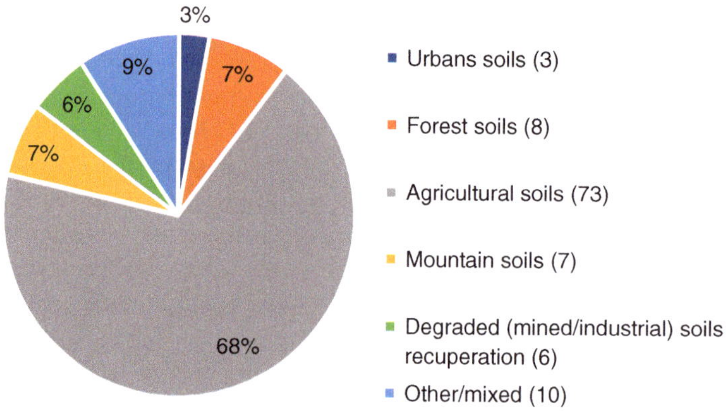

Figure 1.8 Topic categories considered within soil quality publications.

(Rodrigo-Comino et al., 2023), seagrass sediments as a palaeoecological land use and soil quality proxy (López-Merino et al., 2015), and the relationship between soil quality and plant invasion (Novoa et al., 2014).

The most frequently mentioned terms within the soil quality publications (each mentioned more than 15 times, excluding the terms soils, soil quality, and Spain)

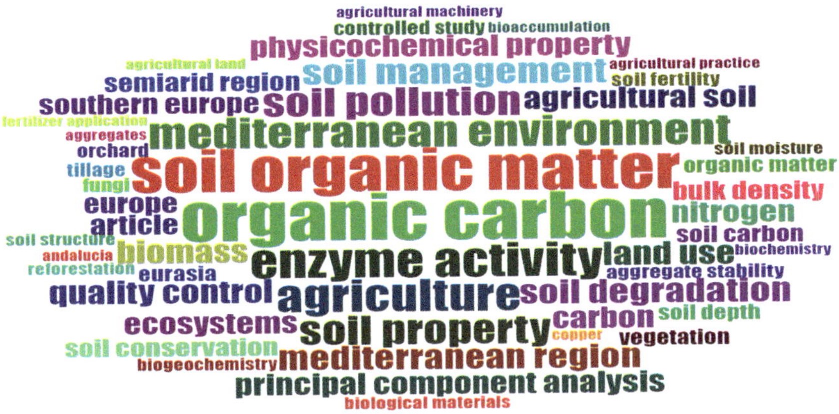

Figure 1.9 Wordcloud with the 50 most frequent terms within soil quality publications.

were the following: organic carbon, soil organic matter, enzyme activity, agriculture, Mediterranean environment, soil pollution, soil properties, soil management, Mediterranean region, and soil degradation (Figure 1.9).

With respect to publications specifically addressing soil health, four focused on soil health assessment in agricultural soils, and two focused on polluted soils, including one on soil in mining area and another one on industrial soils. Finally, two publications focused on soil health assessment in different land uses and land covers (Table 1.1).

Asensio et al. (2013) defined soil ecosystem health as the “ability of a soil to carry out its functions in an appropriate way for the ecosystem operation,” and proposed an integrated soil health assessment strategy based on biomarkers (earthworms) for contaminated industrial soils. In the other publication on polluted soil restoration, Cárdenas-Aguiar et al. (2022) proposed a soil remediation strategy for mining areas using rabbit manure-derived applications combined with plant cover to enhance soil enzyme activities and soil microbial biomass.

Regarding agricultural soils publications, Sánchez-Moreno et al. (2018) studied the effect of organic management in olive groves and vineyards on soil health, based on nematode diversity and other soil properties. The authors highlighted the importance of preserving and improving the health of agroecosystems as a priority for landscape management, especially in southern Europe. Other publications on agricultural soils followed a similar line, evaluating various agricultural strategies to improve soil health. For example, Zuazo et al. (2020) concluded that alternative modifications to conventional production systems for rainfed olive orchards, based on organic, integrated, and conservation agriculture, could improve important soil health features.

Table 1.1 Soil Health References.

	Reference	Study areas	Land/soil use	Topic	Type
1	Visconti et al., 2024	Valencia (Spain)	Agricultural soils	Straw mulching strategies to increase soil health	Article
2	Rutigliano et al., 2023	Alentejo (Portugal), Extremadura (Spain), and Sicily (Italy)	Mixed use	Soil health indicators for different land cover types	Article
3	Zegada-Lizarazu et al., 2022	Cadriano (Italy), Extremadura (Spain), and Aliartos (Greece)	Agricultural soils	Effects of integrated food and bioenergy cropping systems on crop yields, soil health, and biomass quality	Article
4	Cárdenas-Aguiar et al., 2022	Murcia (Spain)	Mining soils	Effects of on soil health and quality in mine soils	Article
5	Zuazo et al., 2020	Granada (Spain)	Agricultural soils	Organic olive rainfed systems to control soil erosion and improve soil health	Article
6	Higueras et al., 2019	Castilla-La Mancha (Spain)	Mixed use	Biogeochemical indicators to assess soil quality and health	Book chapter
7	Sánchez-Moreno et al., 2018	Castilla-La Mancha (Spain)	Agricultural soils	Organic management to improve soil health: microfaunal soil food webs	Article
8	Asensio et al., 2013	Basque Country (Spain)	Industrial soils	Integrative soil health assessment strategy	Article

Higueras et al. (2019) briefly reviewed the evolution of the concepts of soil quality and soil health, updating the definition of soil health as "the continued capacity of the soil to function as a vital living ecosystem that sustains plants, animals, and humans", emphasizing that soil health assessment involves identifying and quantifying the temporal evolution of soil quality. They argued that soil health assessment should focus on the integration and optimization of soil chemical, physical, and biological characteristics, which are fundamental for sustained productivity and environmental quality. This publication, in the form of a book chapter, briefly presents three case studies in Castilla la Mancha (Spain), one on agricultural soils and two on soils in mining areas.

Finally, Rutigliano et al. (2023) integrated the most current definition of soil health proposed by Bünemann et al. (2018) as "the ability of a soil to function

Figure 1.10 Wordcloud with the 50 most frequent terms within soil health publications.

within ecosystems and land use boundaries to sustain biological productivity, maintain or improve environmental quality, and promote plant and animal health," complementing it with the definition suggested by the European Commission (2020): "the continued capacity of soils to support ecosystem services." The authors emphasized in this work that a comprehensive understanding of soil properties and functions is a prerequisite for implementing effective strategies to prevent the risk of desertification. They proposed a set of microbial, physical, and chemical indicators to assess soil health.

The most frequently mentioned terms within the soil health publications (each mentioned more than three times, excluding the terms soils, soil quality, and Spain) were the following: *Eisenia fetida*, biomass, ecosystem health, and soil carbon (Figure 1.10).

In summary, the agricultural use of soils, and the assessment of functions related to the soil organic fraction and biological activity, seem to have been the focus of most of the research conducted in this field. However, other aspects seem also to be recurrent, such as the reclamation of degraded soils (especially by mining activities), and forest soils. In terms of the conceptual approach, in agreement with the general trend (Karlen et al., 2021a), an evolution is observed from the initial soil quality activities and inquiries towards the acceptance, promotion, and adoption of the concept of soil health as such.

Case Studies in This Book

As previously explained, agriculture and forestry have traditionally been the most widespread types of land use in the Iberian Peninsula and have shaped the rural landscape, as in most of Western Europe (Bregt & de Zeeuw, 2002; Virto et al., 2015). The particularities of some Iberian soils in terms of soil health assessment arise from their natural characteristics, climate conditions, and historical land use. For instance, Durán et al. (2024) recently summarized the most relevant indicators to be included in a potential core set for calcareous agricultural soils in the island of Mallorca, observing different responses for rainfed and irrigated soils.

In this book, we have compiled case studies and long-term assessments corresponding to contrasting land uses and soil types in different agroclimatic areas of the peninsula. This focus includes agricultural, forest, and urban soils in a gradient of climates corresponding to the most representative conditions of the peninsula (from the humid northwest to the arid southeast, Figure 1.3).

Agricultural Soils

Cropland covers approximately one-third of the continental surface of Spain and Portugal (Figure 1.6). The two countries together represent 17% of the European Union's farmland area, and in 2020, Spain alone had the highest production among the EU member states of fresh vegetables (15.4 million Mg), citrus fruits (6.0 million tons), and olives (6.0 million tons; Eurostat, 2020).

In agreement with this relevance of agriculture in the territory and the economy, five out of the nine chapters of this book deal with agricultural uses of soil. Four of them (chapters 2, 3, 5, and 8) represent examples of long-term monitoring of soil health linked to the crops and crop management more frequently used and adapted to their areas of study. The fifth (chapter 6) constitutes a particular case study of a very frequent situation of management change within agricultural land in semi-arid and arid Iberia: the transition from rainfed extensive crops to irrigated agriculture.

In chapter 2, Almagro et al. also offer a detailed general introduction on the framework of soil health assessment in the present conception of soil management as a tool to combat global threats that are of special relevance in the Mediterranean region and on semi-arid land: desertification, climate change, and biodiversity loss. They focus on soil organic carbon (SOC) sequestration and stabilization as an indicator of soil health in the particular case of rainfed almond orchards in southeast Spain. These are low-input cropping systems traditionally adapted to areas with a semi-arid Mediterranean climate (Figure 1.3), frequently found on carbonate-rich soils common to this region (Figures 1.4 and 1.5).

This chapter shows the results of a long-term experiment (LTE) located in Murcia (southeast Spain) testing different soil management strategies, some of which include the suppression or reduction of tillage and/or the use of cover crops or spontaneous vegetation or exogenous organic matter incorporation. One of the most important benefits of this type of LTE is that it enables understanding of the consequences of changes locally. In this case, it was observed that although the total suppression of tillage through the use of no-till practices improved soil condition in some ways, it could also have deleterious effects (in particular in terms of compaction and yield decrease). Therefore, practices with shallow or reduced tillage could be more interesting at the local scale. In this sense, the chapter summarizes the present knowledge gaps, research needs, and policy recommendations in Europe. They include a need for revisioning the present eco-schemes proposed by the Common Agricultural Policy of the European Union. Indeed, local adaptation of the strategies to increase SOC in arable land in Europe is crucial, has been acknowledged in previous assessments (Costantini et al., 2020), and arises from the wide diversity in soil and climate conditions in the continent. For instance, this chapter details how pedoclimatic constraints in southeast Spain (e.g., scarce water availability and calcareous soils with low organic matter contents and prone to compaction) make it difficult to seed and grow vegetation covers under rainfed semi-arid conditions without adopting occasional tillage.

This chapter also represents an interesting first example on how crop yields can be compromised when innovative soil management strategies for improving most soil health indicators are implemented. This is the case of no-till in the studied conditions, and it is for instance also reported in chapter 8 for some organic amendments.

Chapter 3 provides information on soil health assessment in other experimental sites in southeast Spain, focusing on the introduction of diversified cropping strategies. In particular, the authors describe the evolution in time of three diversified systems (namely, almond orchards combined with thyme crop and reduced tillage, melon with cowpeas, green manure and reduced tillage, and mandarin orchards combined with a rotation of fava beans and barley or vetch using green manure) compared with the evolution of the conventional monocrop systems for almonds, melons, and mandarins.

After two crop cycles, crop diversification was found to have led to an improvement of the soil structure in all crops studied. Other effects were uneven between crops: diversification increased soil moisture indicators in mandarin and melon crops, whereas soil nutrient contents were improved in the almond crop. At the same time, it was observed in the mandarin crop that the monocrop showed higher availability of nutrients compared with the diversified mandarin. Related to this, crop productivity was enhanced by crop diversification in almond and melon, but not in mandarin. The authors highlight the need for adapting the assessment tools to the particularities of crops, and introduce the idea that changes in soil health following crop diversification may occur at different paces depending on the crop and agrosystem characteristics.

In this sense, water management seems crucial in agriculture in areas such as those described in this study, with low to very low rainfall-to-evaporation ratios.

Indeed, two of the three case studies described in chapter 3 (melon and mandarin crops) include irrigation to enable economically sustainable production. Here and in many other arid and semi-arid areas of Spain and Portugal, the use of water becomes a key strategy for supporting yields and farmers' income. As a result, irrigation has expanded in the past few decades in both countries, making agriculture the largest water-using activity. From the perspective of soil health, the adoption and the use of irrigation has been traditionally associated with some risks related to the quality of irrigation water and the risk of salinization.

With respect to irrigation adoption, chapter 6 focuses on another topic not often considered in this transition, and it offers a multi-scale analysis of its effects on agricultural soils in the region of Navarre (northern Spain). In this chapter, the authors present a comprehensive view on the potential changes in soil health indicators induced by the transformation of rainfed agricultural land to irrigation. The chapter focuses on one particular region (Navarre), where the extension of irrigation has resulted in more than 22,000 ha being converted in the past few decades. In an LTE designed to assess the consequences of the change in water regime and crops on SOC, it was observed after only 8 years that the introduction of irrigation can affect the mechanisms that regulate its stabilization, enhancing SOC storage through increased carbon inputs and a more efficient incorporation of crop residues into SOC. Later, this study was enlarged to the regional scale, using a network of paired plots in which different irrigated agrosystems were compared with their rainfed counterparts on the same soil units and locations. With the use of response ratios as a tool to harmonize comparisons between different sites, it was observed that irrigation can result in enhanced key soil-based ecosystem services such as erosion control and climate regulation, simultaneously enhancing crop yields and net gains to farmers. However, this effect is dependent on the combination of management and location. In addition, the effect on other services such as climate regulation through SOC stabilization can be mitigated by the higher emissions of greenhouse gasses in irrigated systems.

In a different agricultural context—rainfed acid soils developed on the so-called *rañas*, which are continental detritic glacis-piedmont type formations related to quartzite-dominated mountain ranges in the Hesperian or Iberian massif (Figure 1.1)—Hontoria et al. explain in chapter 5 the results of a number of LTEs designed to assess the effects of conservation agriculture coupled with calcium amendments on soil health. Two trials were conducted on soils with serious soil health issues related to a critical organic carbon loss after decades of cultivation.

The response of the soil chemical, physical and microbial soil properties, as well as organic matter storage, greenhouse gas emissions, and yield, was dependent on the type of amendment used and on its combination with soil management strategies. For example, the two LTEs showed that the suppression of tillage and the introduction of pastures improved soil structure and its related parameters (i.e., water retention). However, calcium amendments had very little, if any effect, despite the well-known positive effect of calcium in soil structural stabilization. At the same time, the increase in soil pH resulting from calcium amendments had a major effect on the structure and functioning of the soil microbial populations (increasing the proportion of bacteria vs. fungi, and the populations of ammonia-oxidizing bacteria and ammonia-oxidizing archaea). In terms of productivity, while calcium amendments resulted in yield gains almost every year, the suppression of tillage took longer to show an effect (from the third campaign on) and was especially noticeable in dry years. The use of calcium was also able to reduce N_2O and CH_4 emissions, especially when combined with tillage. In summary, this chapter represents another example on the need for selecting the adequate soil indicators in each condition for long-term monitoring of soil health.

Finally, in chapter 8, Simoes-Mota et al. describe the results of another rainfed LTE in Navarre (northern Spain), representative of cereal production in the region, and focus on one particular strategy: the use of organic amendments as substitutes of mineral fertilizers to improve or enable yields and enhance soil health. As many cultivated soils in the semi-arid and arid areas of the Iberian Peninsula have low organic matter contents, this kind of strategy is seen as an opportunity to improve the productivity using organic sources of different types. In particular, the use of sewage sludge or composted sludge from urban wastewater treatment plants is seen as a win-win strategy because it allows for the disposal of the organic matter–rich fraction separated from wastewater. Their use can be controversial, however, as sewage sludge can contain different components harmful to soil health, especially trace metals and pollutants, and has been identified as a potential risk for soil quality in Europe (Virto et al., 2015). In this chapter, the effect of the application of different doses of sewage sludge, combined with different schedules, on soil health indicators and yield after more than 25 years is described in detail.

Together with yield, soil chemical, physical, and biological indicators, as well as organic matter–related parameters were studied for the different accumulated doses of sewage sludge after 25 years, compared with conventional mineral fertilization and no fertilization at all. As expected, the results showed that the effect was different for the different indicators. While the use of moderate doses (~10–20 $t \cdot ha^{-1}$ per year) supported yields comparable to those from mineral fertilization, the use of excessive doses of sewage sludge (~80 $t \cdot ha^{-1}$) had clear negative effects on soil health, resulting not only in lower productivity, but also worse conditions for soil structure and earthworm activity and the accumulation of some trace metals. However, these higher doses resulted in the greatest increment in the stock of SOC. This chapter thus represents an example of the use of organic amendments not only having variable effects on soil health depending on the dose and frequency of their application, but also that they can differ depending on the soil functions assessed in the agrosystem considered (e.g., productivity vs. climate regulation).

In summary, all these chapters are a collection of representative examples of soil health assessment linked to cultivation strategies aimed at improving soil health while supporting crop yields and the profitability of agriculture in the Iberian Peninsula. As expected, the management techniques tested varied depending on the agro-environmental conditions of each area. In most cases, improving soil health was possible through innovative and conservation practices, but local conditions needed to be considered, which showed that universal solutions are unlikely to work. They are also an example on the need for LTEs for comprehensive assessments of soil health.

Forest Soils

Chapters 7 and 9 describe different approaches to soil health evaluation of forest soils in northern Spain and Portugal. Given that forests and shrublands are the most extensive type of land cover on the peninsula (Figure 1.6), the assessment and preservation of soil health in these ecosystems are of paramount relevance. Forest soils provide essential services such as a clean water supply and the protection of biodiversity, as explained by Gartzia-Bengoetxea et al. in chapter 7. The authors describe how, in northern Spain, many forested areas are actually forest plantations. The techniques used for planting and harvesting these plantations can have a notable impact on soil health, not only because of the physical disturbance, but also because of their effects on the capacity of forest soils to provide ecosystem services beyond the provision of wood and fiber. The authors show how a relatively simple protocol for visually assessing soil condition can be sensitive enough to detect differences between undisturbed areas and areas affected by different operations of mechanical forestry. The validation of this protocol was conducted by measuring different key soil properties in undisturbed and disturbed areas, and then comparing their incidence in plantations managed with different levels of mechanical and manual handling. This assessment seems especially useful in the framework of both new regulations to protect soil health and new certification programs in wood products.

The concern about the consequences of plantations on soil health is also a constant in *Eucalyptus* sp. stands across Portugal, which were extensively established in the last decades of the 20th century in Portugal, as explained by Madeira et al. in chapter 9. The authors explain how the management of the three major land uses related to trees or forests in Portugal (namely eucalypt plantations, *montados* or evergreen oak woodlands, and chestnut groves) can affect soil health, depending on a number of factors.

In eucalypt plantations, the extent of soil disturbance during the preparation for planting, as well as the management of harvest residues, was found to have site- and soil-dependent effects. Whereas deep plowing resulted in poorer soil health than disc harrowing for almost all soil functions considered, the effect of residue management was not comparable between the different sites tested. As such, the removal of harvest residues seemed possible in sites sufficiently resistant and resilient to carbon and nutrient losses, although their incorporation by different means seems to be a better option to support soil health. This example shows how similar systems (e.g., eucalypt plantations) can display different results in terms of soil health depending on their natural soil (and climate) characteristics within a relatively small area.

Permanent agroforestry systems based on evergreen oak trees (*Quercus suber* and/or *Quercus rotundifolia*, called *montados* in Portugal and *dehesas* in Spain) are common to a wide area in the central and west-central area of the peninsula. They represent a traditional agroforestry system in which soils provide a wide range of ecosystem services. These systems combine wood or cork production and grazing-based animal production, while also providing shelter for game and allowing other extensive land uses such as harvesting of wild mushrooms. Their management has been the object of research for decades, especially in relation to cork and animal production. In terms of soil health, different strategies have been assessed by Madeira et al. In a number of mid- to long-term experimental sites located in *montado* systems across Portugal, the effect of the presence of tree islands, the replacement of natural shrubs and pastures by improved pastures, different grazing systems, and shrub encroachment in areas with degraded soil condition have been tested for years. While the existence of tree islands and a widespread presence of sparse trees is acknowledged to result in better soil health compared with bare soil, the consequences of changes in pasture type and management were uneven. Improved pastures can result in better conditions for soil nutrients, but not always in better soil biological indicators compared with natural pastures. The positive effect of improved pastures compared with shrubs can also be modulated by the intensity of grazing, with higher loads or larger animals (i.e., cattle vs. sheep) resulting in poorer soil health. Finally, the colonization of *montado* areas with low soil health by shrubs (usually *Cistus* sp.) can alter soil carbon cycling, without noticeable effects on carbon use efficiency. These observations imply that soil health in these delicate systems needs to be monitored with caution in the long run after introducing new management systems.

Finally, chestnut groves are perennial and traditional staple food crops mostly present in northeastern Portugal on soils developed on schists. The studies conducted in these systems were mostly focused on changing soil management and cover from traditional bare soil with repeated tillage, to a reduction or suppression of tillage, introducing permanent grass cover of different types. In a clear example of how local conditions of soil, climate, and agrosystem can determine the net effect of tillage suppression, and as observed in other systems (Virto et al., 2012), no changes were observed in relation to carbon storage when tillage was suppressed, even after 8 years. No clear trends were observed on the soil physical or biological indicators. However, the net system profitability improved overall when tillage was replaced by permanent grass cover, mostly because of better economic yields and improved conditions for growth of wild edible fungi. Interestingly, when irrigation was introduced in the system, no tillage resulted in better soil condition than conventional management, suggesting that water limitation can be a reason for the poor performance of no tillage in these systems.

Urban Soils

Urban soils, although representing a minor proportion of the total surface in Spain and Portugal (Figure 1.6), usually are intensely and deeply disturbed by anthropic activities, and are therefore exposed to many degradation risks. In addition, these soils are those closest to most of the population (around 80% of the population in Spain and Portugal lives in urban areas, according to the UN Department of Economic and Social Affairs, 2024). Studying soil health in urban environments therefore seems to be strategic both to determine the capacity of these soils to provide basic services and to increase the awareness of the population on soil conservation.

Chapter 4 is devoted to a comprehensive study of the soils of the city of Santiago de Compostela, in northwest Spain, corresponding to the temperate oceanic (Cfb, Figure 1.3) climate area. It is therein shown how natural soil properties (mostly derived from parent material) can result in different baseline levels of many common soil indicators.

Acknowledgments

We acknowledge the E-OBS dataset and the data providers in the ECA&D project (https://www.ecad.eu, Cornes et al., 2018)

References

Aria, M., & Cuccurullo, C. (2017). Bibliometrix: An R-tool for comprehensive science mapping analysis. *Journal of Informetrics*, *11*(4), 959–975. https://doi.org/10.1016/j.joi.2017.08.007

Asensio, V., Rodríguez-Ruiz, A., Garmendia, L., Andre, J., Kille, P., Morgan, A. J., Soto, M., & Marigómez, I. (2013). Towards an integrative soil health assessment strategy: A three tier (integrative biomarker response) approach with *Eisenia fetida* applied to soils subjected to chronic metal pollution. *Science of the Total Environment*, *442*, 344–365. https://doi.org/10.1016/j.scitotenv.2012.09.048

Bregt, A., & de Zeeuw, K. (2002). Agriculture, forestry and nature: Trends and development across Europe. In J. Stillwell & H. Scholten (Eds.), *Land use simulation for Europe* (Vol. 1, pp. 37–44). Kluwer Academic Publishers.

Bünemann, E. K., Bongiorno, G., Bai, Z., Creamer, R. E., De Deyn, G., de Goede, R., Fleskens, L., Geissen, V., Kuyper, T. W., Mäder, P., Pulleman, M., Sukkel, W., van Groenigen, J. W., & Brussaard, L. (2018). Soil quality—A critical review. *Soil Biology and Biochemistry*, *120*, 105–125. https://doi.org/10.1016/j.soilbio.2018.01.030

Calvo de Anta, R., Luís, E., Febrero-Bande, M., Galiñanes, J., Macías, F., Ortíz, R., & Casás, F. (2020). Soil organic carbon in peninsular Spain: Influence of environmental factors and spatial distribution. *Geoderma*, *370*, Article 114365. https://doi.org/10.1016/j.geoderma.2020.114365

Cárdenas-Aguiar, E., Méndez, A., Paz-Ferreiro, J., & Gascó, G. (2022). The effects of rabbit manure-derived biochar on soil health and quality attributes of two mine tailings. *Sustainability*, *14*(3), Article 1866. https://doi.org/10.3390/su14031866

Cornes, R. C., van der Schrier, G., van den Besselaar, E. J. M., & Jones, P. D. (2018). An ensemble version of the E-OBS temperature and precipitation data sets. *Journal of Geophysical Research: Atmospheres*, *123*(17), 9391–9409. https://doi.org/10.1029/2017JD028200

Costantini, E. A. C., Antichi, D., Almagro, M., Hedlund, K., Sarno, G., & Virto, I. (2020). Local adaptation strategies to increase or maintain soil organic carbon content under arable farming in Europe: Inspirational ideas for setting operational groups within the European innovation partnership. *Journal of Rural Studies*, *79*, 102–115. https://doi.org/10.1016/j.jrurstud.2020.08.005

Durán, G. A., Sacristán, D., Farrús, E., & Vadell, J. (2024). Towards defining soil quality of Mediterranean calcareous agricultural soils: Reference values and potential core indicator set. *International Soil and Water Conservation Research*, *12*(1), 145–155. https://doi.org/10.1016/j.iswcr.2023.06.001

ESDAC (European Soil Data Center). (2004). ESDB v2.0: The European Soil Database distribution version 2.0, European Commission and the European Soil Bureau Network, CD-ROM, EUR 19945 EN. Available at https://esdac.jrc.ec.europa.eu/content/european-soil-database-v20-vector-and-attribute-data

European Commission: Directorate-General for Research and Innovation. (2023). *EU missions – Soil deal for Europe – What is the EU mission – A soil deal for Europe*, Publications Office of the European Union. https://data.europa.eu/doi/10.2777/171313

Eurostat. (2020). *Agriculture, forestry and fishery statistics—2020 edition*. Publications Office of the European Union. https://ec.europa.eu/eurostat/documents/3217494/12069644/KS-FK-20-001-EN-N.pdf/

Faber, J. H., Cousin, I., Meurer, K. H. E., Hendriks, C. M. J., Bispo, A., Viketoft, M., ten Damme, L., Montagne, D., Hanegraaf, M. C., Gillikin, A., Kuikman, P., Obiang-Ndong, G. Bengtsson, J., & Taylor, A. (2022). *Stocktaking for agricultural soil quality and ecosystem services indicators and their reference values*. EJP SOIL Internal project SIREN Deliverable 2. Available at https://ejpsoil.eu/fileadmin/projects/ejpsoil/1st_call_projects/SIREN/SIREN_D2_final_report.pdf.

Fick, S. E., & Hijmans, R. J. (2017). WorldClim 2: new 1km spatial resolution climate surfaces for global land areas. *International Journal of Climatology*, *37*(12), 4302–4315. https://doi.org/10.1002/joc.5086

Gallardo, J.F. (ed.) (2016). *The Soils of Spain, World Soils Book Series, Springer International Publishing Switzerland*. https://doi.org/10.1007/978-3-319-20541-0

Gómez-Miguel, V., & Badía-Vilas, D. (2016). *Soil distribution and classification*. In J. Gallardo (Ed.), *The Soils of Spain*. World Soils Book Series. Springer. https://link.springer.com/chapter/10.1007/978-3-319-20541-0_2

Governo do Porrtugal. Ministério da Agricultura. Direção-Geral de Agricultura e Desenvolvimento Rural. 2014. Aproveitamentos hidroagrícolas do grupo II no continente. Áreas e culturas regadas em 2013 [in Portuguese]

Higueras, P., Campos, J. A., Esbrí, J. M., García-Noguero, E. M., & Elmayel, I. (2019). Biogeochemical mapping: A new tool to assess the soil quality and health. In D. M. Doronzo, B. Zoheir, E. Schingaro, & J. S. Armstrong-Altrin (Eds.), *Petrogenesis and exploration of the earth's interior* (pp. 3–5). Springer Nature Switzerland.

IGME (Instituto Geológico y Minero de España), 2015. Mapa Geológico de la Península Ibérica, Baleares y Canarias a escala 1:1.000.000.

Intergovernmental Technical Panel on Soils. (2021). *Towards a definition of soil health*. Food and Agriculture Organization of the United Nations. https://www.fao.org/3/cb1110en/cb1110en.pdf

IUSS Working Group WRB. (2015). *World Reference Base for Soil Resources 2014, update 2015. International soil classification system for naming soils and creating legends for soil maps*. Rome.

IUSS Working Group WRB. (2022). *World Reference Base for Soil Resources. International soil classification system for naming soils and creating legends for soil maps* (4th ed.). International Union of Soil Sciences (IUSS). https://wrb.isric.org/files/WRB_fourth_edition_2022-12-18.pdf

Janzen, H. H., Janzen, D. W., & Gregorich, E. G. (2021). The 'soil health' metaphor: Illuminating or illusory? *Soil Biology and Biochemistry*, *159*, Article 108167. https://doi.org/10.1016/j.soilbio.2021.108167

Karlen, D. L., De, M., McDaniel, M. D., & Stott, D. E. (2021a). Evolution of the Soil Health Movement. In D. L. Karlen, D. E. Stott, & M. M. Mikha (Eds.), *Soil Health*

Series: Approaches to Soil Health Analysis (1st ed., Vol. 1, pp. 21–48). Soil Science Society of America, Inc. Published by John Wiley & Sons, Inc.

Karlen, D. L., Ditzler, C. A., & Andrews, S. S. (2003). Soil quality: Why and how? *Geoderma*, *114*(3–4), 145–156. https://doi.org/10.1016/S0016-7061(03)00039-9

Karlen, D. L., Stott, D. E., Mikha, M. M., & Moebius-Clune, B. (2021b). Soil health: An overview and goals of these volumes. In D. L. Karlen, D. E. Stott, & M. M. Mikha (Eds.), *Soil health series: Approaches to soil health analysis* (Vol. 1, pp. 1–20). Soil Science Society of America, Inc. & Wiley & Sons, Inc.

López-Merino, L., Serrano, O., Adame, M. F., Mateo, M. Á., & Martínez Cortizas, A. (2015). Glomalin accumulated in seagrass sediments reveals past alterations in soil quality due to land-use change. *Global and Planetary Change*, *133*, 87–95. https://doi.org/10.1016/j.gloplacha.2015.08.004

Maestro, A., López-Martínez, J., Llave, E., Bohoyo, F., Acosta, J., Hernández-Molina, F. J., Muñoz, A., & Jané, G. (2013). Geomorphology of the Iberian Continental Margin. *Geomorphology*, *196*, 13–35. https://doi.org/10.1016/j.geomorph.2012.08.022

Martín-Serrano, A., Maestro, A., Nozal, F., Salazar, A., & Suarez, A. (2005). *Mapa Geomorfológico de España y del Margen Continental, Escala 1:1,000,000*. Instituto Geológico y Minero de España.

Ministerio de Agricultura, Alimentación y Medio Ambiente. Servicio General de Estadística. 2022. Estadística anual de superficies y producciones de cultivos 2021. [in Spanish]

Mizuta, K., Grunwald, S., & Cropper, W. P. (2021). Developmental history of soil concepts from a scientific perspective. *Applied Sciences*, *11*, Article 4275

Novoa, A., Rodríguez, R., Richardson, D., & González, L. (2014). Soil quality: A key factor in understanding plant invasion? The case of *Carpobrotus edulis* (L.) N.E.Br. *Biological Invasions*, *16*(2), 429–443. https://doi.org/10.1007/s10530-013-0531-y

Panagos, P., Montanarella, L., Barbero, M., Schneegans, A., & Aguglia, L. (2022). Soil priorities in the European Union. *Geoderma Regional*, *29*, Article e00510. https://doi.org/10.1016/j.geodrs.2022.e00510

Prăvălie, R., Borrelli, P., Panagos, P., Ballabio, C., Lugato, E., Chappell, A., Miguez-Macho, G., Maggi, F., Peng, J., Niculiță, M., Roșca, B., Patriche, C., Dumitrașcu, M., Bandoc, G., Nita, I.-A., & Birsan, M.-V. (2024). A unifying modelling of multiple land degradation pathways in Europe. *Nature Communications*, *15*, Article 3862. https://doi.org/10.1038/s41467-024-48252-x

Rodrigo-Comino, J., Keshavarzi, A., & Senciales-González, J. M. (2023). Evaluating soil quality status of fluvisols at the regional scale: A multidisciplinary approach crossing multiple variables. *River Research and Applications*, *39*(7), 1367–1381. https://doi.org/10.1002/rra.3865

Rutigliano, F. A., Marzaioli, R., Grilli, E., Coppola, E., & Castaldi, S. (2023). Microbial, physical and chemical indicators together reveal soil health changes related to land cover types in the southern European sites under desertification risk. *Pedobiologia*, *99–100*, Article 150894. https://doi.org/10.1016/j.pedobi.2023.150894

Sánchez-Moreno, S., Cano, M., López-Pérez, A., & Rey Benayas, J. M. (2018). Microfaunal soil food webs in Mediterranean semi-arid agroecosystems. Does organic management improve soil health? *Applied Soil Ecology*, *125*, 138–147. https://doi.org/10.1016/j.apsoil.2017.12.020

Soil Atlas of Europe, European Soil Bureau Network. European Commission. (2005). Office for Official Publications of the European Communities, L-2995 Luxembourg. 128 pp.

UN Department of Economic and Social Affairs. (2024). Population Dynamics. World Urbanization Prospects 2018. Country Profiles. Available at https://population.un.org/wup/Country-Profiles/ (accessed Nov 16, 2024).

European Commission: Directorate-General for Research and Innovation:Veerman, C., Pinto Correia, T., & Bastioli, C. (2020). *Caring for soil is caring for life—Ensure 75% of soils are healthy by 2030 for healthy food, people, nature and climate—Interim report of the mission board for soil health and food*. Publications Office of the European Union. https://data.europa.eu/doi/10.2777/918775

Vera, J. A., Ancoceha, E., Barnolas, A., Bea, F., Calvo, J. P., et al. (2004). *Introducción*. In J. A. Vera (Ed.), *Geología de España* (pp. 3–17). Instituto Geológico y Minero de España (IGME).

Virto, I., Barré, P., Burlot, A., & Chenu, C. (2012). Carbon input differences as the main factor explaining the variability in soil organic C storage in no-tilled compared to inversion tilled agrosystems. *Biogeochemistry*, *108*(1–3), 17–26. https://doi.org/10.1007/s10533-011-9600-4

Virto, I., Imaz, M., Fernández-Ugalde, O., Gartzia-Bengoetxea, N., Enrique, A., & Bescansa, P. (2015). Soil degradation and soil quality in Western Europe: Current situation and future perspectives. *Sustainability*, *7*(1), 313–365. https://doi.org/10.3390/su7010313

Visconti, F., Peiró, E., Pesce, S., Balugani, E., Baixauli, C., & de Paz, J. M. (2024). Straw mulching increases soil health in the inter-row of citrus orchards from Mediterranean flat lands. *European Journal of Agronomy*, *155*, Article 127115. https://doi.org/10.1016/j.eja.2024.127115

Zegada-Lizarazu, W., Carvalho, J. L. N., Parenti, A., Tenelli, S., Martín Sastre, C., Ciria, P., Christou, M., Efthymia, A., Bonomi, A., & Monti, A. (2022). The effects of integrated food and bioenergy cropping systems on crop yields, soil health, and biomass quality: The EU and Brazilian experience. *GCB Bioenergy*, *14*(5), 522–538. https://doi.org/10.1111/gcbb.12924

Zuazo, V. H. D., Rodríguez, B. C., García-Tejero, I. F., Ruiz, B. G., & Tavira, S. C. (2020). Benefits of organic olive rainfed systems to control soil erosion and runoff and improve soil health restoration. *Agronomy for Sustainable Development*, *40*(6), Article 41. https://doi.org/10.1007/s13593-020-00644-1

2

Soil Health Evaluation for Permanent Crops Under Semiarid Mediterranean Conditions: Principles, Processes, Practices, and Policy Options

María Almagro, Noelia Garcia-Franco, and María Martínez-Mena

Chapter Overview

This chapter centers on the potential of different sustainable soil management practices for soil organic carbon sequestration and stabilization as indicators of soil health and functioning in organic rainfed almond orchards. These orchards are representative of low-input cropping systems in semiarid Mediterranean regions. The state-of-the-art of the current knowledge on soil organic carbon sequestration mechanisms and soil health indicators is presented, and the research questions still to be answered are identified. Results from this review highlight that reducing tillage, alone or in combination with other practices such as green manuring, offers a promising sustainable soil management practice that may contribute to climate change mitigation and adaptation through enhancing soil health and functions without compromising crop yields. However, yield penalties observed under no-tillage systems suggest that this practice is less beneficial in semiarid calcareous soils even though soil health and functioning indicators were improved. Altogether, our findings highlight that the magnitude and direction of the agro-environmental impacts derived from soil management will depend on prevailing local conditions (soil and crop type, degradation level, microclimate, and soil management history) and that there are no universal solutions for these semiarid regions. Therefore, integrated cost-benefit assessments are needed to guide and identify the best soil management practices according to site-specific environmental and socio-economic conditions.

Soil Health Series: Volume 5 Soil Health and Sustainability in Spain and Portugal, First Edition.
Edited by Iñigo Virto and Rodrigo Antón.

Published 2025 by Soil Science Society of America, Inc. and John Wiley & Sons, Inc.

Soil Relevance and Global Threats

Soils, a non-renewable natural resource and the foundation of terrestrial ecosystems, support a quarter of the world's biodiversity and contribute to many ecosystem functions (Bardgett and van der Putten, 2014; Cameron et al., 2019). In recent decades, the perception of the soil has changed from being considered an "inert" component of the ecosystem to a "living" component crucial for ecosystem functioning (GSBI 2020; Hermoso et al., 2022). Besides providing a broad range of ecosystem services, soils contribute to climate change mitigation and adaptation goals (IPCC, 2019; Pörtner et al., 2022). Examples of soil ecosystem services include (a) the production of food, wood, and pharmaceuticals (Kirjavainen et al., 2019; D. G. Smith et al., 2019; Tamburini et al., 2020; Zhu et al., 2019); (b) climate and water regulation (Martínez-Mena et al., 2021a); (c) erosion and flood prevention (Boix-Fayos et al., 2020; Eekhout & de Vente, 2019); and (d) that they support relevant terrestrial ecosystems that are part of a region's landscape and cultural heritage (Hanaček & Rodríguez-Labajos, 2018), as for example the *dehesa* (in Spain) or *montado* (in Portugal), a silvopastoral system typical of the southern Iberian Peninsula with a tree cover that is predominantly *Quercus suber* with some *Quercus rotundifolia*, where cattle or sheep graze the understory. Soil ecosystems also have several crucial support functions for other services such as soil structure, nutrient cycling, and primary production (Almagro et al., 2023a; Almagro et al., 2021; Martínez-Mena et al., 2021b). Importantly, all these services are interrelated and mutually dependent, such that impacts on one service will have repercussions for the others (Reid et al., 2005; Food and Agriculture Organization & Intergovernmental Technical Panel on Soils [FAO & ITPS], 2021).

According to the European Union Thematic Soil Strategy, soils around the world are subject to several anthropogenic threats, such as erosion, sealing, pollution, organic matter and biodiversity decline, compaction, and degradation, among others (Stolte et al., 2016; Veerman et al., 2020; European Commission [EC], 2006). Twenty-five percent of the European land surface shows erosion rates well above the threshold rate recommended as sustainable (2 t ha^{-1} yr^{-1}), and 6% exhibits severe erosion (11 t ha^{-1} yr^{-1}; Panagos et al., 2020). Intensive agriculture (particularly machinery weight) and urban development are responsible for the reduction of soil porosity, and according to the European Environment Agency, approximately 500 km^2 of soil is sealed every year (van der Putten et al., 2018).

Soil erosion and the associated soil organic carbon (SOC) losses are considered major soil threats in the European Union's Thematic Strategy for Soil Protection (EC, 2006). In Europe, average soil erosion rates of 2.45 t ha^{-1} yr^{-1}, paralleled by reductions of cropland SOC stocks by 28%, have been reported for the past 30 years (Panagos et al., 2020). As a result, crop yields have also decreased (Bakker et al., 2007; Oldfield et al., 2019). Furthermore, according to recent scientific reports, climate change will exacerbate current crop yield loss rates if no adaptation strategies are implemented. This risk is particularly prominent in Mediterranean countries, which are highly vulnerable to soil degradation and erosion. For example, Spain has higher soil erosion rates (4 t ha^{-1} yr^{-1}) than the European average (Feyen et al., 2020). Moreover, and given the climate change projections for some areas, erosion rates are expected to increase because high-intensity extreme rainfall events may become more frequent, especially in the Mediterranean Basin. Likewise, soil compaction and sealing from intensive agriculture and urbanization are responsible for reducing soil porosity and thus the soil's capacity for water storage and infiltration (Shah et al., 2017). In addition, microplastics and nanoplastics have been currently recognized as a new global threat for soils since they affect crucial soil properties such as structure, bulk density, and water holding capacity, with consequences for process rates and net primary production (Rillig & Lehmann, 2020). There are multiple sources and input pathways into soils, such as aerial deposition (Brahney et al., 2020), discarded plastic litter which is further fragmented (Rillig, 2012), agricultural products such as plastic mulching, compost or sewage sludge addition (Bläsing & Amelung, 2018; Weithmann et al., 2018), and tire abrasion in roads (Baensch-Baltruschat et al., 2020). All these microplastics are transported vertically and horizontally into soil not only by living organisms such as earthworms and collembola (Maaß et al., 2017; Rillig et al., 2017b) but also by other means of transportation such as plowing and root channels (Rillig et al., 2017a).

All these threats and the associated negative environmental impacts are due to intensified land use and inadequate land management (Foley et al., 2005; Montanarella & Panagos, 2021). These issues cause the loss of many important soil functions such as primary productivity provision, biodiversity maintenance, uptake of atmospheric CO_2 emissions, and regulation of nutrient and water cycling, ultimately leading to a degenerative process with regional and global negative repercussions (Costantini et al., 2020; Ruiz et al., 2020; FAO & ITPS, 2021; EC, 2006). Some examples of unsustainable agricultural practices causing negative environmental impacts at local, regional, and global scales include the following:

- The loss of fertile topsoils rich in organic matter through erosion and degradation, which is compensated for by applying more fertilizers, thus aggravating the problems of eutrophication of aquatic ecosystems and global warming (Lal, 1998);

- The predominance of woody monocrops without vegetation covers in the inter-tree rows, which makes these agroecosystems highly vulnerable to erosion (Martinez-Mena et al., 2008, 2020) and pathogens and pests such as *Xylella fastidiosa* Wells et al. and *Capnodis tenebrionis* L. (Loureiro et al., 2023);
- The pollution from excessive use of fertilizers, pesticides, and plastics, which pose a threat to soil organisms (Rillig et al., 2019; Rillig & Lehmann, 2020);
- Floods and damage to infrastructure (Cantón et al., 2011).

As a consequence of the above-mentioned problems, around 30% of Earth's land surface is in a state of degradation because of agriculture, affecting 25% of the global population and resulting in annual costs of 300 billion USD (Mäder et al., 2002; Lal, 2010; Wagg et al., 2014; MEA, 2005; Banwart, 2011; Cherlet et al., 2018). Further, climate change is expected to exacerbate the negative impacts. In this regard, sustainable soil management practices are aimed at combating the negative environmental impacts of agriculture, improving soil functions, and thus providing the greatest number of ecosystem services (Almagro et al., 2016; Sanz et al., 2017; FAO & ITPS, 2021; Montanarella & Panagos, 2021). Specifically, the practices are intended to improve soil structure and carbon and nutrient storage, reduce erosion, and improve water availability and use efficiency, without compromising agricultural productivity (Lal, 2010; Mäder et al., 2002; Wagg et al., 2014).

Soil Health, Sustainable Soil Management, and Ecological Intensification to Combat Global Threats: Desertification, Climate Change, and Biodiversity Loss

Soil health is defined as "the capacity of soil to function as a vital living system, within ecosystem and land-use boundaries, to sustain plant and animal productivity, maintain or enhance water and air quality, and promote plant and animal health" (Doran & Zeiss, 2000). According to the U.S.A. Natural Resources Conservation Service (2019), soil health is based on four pillars: (a) minimizing soil disturbance through reduced and smart tillage; (b) maximizing soil cover through crop residue retention and mulching, green manuring, and cover cropping; (c) maximizing biodiversity through crop rotation and intercropping, vegetation buffer strips to attract pollinators, and agroforestry and livestock integration; and (d) maximizing continuous living roots through perennial crops and cover cropping.

According to the voluntary guidelines of sustainable soil management by the FAO (VGSSM, 2015), sustainable soil management refers to a set of activities that maintain or enhance the support, provision, regulation, and cultural services of soil, without significantly impairing the soil functions that enable those services or the biodiversity. Among other goals, the voluntary guidelines for sustainable soil management seek to refine the criteria for determining whether a soil is functioning effectively and being managed sustainably. An indicative and non-exhaustive list of such criteria include the following:

- Loss of nutrients (e.g., by leaching, gaseous emissions, erosion) is low.
- Biological production (e.g., crops, trees, fodder) is high relative to the potential limits set by climate and additional water sources (e.g., irrigation), where appropriate.
- Levels of biodiversity within and above the soil are relatively high.
- SOC stores are maintained or increased.
- Rainfall and supplementary water are efficiently captured and held within the root zone.
- Rates of soil erosion are similar to sedimentation rates.
- Contaminants should not be introduced into the landscape at rates that exceed the filtering capacity of the soil, and existing contaminants are remediated if they are at levels that cause harm.

Both "soil health" and "sustainable soil management" concepts are closely related to the "ecological intensification" concept, which refers to a set of management practices aimed at improving agricultural production by increasing the functional biodiversity of the agroecosystem, thereby reducing external inputs to agriculture (Bommarco et al., 2013; García-Palacios et al., 2019). Each living soil organism performs a variety of essential functions for ecosystems (Bardgett & van der Putten, 2014; Orgiazzi et al., 2016; Erktan et al., 2020). The macroinvertebrates, such as earthworms and ants, are well-known ecosystem "engineers" that improve soil structure and its water infiltration and storage capacity, while controlling surface runoff and erosion (Bouché & Al-Addan, 1997; Bertrand et al., 2015; Orgiazzi et al., 2016). The invertebrates transform plant residues into organic matter and nutrients, making them more accessible to other microorganisms and plants and thus improving primary production while contributing to climate regulation (Naeem et al., 1994; Nielsen, 2019; Wagg et al., 2014). The micro-predators regulate microbial populations, while offering crop protection against pathogens (e.g., by avoiding pests; Altieri & Rogé, 2009; Tamburini et al., 2020). Finally, the microorganisms transform and decompose organic matter, as well as wastes and pollutants; regulate climate, water, and nutrients; offer protection against pathogens; and are a potential source of medicines (Six et al., 2004; Briones et al., 2014; Kirjavainen et al., 2019; Smith et al., 2019; Zhu et al., 2019).

Currently, the scientific community has reached a consensus on the use of several ecosystem response variables, such as soil respiration, organic matter decomposition, and aggregate formation, for assessing soil functional biodiversity and health (Guerra et al., 2021). All these variables are closely related to SOC content, which is considered an excellent indicator of soil functionality and health. Maintaining and increasing SOC content improves essential aspects such as soil structure and aggregation, biotic activity, and water and nutrient retention capacity and availability for plants, thus contributing to multiple ecosystem services (FAO and ITPS, 2021; Lal, 2010; Mäder et al., 2002; Wagg et al., 2014).

In short, to mitigate global problems such as desertification and climate change, sustainable soil management and ecological intensification are aimed at optimizing the carbon and nutrient balances of agroecosystems by increasing the inputs into the system (e.g., through crop residue retention, compost addition, or intercropping), while minimizing greenhouse gas (GHG) emissions and possible losses that may occur in the agroecosystem through erosion, leaching, photodegradation, and other processes (Figure 2.1; Montgomery, 2007; Almagro and Martínez-Mena, 2014; Almagro et al., 2015; Sanz-Cobeña et al., 2017; Costantini et al., 2020; Lassaletta et al., 2021).

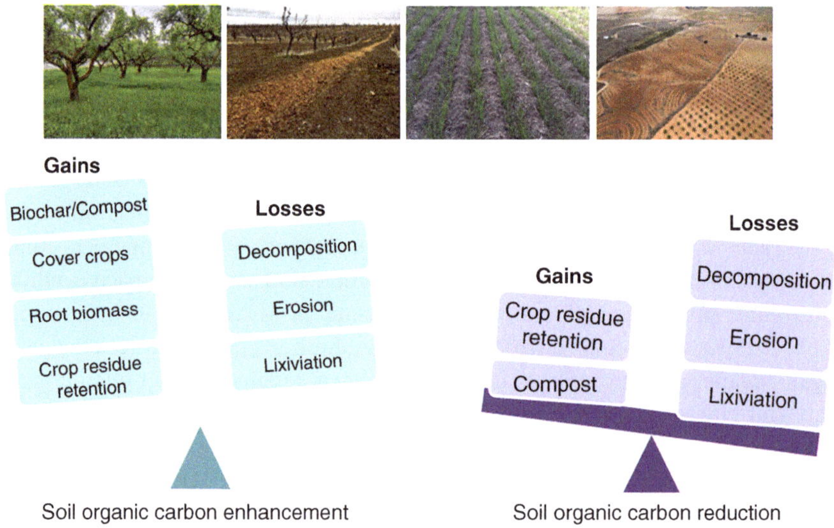

Figure 2.1 Creating Positive Soil Carbon, Nutrient, and Water Balances.The annual carbon gains and losses can be estimated at the agroecosystem level to evaluate whether the net SOC balances are being improved or worsened by the adoption of a particular soil management practice. For example, if the annual soil carbon inputs (derived from addition of organic amendments, cover cropping, and/or crop residue retention) exceed the losses caused by processes such as soil water erosion, leaching, and/or decomposition, the SOC stock would increase. If the opposite occurs, SOC stocks will be reduced with time. Photo by María Almagro Bonmati.

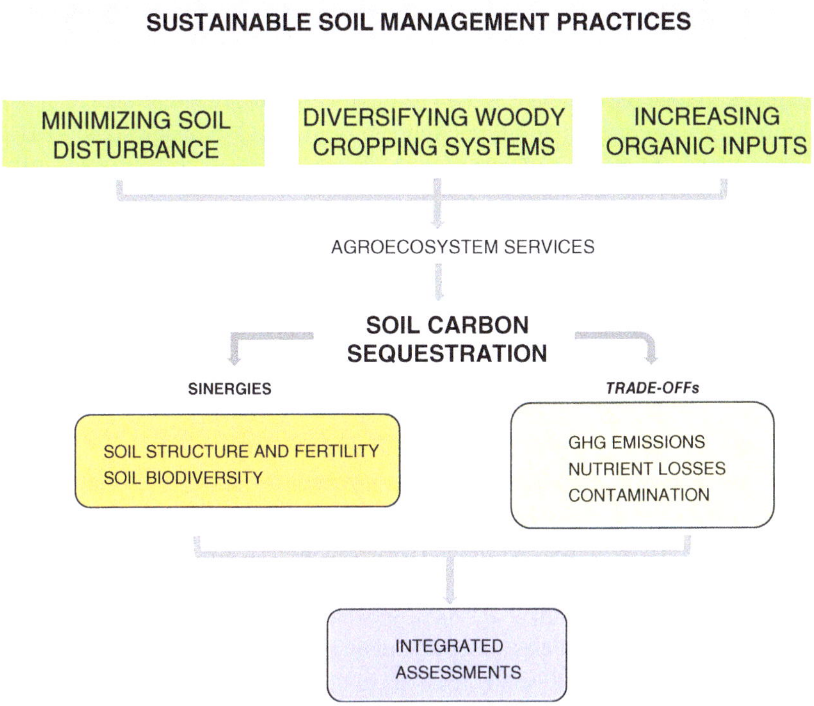

Figure 2.2 Synergies and Trade-offs Resulting from the Adoption of Sustainable Soil Management Practices for Increasing Carbon Sequestration in Agroecosystems.

Despite the well-known benefits of improving the organic carbon (OC) content in agricultural soils, the adoption of sustainable soil management practices, such as conservation agriculture or crop diversification, has been controversial among farmers and advisory services because the practices may compromise agroecosystem productivity to some extent, particularly in semiarid areas (Morugán-Coronado et al., 2020; Schoonhoven & Runhaar, 2018; Zabala et al., 2023). However, a recent meta-analysis showed that increasing biodiversity in agricultural systems through polycultures, diversifying habitats, minimum tillage, or inoculating beneficial microorganisms into the soil, among other examples, promotes the provision of multiple ecosystem services such as carbon sequestration, nutrient cycling, pollination, and pest control, without compromising crop yields (Tamburini et al., 2020). In short, these emerging sustainable soil management practices (Figure 2.2) may accomplish the following:

- Minimize soil structure disturbance by adopting no-tillage, minimum tillage, or strip tillage (Costantini et al., 2020; FAO & ITPS, 2021; Francaviglia et al., 2023)

- Increase agroecosystem biodiversity through polycultures, crop rotations, cover cropping, buffer strips, ponds, agroforestry systems, and livestock integration (Power, 2010; Stavi et al., 2016; Tamburini et al., 2020; Almagro et al., 2023a);
- Increase organic inputs to the system through crop residue retention and green manure and/or compost addition (Aguilera et al., 2013; Almagro et al., 2021; Garcia-Franco et al., 2021; Guenet et al., 2021; Liu et al., 2022; Power, 2010; Sanz-Cobeña et al., 2017).

Through a combination of these sustainable soil management practices, ecosystem services (e.g., climate, water, and nutrient regulation; erosion control; habitat provision for biodiversity; soil regeneration) should be enhanced, mainly through the promotion of carbon sequestration and its synergies, such as improved soil structure, fertility, and biodiversity (Almagro et al., 2021; Guenet et al., 2021; Power, 2010). However, despite all efforts since 2014 having been focused on promoting soil management practices that increase carbon sequestration, this land management solution is not free of controversy because it has limitations and trade-offs (Guenet et al., 2021; Liu et al., 2022; Power, 2010). First, soil carbon sequestration is not infinite given that soils become saturated after a period depending on the pedoclimatic conditions, land use legacy, and management history (Schlesinger, 2000; J. O. Smith et al., 2005; Stewart et al., 2007). In addition, boosting agro-environmental policies aimed at enhancing soil carbon sequestration can have adverse effects on biodiversity itself, as in the case of the huge expansion of oil-palm plantations or the so-called "green deserts" created by eucalyptus plantations or other nonnative species motivated by global carbon markets (Bradshaw et al., 2013). Moreover, the trade-offs have to be assessed when a soil management practice is being adopted, with attention not only to the improvement in the SOC stocks, but also to the GHG emission balances (CO_2, N_2O, and CH_4) after the application of certain fertilizers, amendments, and slurries and to the possible soil contamination and water eutrophication problems caused by such practices (Sanz-Cobeña et al., 2017; Sandén et al., 2018; Liu et al., 2022). Likewise, attention must be paid to local conditions before adoption of a certain soil management practice to avoid yield penalties, which will be an important barrier among farmers. For example, although higher fungi: bacteria ratios mediated by no-tillage are associated with higher rates of soil carbon sequestration (Briones et al., 2014) and biogenic aggregation that protects soil organic matter, which is a positive environmental impact, tillage cessation can sometimes have detrimental effects on crop yields, as in the case of permanent crops in semiarid environments under rainfed conditions (Martínez-Mena et al., 2013, 2021b). Therefore, integrated assessments that include several agro-environmental indicators while also considering the interactions and trade-offs are crucial to support decision-making in soil management based on the desired objectives (Derak et al., 2017).

Soil Health and Sustainable Soil Management: Concept Evolution and Research Progress

Many examples of soil conservation measures practiced by very ancient civilizations (e.g., terraces) exist around the world, with some dating back to the beginning of agriculture (10,000 BCE). The first agronomic trials to study the long-term effects of soil management on soil properties and crop productivity were established in 1843 by John Lawes at the Rothamsted research farm in the United Kingdom. However, the phenomenon known as the Dust Bowl in the United States (1932–1939), one of the worst ecological disasters of the 20th century, was decisive in raising national and international awareness of the problems of soil degradation and over-exploitation of natural resources and the negative consequences for the environment and society. During this period, the concept of "conservation agriculture" emerged when Edward Faulkner questioned the utility of plowing in his book "Plowman's Folly." The idea gained popularity during the 1960s in the US Midwest as an approach for preventing soil degradation, and in parallel, studies on erosion impacts started to be conducted (Schwab et al., 1982; Wischmeier & Smith, 1962). According to the FAO, conservation agriculture is an agricultural management practice that includes continuous minimal soil disturbance (no-tillage) and permanent organic soil cover (mulching) combined with diversification of crops grown in sequence/association, preferably including at least one legume.

The relationship between biodiversity and ecosystem productivity has now been studied for decades. The impacts of soil management on soil biodiversity and its functions began to be studied by Hendrix et al. (1986), who developed a conceptual model to describe how soil management modulates the soil biotic structure, food change, and functions. Their model was based on a comparison of a conventional intensive tillage scenario under which crop residues were buried into soil and a no-tillage scenario under which crop residues were left on the soil surface as mulching (Figure 2.3). Next, Blum (1990) identified the concept of soil functionality as an attribute beyond plant biomass production. These changes in the perception of the soil system and how to study it occurred in the historical context of the Rio de Janeiro Earth Summit (1992) and the United Nations Convention to Combat Desertification (UNCCD), which promoted the sustainable use of natural resources. In parallel, the European Common Agricultural Policy (CAP) began to support actions oriented to soil and water conservation in agricultural areas. In a study based on building plant and invertebrate communities with decreasing levels of diversity, Naeem et al. (1994) were the first researchers to demonstrate experimentally that more diverse biological communities are more productive in terms of biomass and better able to resist environmental changes.

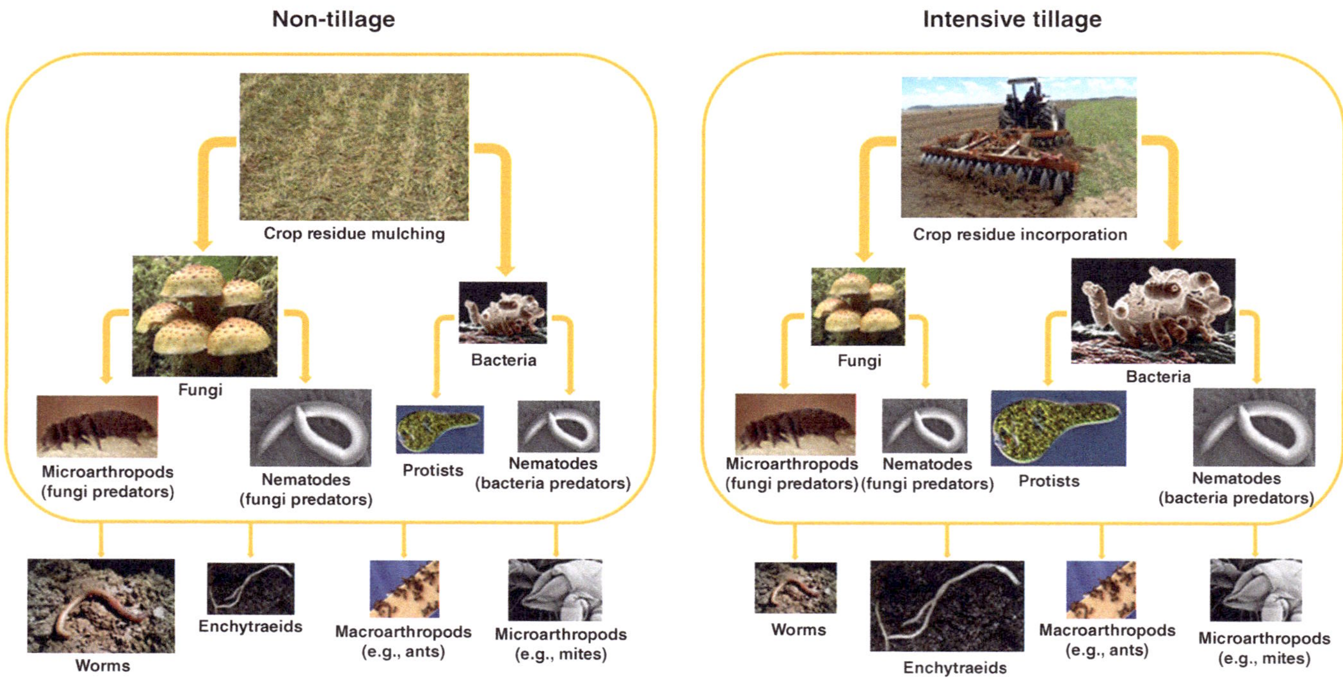

Non-tillage
Crop residue mulching
Fungi
Bacteria
Microarthropods (fungi predators)
Nematodes (fungi predators)
Protists
Nematodes (bacteria predators)
Worms
Enchytraeids
Macroarthropods (e.g., ants)
Microarthropods (e.g., mites)
Intensive tillage
Crop residue incorporation
Fungi
Bacteria
Microarthropods (fungi predators)
Nematodes (fungi predators)
Protists
Nematodes (bacteria predators)
Worms
Enchytraeids
Macroarthropods (e.g., ants)
Microarthropods (e.g., mites)

Figure 2.3 Impact of Agricultural Intensification on Soil Functional Biodiversity and the Trophic Chain Diagram adapted from (Hendrix et al., 1986) showing two simplified models of how soil management determines the importance and abundance of each taxonomic group and its function within the soil trophic chain. Intensive tillage physically disrupts the soil structure, destroying many arthropods and insect nests, while the weight of machinery causes soil compaction and reduces its porosity. Moreover, under intensive tillage systems the vertical distribution of plant residues in the topsoil makes bacterial activity predominate over fungal activity because bacteria decompose buried plant residues in contact with the soil better. As a result, mineralization rates are relatively faster and the abundance of protozoa that feed on bacteria is favored, which in turn determines changes in the food chain. In contrast, fungi are more abundant when crop residues are left as mulching, particularly in semiarid systems, as they are able to translocate nutrients from the soil (which causes nutrients to be immobilized in the crop residue itself) and better resist drought. As a consequence, fungal feeding communities predominate, and organic matter and nutrients mineralization rates are also relatively slower. In addition, higher fungi: bacteria ratios are associated with higher rates of soil carbon sequestration (Briones et al., 2014), and biogenic aggregation protects soil organic matter. Similarly, nematodes feed on bacteria and others on fungi, so their presence and abundance will be determined by the fungi: bacteria ratios. Generally, there are more microarthropods, macroarthropods, and annelids in no-tillage than in conventional tillage systems. Microarthropods can act as predators or prey while controlling detritus decomposition directly through physical fragmentation and indirectly when acting as predators and thus produce changes in the structure of the microbial community. Macroarthropods such as crickets, beetles, and earthworms fragment and distribute plant residues, defecate nutrient-rich feces, and provide a favorable habitat for microbial activity, which helps accelerate nutrient turnover rates and stimulate crop production. The macroarthropods also feed on fungi, bacteria, protozoa, and nematodes, which influences the distribution and abundance of these organisms and their potential to mineralize nutrients.

The role of soil as a sink for atmospheric CO_2 was soon recognized (Schlesinger, 2000), and initiatives such as the Millennium Ecosystem Assessment (MEA, 2005) and the European Soil Protection Strategy were launched. These developments contributed to sustainable soil management being considered from a more holistic approach, including biophysical, socio-economic, and cultural aspects. Next, Austin and Vivanco (2006) and Lee et al. (2012) demonstrated the importance of photodegradation in semiarid ecosystems. Through this abiotic mechanism, solar UV radiation oxidizes the carbon and nitrogen in organic matter, causing direct CO_2 and N_2O emissions to the atmosphere and thereby negatively affecting carbon sequestration and fertility of semiarid soils. This mechanism could have serious implications for cover crop management in semiarid agroecosystems under climate change scenarios, as demonstrated by Almagro et al. (2015) in a manipulative climate change and solar UV radiation exposure experiment. These authors proved that plant residue decomposition rates by photodegradation were exacerbated by increasing temperatures, which ultimately hampered carbon and nutrient storage in semiarid soils under warming conditions. Results from this study raised the question of whether under drier and warmer conditions, it would be better to incorporate cover crop residues into the soil by shallow tillage, instead of cutting them back with a brush cutter and leaving them as mulching on the soil surface, to avoid carbon and nutrient losses to the atmosphere by photodegradation. Concurrent studies under semiarid conditions also highlighted the important role of plant litter quality and the interaction with soil erosion to explain organic matter decomposition dynamics and subsequent carbon and nitrogen storage along a land use intensification gradient under semiarid Mediterranean conditions (Almagro & Martínez-Mena, 2012, 2014).

Since 2000, the concepts of soil health and sustainable soil management emerged and the World Overview of Conservation Approaches and Technologies (WOCAT) initiative was consolidated as a global reference database of sustainable soil management practices (Liniger et al., 2007). Likewise, international initiatives such as 4 per mil, the Sustainable Development Goals, the Paris Agreement, and the Global Soil Alliance were created and now facilitate the transference of knowledge on the benefits of sustainable soil management between science and policy.

In the early 2020s, soil started being considered as a cross-cutting issue in international and European agro-environmental policies (Figure 2.4), and international organizations such as the UNCCD and the FAO recognized the importance of adapting soil management to local conditions as well as of standardizing SOC monitoring methodologies (Sanz et al., 2017; FAO, 2020; FAO & ITPS, 2021). Current new initiatives such as the European Green Deal (European Commission, 2019), the EU Mission: A Soil Deal for Europe (Figure 2.5), the "From Farm to Fork" strategy (European Commission, 2020a), the EU Biodiversity Strategy

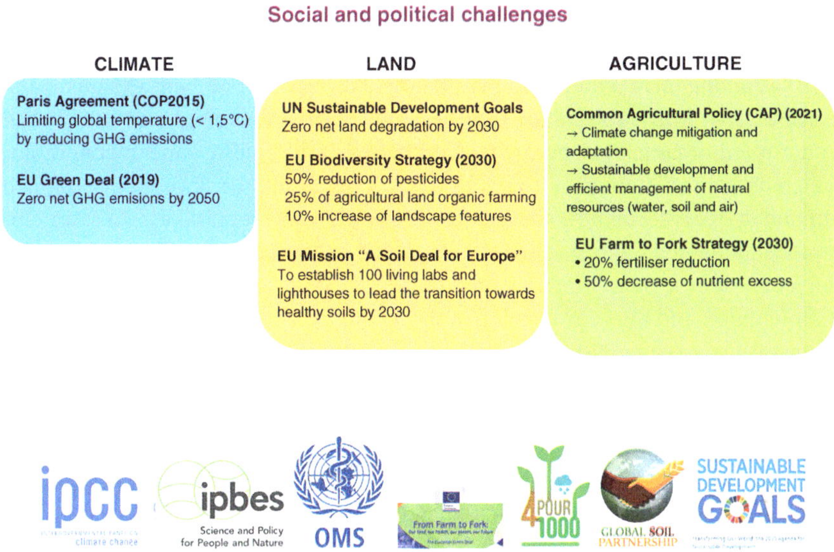

Figure 2.4 Climate and Agro-environmental Policy Framework to Cope with Emerging Social and Political Challenges.

(European Commission, 2020b), and the EJP Soil Joint Programme clearly highlight that the soil is humanity's best ally in the fight against climate change and biodiversity loss. However, and despite a proposal for a European Directive on Soil Monitoring and Resilience (European Commission, 2023) being currently under consultation, policy instruments to explicitly protect soil biodiversity and its functions are still lacking.

1. Reduce desertification
2. Conserve soil organic carbon stocks
3. Stop soil sealing and increase re-use of urban soils
4. Reduce soil pollution and enhance restoration
5. Prevent erosion
6. Improve soil structure to enhance soil biodiversity
7. Reduce the EU global footprint on soils
8. Improve soil literacy in society

Figure 2.5 Soil Health Objectives According to the EU Mission "A Soil Deal for Europe" (Veerman, 2023).

Lessons Learned from a Case Study

In 2008, a long-term agriculture research field called "Burete" was established in northwestern Murcia Province (southeastern Spain). This field was established in a rainfed organic almond orchard under semiarid Mediterranean conditions, which is representative of the traditional, low-input, less-productive rainfed almond systems located in southeastern Spain. These cropping systems are currently endangered owing to the lack of specific policy support, despite the important ecosystem services that they provide (e.g., food provisioning, local population maintenance, and the guardianship of the territory). The annual precipitation and temperature average 370 mm and 16 °C, respectively, and the mean potential evapotranspiration reaches 1,200 mm yr^{-1}, leading to the mean annual water deficit being approximately 900 mm. The soils in the study area are relatively shallow with moderate southeast-facing slopes (~7%), are classified as Petric Calcisols (IUSS Working Group WRB, 2015), and have a silt-loam texture and a high level of $CaCO_3$ (~45%).

The traditional soil management in the study site consists of three to five moldboard passes per year after important rainfall events to control weeds, and this management was used as our reference (control) treatment. In October 2008, four different soil management practices were implemented: (a) reduced tillage (RT), (b) reduced tillage plus green manure (RTG), (c) reduced tillage plus organic manure (RTOM), and (d) no-tillage (NT). The reduced tillage consisted of chisel plowing to 15-cm depth using a cultivator twice a year (in fall and spring) to control weeds. The green manure consisted of seeding a mixture of common vetch (*Vicia sativa* L.) and common oat (*Avena sativa* L.) at 150 kg ha^{-1} in early fall to provide a cover crop during winter. The organic manure consisted of goat manure (16% OC; 2.3% N; C : N = 6.95), added at 400 and 1,500 kg/ha during 2009–2011 and 2012–2016, respectively, in early fall. In the NT treatment, tillage was suppressed from the beginning of the experiment and no fertilizers or herbicides were applied from that time onward. Owing to the changes in soil management, different plant covers were established for each tillage treatment from the beginning of the experiment: (a) a mix of spontaneous herbaceous plants [*Hordeum murinum* L., *Lolium perenne* L., and *Eruca vesicaria* (L.) Cav.] in the RT and RTOM treatments; (b) a plant cover dominated by the shrub *Teucrium capitatum* L. in the NT treatment; and (c) a mix of *V. sativa* and *A. sativa* in the RTG treatment (Figure 2.6). In all treatments, the plant cover was allowed to grow from mid-fall to early spring, when mowing was done to avoid competition for water with the main crop given the Mediterranean climate with warm, dry summers.

Monitoring a series of variables in the field that serve as indicators of improved function and health of the agroecosystem enabled demonstrating the agro-environmental and economic impacts of agricultural practices such as reduced tillage in rainfed organic almond orchards under semiarid conditions

Figure 2.6 Aspect of different soil management practices and vegetation covers in late spring in organic rainfed almond orchards under semiarid Mediterranean conditions. Partial views of the different soil management practices showing the aspect of the plant cover in the rows at the end of the growing season (late spring): (a) traditional tillage (TT), (b) non-tillage (NT), (c) reduced tillage (RT), and (d) reduced tillage plus green manure (RTG). In the TT treatment, the lack of plant cover in the rows is noticeable year-round because these almond fields are generally plowed between three and five times per year, following important rainfall events. However, for both reduced tillage treatments (RT and RTG) and the NT treatment, significant plant cover is noticeable at the end of the growing season since these fields are plowed only twice a year or are not tilled at all, respectively. Photos by María Almagro Bonmati.

(Almagro et al., 2017, 2021; Garcia-Franco et al., 2015; Martínez-Mena et al. 2021b). As shown in Table 2.1, all soil management practices had positive impacts on most of the variables, indicating improvements in soil health and functioning, such as enhancements in the aggregate stability, its water retention capacity, and OC sequestration. However, the NT treatment did not have the best results for some indicators of soil structure improvement, and in particular, it had a very negative impact on almond tree productivity and the carbon footprint (Martin-Gorriz et al., 2020).

Our results from long-term observations provide scientific evidence supporting the adoption of reduced tillage instead of no-tillage in rainfed organic almond orchards in semi-arid calcareous soils to avoid yield penalties despite

Table 2.1 Main Findings on Beneficial and Detrimental Agro-environmental and Economic Effects of Different Sustainable Soil Management Practices in rainfed organic almond orchards under semiarid Mediterranean conditions.

	Soil Management Practice			
Soil functions and indicators	**No-tillage**	**Reduced tillage**	**Reduced tillage + green manure**	**References**
Soil water erosion	↓↓↓ (+)	↓↓↓ (+)	↓↓↓ (+)	Almagro et al., 2016; Martínez-Mena et., al 2020; 2021a
Eroded organic carbon (OC), nitrogen (N), and phosphorous (P)	↓↓↓ (+)	↓↓↓ (+)	↓↓↓ (+)	Martínez-Mena et al., 2020; 2021a
Runoff	↓↓↓ (+)	↓↓↓ (+)	↓↓↓ (+)	Martínez-Mena et al., 2020
Soil water retention capacity	↑↑	↑	↑↑	Almagro et al., 2016; 2023a; Martínez-Mena et al., 2021b
Bulk density	↑ (−)	↓ (+)	↓ (+)	Almagro et al., 2017; Martínez-Mena et al., 2021b
Soil OC stock	↑ (+)	↑↑ (+)	↑↑↑ (+)	Almagro et al., 2017; 2021; 2023b
Particulate OC stock	↑ (+)	↑ (+)	↑↑ (+)	Martínez-Mena et al., 2021a; Fernández-Soler et al., 2024
Soil CO2 emissions	↓ (+)	=	=	Almagro et al., 2016; 2017; Martínez-Mena et al., 2021a
OC and N in aggregates	↑↑↑ (+)	↑↑↑ (+)	↑↑↑ (+)	Garcia-Franco et al., 2015; 2021; Almagro et al., 2017; 2021; 2023b; Fernández-Soler et al., 2024
Occluded: free OC fraction	↑↑↑ (+)	↑↑ (+)	↑↑ (+)	Garcia-Franco et al., 2015; 2021; Almagro et al., 2021; 2023b; Fernández-Soler et al., 2024
Aggregate stability	↑↑↑ (+)	↑ (+)	↑↑ (+)	Garcia-Franco et al., 2015; 2021; Luján-Soto et al., 2021; Almagro et al., 2023a; Fernández-Soler et al 2024

Table 2.1 (Continued)

	Soil Management Practice			
Soil functions and indicators	No-tillage	Reduced tillage	Reduced tillage + green manure	References
Microbial biomass and activity	↑ (+)	↑ (+)	↑ (+)	Almagro et al., 2021; 2023b; Luján-Soto et al., 2021
Crop yield	↓↓↓ (−)	=	=	Martínez-Mena et al., 2013; 2021b; Almagro et al., 2023a
Yield costs	↓↓ (+)	↓ (+)	↑ (−)	Martín-Gorriz et al., 2020
Carbon footprint (greenhouse gas emissions: yield)	↑↑↑ (−)	↓ (+)	=	Martín-Gorriz et al., 2020

Note. Most of the outcomes summarized in Table 2.1 come from different studies conducted in the long-term agriculture research "Burete," where different sustainable soil management practices were adopted in 2008 to be compared with the traditional soil management, but also from other research conducted in other almond fields under similar climatic and management conditions. The magnitude and direction of the effects are denoted by the number and orientation of arrows (i.e., up arrow symbols indicate enhancements while down arrow symbols indicate the opposite), respectively. Positive, neutral, and negative impacts of each sustainable soil management practice compared with the traditional soil management, under which three to five passes per year are conducted with a cultivator to maintain bare soils year round, are also indicated by symbols +, = and –, respectively.

improvement of the soil structure (Martínez-Mena et al., 2013, 2021b). These findings raise questions about what is currently being proposed by the new eco-schemes for permanent crops in semi-arid regions like South-Eastern Spain. The results also demonstrated that reducing tillage in this semiarid woody crop system, especially in combination with adding green manure, improves the soil structure (owing to the occlusion of rich-OC micro-aggregates within macro-aggregates) and the stabilization of OC from the labile to the passive pools (Garcia-Franco et al., 2015), preserving SOC against erosion or tillage operations, while improving almond production in the long term (Martínez-Mena et al., 2021b).

A parallel study demonstrated that adopting green manure in semiarid permanent crops such as olive groves would compensate soil carbon losses driven by erosion or even those caused by land-use change after 20 years from adoption (Almagro & Martínez-Mena, 2014). Moreover, it was demonstrated that soil under vegetation cover, in contrast to bare soil, will contribute to climate change mitigation because it emits less CO_2 to the atmosphere at higher temperatures and after rainfall events (Almagro et al., 2017). Another parallel long-term study demonstrated that vegetation cover is also a cost-effective soil management practice for erosion control and nutrient regulation, particularly because carbon losses by erosion after extreme rainfall events were significantly reduced compared to those observed under bare soils (Martínez-Mena et al., 2020).

Altogether, from a climate change mitigation and adaptation perspective, our results highlight the importance of building resilient soils through the adoption of vegetation covers in semiarid rainfed permanent crops. Likewise, our results underline the need for integrated assessments that include multiple agro-environmental and socioeconomic indicators to assess the improvement of soil and agroecosystem function and health, such as soil aggregate stability and water retention capacity, vegetation cover biomass production, and main crop yields, which are then translated into agroecosystem services. This outcome is based on small-scale indicators such as soil aggregate stability and OC content within aggregates providing information on the benefits of sustainable soil management at local scales (such as improved soil structure and water-holding capacity), which have regional implications in regulating services (such as erosion control and flood prevention), and even at global scales (such as climate change mitigation through soil carbon sequestration).

Scientific Knowledge Gaps, Research Needs, and Policy Recommendations

Despite the many recent studies, including meta-analyses studies, addressing the agro-environmental and socioeconomic impacts of soil management practices in Mediterranean permanent crops (Aguilera et al., 2013; Bolinder et al., 2020;

Costantini et al., 2020; Kassam et al., 2012; Morugán-Coronado et al., 2020; Vicente-Vicente et al., 2016), no robust conclusions have been drawn to date in relation to the best management options for semiarid and Mediterranean regions. In particular, and despite several international organizations having acknowledged the environmental benefits of adopting vegetation covers in woody crop systems (Sanz et al., 2017; FAO & ITPS, 2021), the adoption of such cover under semiarid rainfed conditions is still in its infancy, largely because its management is particularly complex owing to competition with the main crop for water and/or nutrients (Martínez-Mena et al., 2013; Aznar-Sánchez et al., 2020). In this regard, contradictory results have been reported on the impacts of vegetation covers on the main crop yields in fruit orchards (Morugán-Coronado et al., 2020), highlighting the need for cautious vegetation cover management in rainfed woody crop systems of semiarid regions to avoid yield penalties, particularly if the fields are no tilled (Martínez-Mena et al., 2021b).

It remains unclear which options are the best for growing vegetation covers (spontaneous or seeded; annuals or perennials) and their termination timing and method (i.e., by using a roller-crimper or a cut-brusher and leaving them on the soil surface as mulch or incorporating them into the soil using a cultivator) to boost the restoration of soil health and functioning in degraded agricultural soils, while avoiding yield penalties under current and future climatic conditions (Almagro et al. 2023a; Almagro et al., 2023b). Each vegetation cover type and management practice have pros and cons, and the final outcomes heavily depend on the specific local context. For example, spontaneous vegetation cover increases agroecosystem functional biodiversity and are more cost-effective, easy to manage, and better adapted to the environmental local conditions than seeded vegetation cover (Jiménez-Alfaro et al., 2020). However, seeded vegetation covers could be more appropriate during the first years if soils are particularly degraded and/or local soil seed banks are unsuited for regenerative agriculture purposes (Almagro et al., 2021; Fernández-Soler et al., 2024). In this regard, selecting a mixture of suitable annual varieties (including legumes) to be used in seeded vegetation covers can restore agroecosystem functions such as soil regeneration, fertility maintenance, and water regulation. On the other hand, managing vegetation covers using a cut-brusher in no-tillage systems improves soil water balance by reducing losses through evaporation and runoff, while avoiding soil biota damage and aggregate disruption by plowing (Dahiya et al., 2007; Erktan et al., 2020). However, incorporating seeded vegetation covers into the soil by shallow tillage in rainfed woody crop systems might be a more effective means of reversing soil degradation in flat semiarid soils. First, because mixing shoots and roots has a synergistic effect by enhancing soil aggregation and carbon and nitrogen stabilization mechanism in semiarid environments (Almagro et al. 2021); and second, because carbon and nitrogen losses by photodegradation could be avoided in semiarid conditions (Almagro et al. 2015).

Although both annual and perennial cover crops can in principle be adopted in permanent crops, in the case of organic rainfed almond orchards under semiarid conditions, establishing annual cover crops is preferable since they can be more effectively terminated to avoid competition with the main crop for water and/or nutrients in spring. Nevertheless, perennial crops can also be an option if they are wisely managed. For example, they can be established in narrow strips between the almond tree rows to (a) reduce competition for water and/or nutrients with the main crop, (b) facilitate harvest and reduce harvest operation costs if the aim is to establish a secondary crop, and (c) allow occasional tillage operations near the almond tree trunks if necessary (e.g., to avoid pest attacks such as *C. tenebrionis*), and/or (d) be managed by livestock without damage to almond trees. Nevertheless, caution is needed when selecting species and varieties for perennial cover crops in permanent crops under rainfed semiarid conditions. They should be well established in the first year of implementation to protect the soil against erosion, and species that produce seed in the late spring or beginning of the summer, as it is the case with *Onobrychis viciifolia*, should be avoided to better optimize their termination period, particularly under rainfed semiarid conditions.

The lack of scientific evidence to date is explained by (a) the absence of comparable data to assess the impacts of different mixtures of vegetation covers, and their management, on soil carbon, nutrient, and water balances under the same pedoclimatic semiarid conditions; (b) limited knowledge about recently identified mechanisms such as photodegradation that affect the soil carbon and nutrient dynamics of semiarid soils; and (c) the lack of integrated field assessments that include integrated cost-benefit analyses for farmers and the administration. In this regard, more research and monitoring efforts are needed under semiarid and Mediterranean conditions to fill the current knowledge gaps and to design robust metrics that support the decision-making process. In particular, and given the great variability of results, long-term observations and multi-factorial experiments are needed to improve the understanding of biotic and abiotic processes and mechanisms regulating carbon, nutrient, water, and GHG emission balances of agricultural soils under current and future climate conditions (Rillig et al., 2019). In addition, we need to identify the effect of local factors (soil type, slope and orientation, microclimate, local soil seed bank, etc.) on modulating the potential of sustainable soil management for soil carbon sequestration, considering its synergies (i.e., ecosystem services) and trade-offs (GHG emissions, nitrogen losses, pollution, plant water availability). Notably, it is also crucial to go deep into socioeconomic aspects by (a) quantifying the production costs and benefits for farmers and designing robust metrics to assess the cost-effectiveness (mitigation potential and profitability) of a given soil management practice (Colombo et al., 2024); (b) identifying and analyzing technical and economic barriers to adoption and how to address them; and (c) incorporating externalities into economic analyses to improve assessments and to propose new policy incentives that are fairer.

Regarding the European agro-environmental policy instruments, the eco-scheme P6 "spontaneous or sown cover crops in woody crops" proposed for permanent crops should be revisited and be more flexible for regions under harsh environmental conditions, as in the case of southeastern Spain, where pedoclimatic constraints (e.g., scarce water availability and calcareous soils with low organic matter content and a tendency to compaction) make seeding and growing vegetation covers under rainfed semiarid conditions difficult without the use of occasional tillage. Indeed, several authors have claimed that under such specific limiting environmental conditions, occasional tillage might be advisable to stimulate carbon and nitrogen mineralization (Çelik et al., 2019; Quincke et al., 2007), which will ultimately improve nutrient availability for the main crop and result in higher, more stable agricultural yields (Martínez-Mena et al., 2021b). Therefore, the beneficial and detrimental effects must be balanced with the resulting economic costs and benefits to determine the optimum sustainable agricultural management system to include in current and future policy guidelines.

Acknowledgments

María Almagro acknowledges the financial support from the Spanish Ministry of Science, Innovation and Universities through the grant RYC2020-029181-I funded by MICIU/AEI/10.13039/501100011033 and by ESF Investing in your future, UE. This research was performed within the framework of different Spanish national and regional projects funded by the Spanish Ministry of Economy and Competitiveness (projects ERHIBAC IMPASEC, AGL2011-25069, AGRISER, SUSTAINARIDCOVER [Grant PID2022-136960OR-I00 funded by MICIU/AEI/ 10.13039/501100011033 and by ERDF, UE], OSIRIS Acción Complementaria), the Fundación Séneca of the Murcia Region (projects 03027/PI/05, 08757/PI/08, EMISEMUR 19350/PI/14, ALGROALNEXT), and the European Union (Next Generation EU [PRTR-C17.I1]). This book chapter is dedicated to my son, Daniel, and his grandfather, Manolo.

References

Aguilera, E., Lassaletta, L., Gattinger, A., & Gimeno, B. S. (2013). Managing soil carbon for climate change mitigation and adaptation in Mediterranean cropping systems: A meta-analysis. *Agriculture, Ecosystems and Environment*, *168*, 25–36. https://doi.org/10.1016/j.agee.2013.02.003

Almagro, M., de Vente, J., Boix-Fayos, C., García-Franco, N., Melgares de Aguilar, J., González, D., Solé-Benet, A., & Martínez-Mena, M. (2016). Sustainable land management practices as providers of several ecosystem services under rainfed Mediterranean agroecosystems. *Mitigation and Adaptation Strategies for Global Change*, *21*(7), 1029–1043. https://doi.org/10.1007/s11027-013-9535-2

Almagro, M., Díaz-Pereira, E., Boix-Fayos, C., Zornoza, R., Sánchez-Navarro, V., Re, P., Fernández, C., & Martínez-Mena, M. (2023a). The combination of crop diversification and no tillage enhances key soil quality parameters related to soil functioning without compromising crop yields in a low-input rainfed almond orchard under semiarid Mediterranean conditions. *Agriculture, Ecosystems & Environment*, *345*, 108320.

Almagro, M., Garcia-Franco, N., & Martínez-Mena, M. (2017). The potential of reducing tillage frequency and incorporating plant residues as a strategy for climate change mitigation in semiarid Mediterranean agroecosystems. *Agriculture, Ecosystems and Environment*, *246*, 210–220. https://doi.org/10.1016/j.agee.2017.05.016

Almagro, M., Maestre, F. T., Martínez-López, J., Valencia, E., & Rey, A. (2015). Climate change may reduce litter decomposition while enhancing the contribution of photodegradation in dry perennial Mediterranean grasslands. *Soil Biology and Biochemistry*, *90*, 214–223. https://doi.org/10.1016/j.soilbio.2015.08.006

Almagro, M., & Martínez-Mena, M. (2012). Exploring short-term leaf-litter decomposition dynamics in a Mediterranean ecosystem: Dependence on litter type and site conditions. *Plant and Soil*, *358*(1–2), 323–335. https://doi.org/10.1007/s11104-012-1187-6

Almagro, M., & Martínez-Mena, M. (2014). Litter decomposition rates of green manure as affected by soil erosion, transport and deposition processes, and the implications for the soil carbon balance of a rainfed olive grove under a dry Mediterranean climate. *Agriculture, Ecosystems and Environment*, *196*(1), 167–177. https://doi.org/10.1016/j.agee.2014.06.027

Almagro, M., Re, P., Díaz-Pereira, E., Boix-Fayos, C., Sánchez-Navarro, V., Zornoza, R., & Martínez-Mena, M. (2023b). Crop diversification effects on soil organic carbon and nitrogen storage and stabilization is mediated by soil management practices in semiarid woody crops. In *Soil and Tillage Research* (Vol. 233). https://doi.org/10.1016/j.still.2023.105815

Almagro, M., Ruiz-Navarro, A., Díaz-Pereira, E., Albaladejo, J., & Martínez-Mena, M. (2021). Plant residue chemical quality modulates the soil microbial response related to decomposition and soil organic carbon and nitrogen stabilization in a rainfed Mediterranean agroecosystem. *Soil Biology and Biochemistry*, *156*, 108198. https://doi.org/10.1016/j.soilbio.2021.108198

Altieri, M. A., & Rogé, P. (2009). The ecological role and enhancement of biodiversity in agriculture. In *Agriculture, biodiversity and markets* (pp. 15–32). Routledge.

Austin, A. T., & Vivanco, L. (2006). Plant litter decomposition in a semi-arid ecosystem controlled by photodegradation. *Nature*, *442*(7102), 555–558. https://doi.org/10.1038/nature05038

Aznar-Sánchez, J. A., Velasco-Muñoz, J. F., López-Felices, B., & del Moral-Torres, F. (2020). Barriers and facilitators for adopting sustainable soil management practices in Mediterranean Olive Groves. *Agronomy*, *10*(4), 506.

Baensch-Baltruschat, B., Kocher, B., Stock, F., & Reifferscheid, G. (2020). Tyre and road wear particles (TRWP)-A review of generation, properties, emissions, human health risk, ecotoxicity, and fate in the environment. *Science of the Total Environment*, *733*, 137823.

Bakker, M. M., Govers, G., Jones, R. A., & Rounsevell, M. D. (2007). The effect of soil erosion on Europe's crop yields. *Ecosystems*, *10*, 1209–1219.

Banwart, S., 2011. Save our soils. Nature *474*, 151152.

Bardgett, R. D., & van der Putten, W. H. (2014). Belowground biodiversity and ecosystem functioning. *Nature*, *515*(7528), 505–511. https://doi.org/10.1038/nature13855

Bertrand, M., Barot, S., Blouin, M., Whalen, J., de Oliveira, T., & Roger-Estrade, J. (2015). Earthworm services for cropping systems. A review. *Agronomy for Sustainable Development*, *35*, 553–567.

Bläsing, M., & Amelung, W. (2018). Plastics in soil: Analytical methods and possible sources. *Science of the Total Environment*, *612*, 422–435.

Blum, W. E. (1990). The challenge of soil protection in Europe. *Environmental Conservation*, *17*(1), 72–74.

Boix-Fayos, C., Boerboom, L. G. J., Janssen, R., Martínez-Mena, M., Almagro, M., Pérez-Cutillas, P., Eekhout, J. P. C., Castillo, V., & de Vente, J. (2020). Mountain ecosystem services affected by land use changes and hydrological control works in Mediterranean catchments. *Ecosystem Services*, *44*. https://doi.org/10.1016/j.ecoser.2020.101136

Bolinder, M. A., Crotty, F., Elsen, A., Frac, M., Kismányoky, T., Lipiec, J., Tits, M., Tóth, Z., & Kätterer, T. (2020). The effect of crop residues, cover crops, manures and nitrogen fertilization on soil organic carbon changes in agroecosystems: A synthesis of reviews. *Mitigation and Adaptation Strategies for Global Change*, *25*, 929–952.

Bommarco, R., Kleijn, D., & Potts, S. G. (2013). Ecological intensification: Harnessing ecosystem services for food security. *Trends in Ecology & Evolution*, *28*(4), 230–238.

Bouché, M. B., & Al-Addan, F. (1997). Earthworms, water infiltration and soil stability: Some new assessments. *Soil Biology and Biochemistry*, *29*(3–4), 441–452.

Bradshaw, C. J., Bowman, D. M., Bond, N. R., Murphy, B. P., Moore, A. D., Fordham, D. A., Thackway, R., Lawes, M. J., McCallum, H., & Gregory, S. D. (2013). Brave new green world–consequences of a carbon economy for the conservation of Australian biodiversity. *Biological Conservation*, *161*, 71–90.

Brahney, J., Hallerud, M., Heim, E., Hahnenberger, M., & Sukumaran, S. (2020). Plastic rain in protected areas of the United States. *Science*, *368*(6496), 1257–1260.

Briones, M. J. I., McNamara, N. P., Poskitt, J., Crow, S. E., & Ostle, N. J. (2014). Interactive biotic and abiotic regulators of soil carbon cycling: Evidence from controlled climate experiments on peatland and boreal soils. *Global Change Biology*, *20*(9), 2971–2982.

Cameron, E. K., Martins, I. S., Lavelle, P., Mathieu, J., Tedersoo, L., Bahram, M., Gottschall, F., Guerra, C. A., Hines, J., & Patoine, G. (2019). Global mismatches in aboveground and belowground biodiversity. *Conservation Biology*, *33*(5), 1187–1192.

Cantón, Y., Solé-Benet, A., De Vente, J., Boix-Fayos, C., Calvo-Cases, A., Asensio, C., & Puigdefábregas, J. (2011). A review of runoff generation and soil erosion across scales in semiarid south-eastern Spain. *Journal of Arid Environments*, *75*(12), 1254–1261.

Çelik, İ., Günal, H., Acar, M., Acir, N., Barut, Z. B., & Budak, M. (2019). Strategic tillage may sustain the benefits of long-term no-till in a Vertisol under Mediterranean climate. *Soil and Tillage Research*, *185*, 17–28.

Cherlet, M., Hutchinson, C., Reynolds, J., Hill, J., Sommer, S., & VON, M. G. (2018). *World atlas of desertification*.

Colombo, S., Castro-Rodríguez, J., Peréz-Pérez, D., & Almagro, M. (2024). Analysis of the environmental and economic performance of common agricultural policy eco-schemes in soil organic carbon sequestration. *Ecological Economics*, *220*. https://doi.org/10.1016/j.ecolecon.2024.108183

Costantini, E. A. C., Antichi, D., Almagro, M., Hedlund, K., Sarno, G., & Virto, I. (2020). Local adaptation strategies to increase or maintain soil organic carbon content under arable farming in Europe: Inspirational ideas for setting operational groups within the European innovation partnership. *Journal of Rural Studies*, *79*, 102–115. https://doi.org/10.1016/j.jrurstud.2020.08.005

Dahiya, R., Ingwersen, J., & Streck, T. (2007). The effect of mulching and tillage on the water and temperature regimes of a loess soil: Experimental findings and modeling. *Soil and Tillage Research*, *96*(1–2), 52–63.

Derak, M., Cortina, J., & Taiqui, L. (2017). Integration of stakeholder choices and multi-criteria analysis to support land use planning in semiarid areas. *Land Use Policy*, *64*, 414–428.

Doran, J. W., & Zeiss, M. R. (2000). Soil health and sustainability: Managing the biotic component of soil quality. *Special Issue: Managing the Biotic Component of Soil Quality*, *15*(1), 3–11. https://doi.org/10.1016/S0929-1393(00)00067-6

EC. (2006). *Thematic Strategy for Soil protection. Available at:* https://ec.europa.eu/environment/soil/three_en.htm *(Accessed on October 12, 2021).*

Eekhout, J. P. C., & de Vente, J. (2019). Assessing the effectiveness of Sustainable Land Management for large-scale climate change adaptation. *Science of the Total Environment*, *654*, 85–93. https://doi.org/10.1016/j.scitotenv.2018.10.350

Erktan, A., Or, D., & Scheu, S. (2020). The physical structure of soil: Determinant and consequence of trophic interactions. *Soil Biology and Biochemistry*, *148*, 107876.

European Commission. (2019). *Communication from the Commission to the European Parliament, the European Council, the Council, The European Economic and Social Committee and the Committee of the Regions. In The European Green Deal COM/ 2019/640 Final; European Commission: Brussel, Belgium, 2019.*

European Commission: Joint Research Centre, Feyen, L., Ciscar, J., Gosling, S., Ibarreta, D., & Soria, A. (2020). Climate change impacts and adaptation in Europe. In D. Ibarreta, A. Soria, (Eds.), *JRC PESETA IV final report*, Publications Office of the European Union. https://data.europa.eu/doi/10.2760/171121

European Commission. (2020a). *From Farm to Fork Strategy: For a Fair, Healthy and Environmentally-Friendly Food System. European Union. 2020*

European Commission. (2020b). *Communication from the Commission to the European Parliament, the Council, The European Economic and Social Committee and the Committee of the Regions. In EU Biodiversity Strategy for 2030: Bringing Nature Back into Our Lives COM/2020/380 Final; European Commission: Brussel, Belgium, 2020.* [Graphic].

European Commission. (2023). *Proposal for a European Directive on Soil Monitoring and Resilience.* https://www.europarl.europa.eu/RegData/etudes/BRIE/2024/757627/EPRS_BRI(2024)757627_EN.pdf (accessed 28 November 2024).

FAO. (2020). *A protocol for measurement, monitoring, reporting and verification of soil organic carbon in agricultural landscapes.*

FAO and ITPS. (2021). *Recarbonizing Global Soils—A technical manual of recommended sustainable soil management. Volume 3: Cropland, Grassland, Integrated systems, and farming approaches—Practices Overview. Rome.*

Fernández-Soler, C., Garcia-Franco, N., Almagro, M., Díaz-Pereira, E., Luján, R., García, E., & Martínez-Mena, M. (2024). Cover crops improve the long-term stabilization of soil organic carbon and total nitrogen through physico-chemical protection in rainfed semiarid Mediterranean woody crop systems. *Soil Use and Management*, *40*(2). https://doi.org/10.1111/sum.13066

Foley, J. A., DeFries, R., Asner, G. P., Barford, C., Bonan, G., Carpenter, S. R., Chapin, F. S., Coe, M. T., Daily, G. C., Gibbs, H. K., Helkowski, J. H., Holloway, T., Howard, E. A., Kucharik, C. J., Monfreda, C., Patz, J. A., Prentice, I. C., Ramankutty, N., & Snyder, P. K. (2005). Global consequences of land use. *Science*, *309*(5734), 570–574. https://doi.org/10.1126/science.1111772

Francaviglia, R., Almagro, M., & Vicente-Vicente, J. L. (2023). Conservation Agriculture and Soil Organic Carbon: Principles, Processes, Practices and Policy Options. *Soil Systems*, *7*(1). https://doi.org/10.3390/soilsystems7010017

Garcia-Franco, N., Albaladejo, J., Almagro, M., & Martínez-Mena, M. (2015). Beneficial effects of reduced tillage and green manure on soil aggregation and stabilization of organic carbon in a Mediterranean agroecosystem. *Soil and Tillage Research*, *153*, 66–75.

Garcia-Franco, N., Wiesmeier, M., Colocho Hurtarte, L. C., Fella, F., Martínez-Mena, M., Almagro, M., Martínez, E. G., & Kögel-Knabner, I. (2021). Pruning residues incorporation and reduced tillage improve soil organic matter stabilization and structure of salt-affected soils in a semi-arid Citrus tree orchard. *Soil and Tillage Research, 213*. https://doi.org/10.1016/j.still.2021.105129

García-Palacios, P., Alarcón, M. R., Tenorio, J. L., & Moreno, S. S. (2019). Ecological intensification of agriculture in drylands. *Journal of Arid Environments, 167*, 101–105.

GSBI, S. (2020). *State of knowledge of soil biodiversity–Status, challenges and potentialities. Summary for policy makers.*

Guenet, B., Gabrielle, B., Chenu, C., Arrouays, D., Balesdent, J., Bernoux, M., Bruni, E., Caliman, J.-P., Cardinael, R., & Chen, S. (2021). Can N2O emissions offset the benefits from soil organic carbon storage? *Global Change Biology, 27*(2), 237–256.

Guerra, C. A., Bardgett, R. D., Caon, L., Crowther, T. W., Delgado-Baquerizo, M., Montanarella, L., Navarro, L. M., Orgiazzi, A., Singh, B. K., & Tedersoo, L. (2021). Tracking, targeting, and conserving soil biodiversity. *Science, 371*(6526), 239–241.

Hanaček, K., & Rodríguez-Labajos, B. (2018). Impacts of land-use and management changes on cultural agroecosystem services and environmental conflicts—A global review. *Global Environmental Change, 50*, 41–59.

Hendrix, P. F., Parmelee, R. W., Crossley, D. A., Coleman, D. C., Odum, E. P., & Groffman, P. M. (1986). Detritus Food Webs in Conventional and No-Tillage Agroecosystems. *BioScience, 36*(6), 374–380. https://doi.org/10.2307/1310259

Hermoso, V., Carvalho, S. B., Giakoumi, S., Goldsborough, D., Katsanevakis, S., Leontiou, S., Markantonatou, V., Rumes, B., Vogiatzakis, I. N., & Yates, K. L. (2022). The EU Biodiversity Strategy for 2030: Opportunities and challenges on the path towards biodiversity recovery. *Environmental Science & Policy, 127*, 263–271.

IPCC. (2019). *IPCC, 2019: Summary for Policymakers. In: Climate Change and Land: An IPCC special report on climate change, desertification, land degradation, sustainable land management, food security, and greenhouse gas fluxes in terrestrial ecosystems.*

IUSS Working Group WRB. (2015). *World Soil Resources IUSS Working Group WRB (2015) World reference base for soil resources 2014, Update 2015 international soil classification system for naming soils and creating legends for soil maps. World Soil Resources Reports No. 106. FAO, RomeReports No. 103.* FAO.

Jiménez-Alfaro, B., Frischie, S., Stolz, J., & Gálvez-Ramírez, C. (2020). Native plants for greening Mediterranean agroecosystems. *Nature Plants, 6*(3), 209–214. https://doi.org/10.1038/s41477-020-0617-3

Kassam, A., Friedrich, T., Derpsch, R., Lahmar, R., Mrabet, R., Basch, G., González-Sánchez, E. J., & Serraj, R. (2012). Conservation agriculture in the dry Mediterranean climate. *Field Crops Research, 132*, 7–17.

Kirjavainen, P. V., Karvonen, A. M., Adams, R. I., Täubel, M., Roponen, M., Tuoresmäki, P., Loss, G., Jayaprakash, B., Depner, M., & Ege, M. J. (2019). Farm-like indoor microbiota in non-farm homes protects children from asthma development. *Nature Medicine*, *25*(7), 1089–1095.

Lal, R. (1998). Soil erosion impact on agronomic productivity and environment quality. *Critical Reviews in Plant Sciences*, *17*(4), 319–464.

Lal, R. (2010). Beyond Copenhagen: Mitigating climate change and achieving food security through soil carbon sequestration. *Food Security*, *2*(2), 169–177. https://doi.org/10.1007/s12571-010-0060-9

Lassaletta, L., Sanz-Cobeña, A., Aguilera, E., Quemada, M., Billen, G., Bondeau, A., Cayuela, M. L., Cramer, W., Eekhout, J. P., & Garnier, J. (2021). Nitrogen dynamics in cropping systems under Mediterranean climate: A systemic analysis. *Environmental Research Letters*, *16*(7), 073002.

Lee, H., Rahn, T., & Throop, H. (2012). An accounting of C-based trace gas release during abiotic plant litter degradation. *Global Change Biology*, *18*, 1185–1195.

Liniger, H., Critchley, W., Gurtner, M., Schwilch, G., & Mekdaschi Studer, R. (2007). *Where the land is greener: Case studies and analysis of soil and water conservation initiatives worldwide*. World Overview of Conservation Approaches and Technologies (WOCAT).

Liu, H., Zheng, X., Li, Y., Yu, J., Ding, H., Sveen, T. R., & Zhang, Y. (2022). Soil moisture determines nitrous oxide emission and uptake. *Science of The Total Environment*, *822*, 153566.

Loureiro, T., Mesquita, M. M., Dapkevicius, M. de L. E., Serra, L., Martins, Â., Cortez, I., & Poeta, P. (2023). *Xylella fastidiosa: An Updated Review*.

Maaß, S., Daphi, D., Lehmann, A., & Rillig, M. C. (2017). Transport of microplastics by two collembolan species. *Environmental Pollution*, *225*, 456–459.

Mäder, P., Fliessbach, A., Dubois, D., Gunst, L., Fried, P., & Niggli, U. (2002). Soil fertility and biodiversity in organic farming. *Science*, *296*(5573), 1694–1697.

Martínez-Mena, M., Boix-Fayos, C., Carrillo-López, E., Díaz-Pereira, E., Zornoza, R., Sánchez-Navarro, V., Acosta, J. A., Martínez-Martínez, S., & Almagro, M. (2021a). Short-term impact of crop diversification on soil carbon fluxes and balance in rainfed and irrigated woody cropping systems under semiarid Mediterranean conditions. *Plant and Soil*, *467*(1–2), 499–514. https://doi.org/10.1007/s11104-021-05101-w

Martínez-Mena, M., Carrillo-López, E., Boix-Fayos, C., Almagro, M., García Franco, N., Díaz-Pereira, E., Montoya, I., & de Vente, J. (2020). Long-term effectiveness of sustainable land management practices to control runoff, soil erosion, and nutrient loss and the role of rainfall intensity in Mediterranean rainfed agroecosystems. *Catena*, *187*. https://doi.org/10.1016/j.catena.2019.104352

Martínez-Mena, M., Garcia-Franco, N., Almagro, M., Ruiz-Navarro, A., Albaladejo, J., de Aguilar, J. M., Gonzalez, D., & Querejeta, J. I. (2013). Decreased foliar nitrogen and crop yield in organic rainfed almond trees during transition from reduced tillage to no-tillage in a dryland farming system. *European Journal of Agronomy, 49*, 149–157. https://doi.org/10.1016/j.eja.2013.04.006

Martinez-Mena, M., Lopez, J., Almagro, M., Boix-Fayos, C., & Albaladejo, J. (2008). Effect of water erosion and cultivation on the soil carbon stock in a semiarid area of South-East Spain. *Soil and Tillage Research*, *99*(1), 119–129. https://doi.org/10.1016/j.still.2008.01.009

Martínez-Mena, M., Perez, M., Almagro, M., Garcia-Franco, N., & Díaz-Pereira, E. (2021b). Long-term effects of sustainable management practices on soil properties and crop yields in rainfed Mediterranean almond agroecosystems. *European Journal of Agronomy*, *123*. https://doi.org/10.1016/j.eja.2020.126207

Martin-Gorriz, B., Maestre-Valero, J. F., Almagro, M., Boix-Fayos, C., & Martínez-Mena, M. (2020). Carbon emissions and economic assessment of farm operations under different tillage practices in organic rainfed almond orchards in semiarid Mediterranean conditions. *Scientia Horticulturae, 261*. https://doi.org/10.1016/j.scienta.2019.108978

MEA, 2005. Millennium ecosystem assessment: Ecosystems and human well-being synthesis. World Resources Institute, Washington, DC.

Montanarella, L., & Panagos, P. (2021). The relevance of sustainable soil management within the European Green Deal. *Land Use Policy*, *100*, 104950.

Montgomery, D. R. (2007). Soil erosion and agricultural sustainability. *Proceedings of the National Academy of Sciences of the United States of America*, *104*(33), 13268–13272. https://doi.org/10.1073/pnas.0611508104

Morugán-Coronado, A., Linares, C., Gómez-López, M. D., Faz, Á., & Zornoza, R. (2020). The impact of intercropping, tillage and fertilizer type on soil and crop yield in fruit orchards under Mediterranean conditions: A meta-analysis of field studies. *Agricultural Systems*, *178*, 102736.

Naeem, S., Thompson, L. J., Lawler, S. P., Lawton, J. H., & Woodfin, R. M. (1994). Declining biodiversity can alter the performance of ecosystems. *Nature*, *368*(6473), 734–737.

Natural Resources Conservation Service. Soil Health Technical Note No. 450-04, Nov 2019.

Nielsen, U. N. (2019). *Soil Fauna Assemblages: Global to Local Scales*. Cambridge University Press. https://doi.org/10.1017/9781108123518

Oldfield, E. E., Bradford, M. A., & Wood, S. A. (2019). Global meta-analysis of the relationship between soil organic matter and crop yields. *Soil*, *5*(1), 15–32.

Orgiazzi, A., Bardgett, R. D., & Barrios, E. (2016). *Global soil biodiversity atlas*. European Commission.

Panagos, P., Ballabio, C., Scarpa, S., Borelli, P., Lugato, E., & Montanarella, L. (2020). *Soil Related Indicators to Support Agro-Environmental Policies. Publications Office of the European Union.*

Pörtner, H. O., et al. (2022). *Climate change 2022: Impacts, adaptation and vulnerability. IPCC.*

Power, A. G. (2010). Ecosystem services and agriculture: Tradeoffs and synergies. *Philosophical Transactions of the Royal Society B: Biological Sciences*, *365*(1554), 2959–2971.

Quincke, J. A., Wortmann, C. S., Mamo, M., Franti, T., & Drijber, R. A. (2007). Occasional tillage of no-till systems: Carbon dioxide flux and changes in total and labile soil organic carbon. *Agronomy Journal*, *99*(4), 1158–1168.

Reid, W., Mooney, H., Cropper, A., Capistrano, D., Carpenter, S., & Chopra, K. (2005). *Millennium ecosystem assessment. Ecosystems and human well-being: synthesis.* Island Press.

Rillig, M. C. (2012). Microplastic in Terrestrial Ecosystems and the Soil? *Environmental Science & Technology*, *46*(12), 6453–6454. https://doi.org/10.1021/es302011r

Rillig, M. C., Ingraffia, R., & de Souza Machado, A. A. (2017a). Microplastic incorporation into soil in agroecosystems. *Frontiers in Plant Science*, *8*, 1805.

Rillig, M. C., & Lehmann, A. (2020). Microplastic in terrestrial ecosystems. *Science*, *368*(6498), 1430–1431.

Rillig, M. C., Ryo, M., Lehmann, A., Aguilar-Trigueros, C. A., Buchert, S., Wulf, A., Iwasaki, A., Roy, J., & Yang, G. (2019). The role of multiple global change factors in driving soil functions and microbial biodiversity. *Science*, *366*(6467), 886–890.

Rillig, M. C., Ziersch, L., & Hempel, S. (2017b). Microplastic transport in soil by earthworms. *Scientific Reports*, *7*(1), 1362.

Ruiz, I., Almagro, M., García de Jalón, S., Solà, M. D. M., Sanz, M. J., 2020. Assessment of sustainable land management practices in Mediterranean rural regions. *Journal of Environmental Management*, *276*, 111293.

Sandén, T., Spiegel, H., Stüger, H.-P., Schlatter, N., Haslmayr, H.-P., Zavattaro, L., Grignani, C., Bechini, L., D′ hose, T., & Molendijk, L. (2018). European long-term field experiments: Knowledge gained about alternative management practices. *Soil Use and Management*, *34*(2), 167–176.

Sanz, M. J., De Vente, J. L., Chotte, J.-L., Bernoux, M., Kust, G., Ruiz, I., Almagro, M., Alloza, J. A., Vallejo, R., & Castillo, V. (2017). *Sustainable land management contribution to successful land-based climate change adaptation and mitigation: A report of the Science-Policy Interface.*

Sanz-Cobeña, A., Lassaletta, L., Aguilera, E., Del Prado, A., Garnier, J., Billen, G., Iglesias, A., Sanchez, B., Guardia, G., & Abalos, D. (2017). Strategies for greenhouse gas emissions mitigation in Mediterranean agriculture: A review. *Agriculture, Ecosystems & Environment*, *238*, 5–24.

Schlesinger, W. H. (2000). Carbon sequestration in soils: Some cautions amidst optimism. *Agriculture, Ecosystems & Environment*, *82*(1–3), 121–127.

Schoonhoven, Y., & Runhaar, H. (2018). Conditions for the adoption of agro-ecological farming practices: A holistic framework illustrated with the case of almond farming in Andalusia. *International Journal of Agricultural Sustainability*, *16*(6), 442–454. https://doi.org/10.1080/14735903.2018.1537664

Schwab, G. O., Frevert, R. K., Edminster, T. W., & Barnes, K. K. (1982). Soil and water conservation engineering. *Soil Science*, *134*(2), 146.

Shah, A. N., Tanveer, M., Shahzad, B., Yang, G., Fahad, S., Ali, S., Bukhari, M. A., Tung, S. A., Hafeez, A., & Souliyanonh, B. (2017). Soil compaction effects on soil health and cropproductivity: An overview. *Environmental Science and Pollution Research*, *24*, 10056–10067.

Six, J., Bossuyt, H., Degryze, S., & Denef, K. (2004). A history of research on the link between (micro) aggregates, soil biota, and soil organic matter dynamics. *Soil and tillage research*, *79*(1), 7–31.

Smith, D. G., Martinelli, R., Besra, G. S., Illarionov, P. A., Szatmari, I., Brazda, P., Allen, M. A., Xu, W., Wang, X., & Nagy, L. (2019). Identification and characterization of a novel anti-inflammatory lipid isolated from Mycobacterium vaccae, a soil-derived bacterium with immunoregulatory and stress resilience properties. *Psychopharmacology*, *236*, 1653–1670.

Smith, J. O., Smith, P., Wattenbach, M., Zaehle, S., Hiederer, R., Jones, R. J., Montanarella, L., Rounsevell, M. D., Reginster, I., & Ewert, F. (2005). Projected changes in mineral soil carbon of European croplands and grasslands, 1990–2080. *Global Change Biology*, *11*(12), 2141–2152.

Soto, R. L., Martínez-Mena, M., Padilla, M. C., & de Vente, J. (2021). Restoring soil quality of woody agroecosystems in Mediterranean drylands through regenerative agriculture. *Agriculture, Ecosystems & Environment*, *306*, 107191.

Stavi, I., Bel, G., & Zaady, E. (2016). Soil functions and ecosystem services in conventional, conservation, and integrated agricultural systems. *A review. Agronomy for Sustainable Development*, *36*(2), 1–12.

Stewart, C. E., Paustian, K., Conant, R. T., Plante, A. F., & Six, J. (2007). Soil carbon saturation: Concept, evidence and evaluation. *Biogeochemistry*, *86*, 19–31.

Stolte, J., Tesfai, M., Oygarden, L., Kvaerno, S., Keizer, J., Verheijen, F., Panagos, P., Ballabio, C., & Hessel, R. (2016). *Soil threats in Europe: Status, methods, drivers and effects on ecosystem services: Deliverable 2.1 RECARE project*. European Commission DG Joint Research Centre.

Tamburini, G., Bommarco, R., Wanger, T. C., Kremen, C., van der Heijden, M. G. A., Liebman, M., & Hallin, S. (2020). Agricultural diversification promotes multiple ecosystem services without compromising yield. *Science Advances*, *6*(45), eaba1715.

van der Putten, W. H., Ramirez, K. S., Poesen, J., Winding, A., Lemanceau, P., Lisa, L., Simek, M., Moora, M., Setala, H., & Zaitsev, A. (2018). *Opportunities for soil sustainability in Europe.*

Veerman, C., Correia, T. P., Bastioli, C., Biro, B., Bouma, J., Cienciala, E., Emmett, B., Frison, E. A., Grand, A., & Filchev, L. H. (2020). *Caring for soil is caring for life: Ensure 75% of soils are healthy by 2030 for healthy food, people, nature and climate: Interim report of the mission board for soil health and food: Study.*

Veerman, C. P. (2023). Activity update of the Mission Board of European Union on Soil Health and Food. In *Soil & Environmental Health* (pp. 100018). Elsevier.

VGSSM. (2015). *Voluntary Guidelines on Sustainable Soil Management.*

Vicente-Vicente, J. L., García-Ruiz, R., Francaviglia, R., Aguilera, E., & Smith, P. (2016). Soil carbon sequestration rates under Mediterranean woody crops using recommended management practices: A meta-analysis. *Agriculture, Ecosystems & Environment, 235*, 204–214.

Wagg, C., Bender, S. F., Widmer, F., & van der Heijden, M. G. A. (2014). Soil biodiversity and soil community composition determine ecosystem multifunctionality. *Proceedings of the National Academy of Sciences, 111*(14), 5266. https://doi.org/10.1073/pnas.1320054111

Weithmann, N., Möller, J. N., Löder, M. G., Piehl, S., Laforsch, C., & Freitag, R. (2018). Organic fertilizer as a vehicle for the entry of microplastic into the environment. *Science Advances, 4*(4), eaap8060.

Wischmeier, W. H., & Smith, D. D. (1962). Soil loss estimation as a tool in soil and water management planning. *International Association of Scientific Hydrology Publication, 59*, 148–159.

Zabala, J. A., Martínez-García, V., Martínez-Paz, J. M., López-Becerra, E. I., Nasso, M., Díaz-Pereira, E., Sánchez-Navarro, V., Álvaro-Fuentes, J., González-Rosado, M., & Farina, R. (2023). Crop diversification practices in Europe: An economic cross-case study comparison. *Sustainability Science*, 1–16.

Zhu, Y.-G., Zhao, Y., Zhu, D., Gillings, M., Penuelas, J., Ok, Y. S., Capon, A., & Banwart, S. (2019). Soil biota, antimicrobial resistance and planetary health. *Environment International, 131*, 105059.

3

Soil Health and Agricultural Production of Diversified Crops in Southeastern Spain

Virginia Sánchez-Navarro, Jessica Cuartero, Onurcan Özbolat, Carolina Boix-Fayos, María Martínez-Mena, María Almagro, Elvira Díaz-Pereira, Margarita Ros, Jose A. Pascual, Juan A. Fernández, and Raúl Zornoza

Chapter Overview

Combining soil management strategies such as crop diversification, reduced tillage, and organic amendments can improve not only soil health but also the sustainability of agroecosystems. This approach can contribute to improved ecosystem services related to supporting and regulating features such as biodiversity, nutrient cycling, water infiltration and retention, and carbon storage, among others, that influence crop production. This study encompassed three different case studies: (a) an almond crop in monocrop and diversified with thyme combined with reduced tillage (CS1), (b) a mandarin crop in monocrop and diversified with the rotation of fava bean and barley/vetch combined with green manure (CS2), and a melon crop in monocrop and diversified with cowpea combined with reduced tillage, green manure, and reduced fertilizer application (CS3). The main goal of this study was to assess the effect of crop diversification combined with soil conservation management practices in woody (almond and mandarin) and fruit (melon) crops on soil physical and chemical properties, soil bacterial communities, and their relationships with crop yield. The combined effect of the diversified crops and the soil conservation practices was jointly assessed in the individual case studies, although these were not equivalent. After two crop cycles, crop diversification paired with soil conservation management led to improved soil structure in all crops studied (almond, mandarin, and melon). In addition,

Soil Health Series: Volume 5 Soil Health and Sustainability in Spain and Portugal, First Edition.
Edited by Iñigo Virto and Rodrigo Antón.

Published 2025 by Soil Science Society of America, Inc. and John Wiley & Sons, Inc.

crop diversification also increased soil moisture content in mandarin and melon crops and soil nutrient content in the almond crop. However, the mandarin monocrop showed higher availability of nutrients compared with the diversified mandarin, likely owing to higher uptake by alley crops. The combination of diversified crops and soil conservation practices also increased soil microbial content in all crops. Hence, crop diversification combined with soil conservation practices resulted in the best alternative for improving soil health in almond and melon crops, but not for the mandarin crop. Thus, the choice of the crops involved in the diversification and the soil conservation practices selected are key factors for improving soil fertility, owing to nutrients from crop residues and root exudation, soil organic matter mineralization, and solubilization of minerals related to higher microbial content. The opposite effect in the mandarin crop suggests a need for modifying the diversification strategy and/or using a longer monitoring period to discern the complementarity between the different crops involved in the diversification. In almond and melon crops under diversification systems, crop yield and overall crop yield were related to increases in soil organic carbon, microbial richness, and nutrient availability and to decreases in bulk density. This outcome confirms that practices that enhance soil health are related to improvements in crop production under Mediterranean conditions, both under rainfed and irrigated systems.

Introduction

Agricultural intensification as a result of the increased demand for food from the growing world population has been accompanied by a decline in soil health (Sünnemann et al., 2021), primarily because of higher pressures on natural resources such as land, water, and energy (Food and Agriculture Organization [FAO], 2011; OECD/FAO, 2018). In Spain, Mediterranean agriculture is generally intensive, with high variety in terms of crop types and specialization. This kind of agriculture negatively affects soil health, mainly owing to soil organic matter losses, decreased biodiversity, and soil pollution. In addition, the excessive use of fertilizers and pesticides in this type of agriculture increases the risk of nutrients being leached into the groundwater, contributing to water pollution (Eurostat, 2015). Woody and horticultural crops are the predominant agricultural land uses in the Spanish Mediterranean area (Ministerio de Agricultura Alimentación y Medio Ambiente [MAGRAMA], 2013). With regard to woody crops, Spain is the world's second largest producer of almond [*Prunus dulcis* (Mill.) D.A. Webb] and the largest producer of mandarin (*Citrus reticulata* Blanco), with an average production of 421,000 tons per year of almonds and an average production of 2 million tons per year of mandarins (MAGRAMA, 2023). With regard to horticultural crops, Spain represents approximately 22% of the total European production, with an average production of 722,000 tons per year of melons (MAGRAMA, 2023).

Owing to intensive agriculture, soil conservation agricultural practices need to be developed, with the aim of enhancing soil health while maintaining high crop production. Soil conservation practices include no-tillage, the introduction of crop rotations, cover crops, mulching, application of manure or compost, and incorporation of crop residues (Lal, 2004). These soil conservation practices provide multiple benefits, such as the reduction of greenhouse gas emissions (Alam et al., 2019), the increase of soil biological activity (Naresh et al., 2016), and the increase of the crop productivity in the long term (Sharma et al., 2012). In this context, establishing crop diversification can optimize crop production and increase soil health through improving soil structure, enhancing nutrient availability, and increasing the function and diversity of soil microorganisms (Abid & Lal, 2008; Maron et al., 2011; Bardgett & van der Putten, 2014; Congreves et al., 2015). These improvements are derived from the various attributes of different crop species included in a diversified system, such as different root structures (depth and branching), the root exudation (turnover rates, composition and quantity of root exudates), and aboveground crop residues (Rovira, 1969; Congreves et al., 2014).

Diversified cropping systems may also promote the delivery of ecosystem services such as (a) regulation of the temperature at global or local levels and the reduction of greenhouse gas emissions, mainly owing to the inclusion of secondary crops such as legumes that enable reduced application of fertilizers for later crops because they fix nitrogen and stimulate soil microbial communities (Preissel et al., 2015); (b) reduced soil erosion, as the denser vegetation cover of diversification systems may prevent soil losses caused by wind, runoff, or other removal processes (Adekalu et al., 2006); (c) enhanced soil formation processes owing to the accumulation of organic material and reduced erosion (Myaka et al., 2006; Van Kessel & Hartley, 2000); (d) support for pollinators and natural enemies from increased plant biodiversity, with decreased disease/pest incidence (Haaland et al., 2011; Campbell et al., 2017; Bertrand et al., 2019; Whitby et al., 2020); and (e) increased primary production of crops and fruits from optimized use of time and space for growing crops (Li et al., 2019; Kumari et al., 2024).

In addition to the diversification systems, management practices and pedoclimatic characteristics also promote the provision of ecosystem services (Chabert

& Sarthou, 2020). Palomo-Campesino et al. (2022) concluded that using more agroecological practices than conventional farming practices increased the potential for providing and regulating ecosystem services such as erosion control, pollination, pest control, and food diversity. The inclusion of crop diversification (rotations, multiple cropping, or intercropping), reduced tillage, and the use of organic fertilizers within a diversification system are associated with erosion control, pollination, pest control, and soil fertility (Holzschuh et al., 2008; Diekötter et al., 2010; Kuntz et al., 2013; Andersson et al., 2014; Forrest et al., 2015; Bonanomi et al., 2016; Isaacs et al., 2016; Cerda et al., 2017; Degrune et al., 2017).

Soil health is evaluated through different soil physical, chemical, and biological properties (Williams et al., 2020). Soil physicochemical properties are related to the availability of nutrients, carbon sequestration and storage, and soil structure and compaction (Blanco-Canqui et al., 2015, Haruna & Nkongolo, 2019; Sainju et al., 2021). In turn, nutrient cycling, soil organic matter, carbon sequestration, and biodiversity are related to biological properties (Sainju et al., 2021). Agricultural soil conservation practices have a positive impact on soil health through soil quality, soil carbon sequestration, and crop yield (Bonetti et al., 2023). For example, Sainju et al. (2021) observed that soil biological properties, which are sensitive to management, are key soil health indicators directly associated with crop yield. This association is based on the decomposition of organic matter by microorganisms, which increases the availability of nutrients and thus helps to maintain crop yields.

Considering the expected benefits to soil health and the associated improved crop productivity from crop diversification, we assessed and compared different soil physicochemical and biological properties and crop yield in different woody and horticultural crops (almond, mandarin, and melon). Our objective was to assess whether crop diversification and soil conservation management practices such as reduced tillage and green manure or reduced fertilizer application could improve soil health and crop productivity. This study encompassed three case studies: (a) an almond crop in monocrop and diversified with thyme (*Thymus hyemalis* Lange) combined with reduced tillage (CS1), (b) a mandarin crop in monocrop and diversified with the rotation of fava bean (*Vicia faba* L.) and barley/vetch (*Hordeum vulgare* L./*Vicia sativa* L.) combined with green manure (CS2); and a melon crop in monocrop and diversified with cowpea [*Vigna unguiculata* (L.) Walp] combined with reduced tillage, green manure and reduced fertilizer application (CS3). Details of these case studies are provided in (Table 3.1). We hypothesized that crop diversification and soil conservation management could promote improved soil health, while reducing the use of external inputs and increasing overall production.

Table 3.1 Key Aspects of the Different Treatments in Each Case Study.

Aspect	CS1	CS2	CS3
Establishment of the trial (date)	November 2018	March 2018	May 2018
Main crop	Almond	Mandarin	Melon
Secondary crop	Thyme	Rotation of fava bean and vetch/barley	Cowpea
Crop type	Rainfed	Irrigated	Irrigated
Weed control	Tillage (only in monocrop)	Tillage (only in monocrop)	Hand-hoeing
Fertilizers applied	Incorporation of the dry outer green shell cover of the almond rind after harvest	Conventional fertilizers	Organic fertilizers
Tillage	Reduced in diversification system	—	Reduced in diversification system
Crop residues	Incorporation of the dry outer green shell cover of the almond rind	Incorporation of pruning residues and secondary crop residues	Incorporation of the main and secondary crop residues in the diversification system

Material and Methods

Study Site and Experimental Design

The study was conducted at three different farms dedicated to almond, mandarin, and melon production that have different diversification strategies and management practices. However, while the case studies are not equivalent, the combined effect of diversified crops and soil conservation practices was not assessed separately. The following diversification strategies and management practices were established on the farms.

Case study 1

(CS1; Figure 3.1) involved a rainfed organic almond orchard covering 2.63 ha, with 540 trees planted in 1950, cultivated on terraces with 7 m × 7 m spacing. This farm is in the Region of Murcia, southeastern Spain (37°57′31″N, 0°56′17″W). The climate is semiarid Mediterranean with a mean annual precipitation and air temperature of 231 mm and 17.5 °C, respectively.

Figure 3.1 Almond monocrop (left) and almond diversified with thyme (right) from CS1. Photo courtesy of Raul Zornoza.

The mean potential evapotranspiration reaches 1,300 mm/yr (Agricultural Information System of Murcia, http://siam.imida.es). The soil is classified as Calcaric Eutric Regosols (International Union of Soil Sciences [IUSS], 2014) and has a silt-loam texture. Two different treatments were established with a randomized block design with three replicates starting on 5 November 2018. Plots of 210 m^2 were established, with the long side of each one following the direction of the maximum slope, including rows of five trees. The average plot slope was 8%. Treatments were (a) almond monocrop with tillage of the entire plot surface (chisel plowing twice per year at 20-cm depth) and (b) almond plantation with reduced tillage (rototiller [Lander 180, Spain] twice per year at 20 cm only 1.5 m around each tree trunk), with no till in the rest of the alley, and diversified with thyme as alley cropping, at a spacing of 0.5 m (between rows) × 1 m (between individuals within the same row). The dry outer green shell cover of almond rinds was incorporated in the soil after harvest in all plots regardless the treatment (Sáez, 1995). The orchard was rainfed, but the thyme was irrigated on four occasions to ensure proper establishment, with 12 L of water added per plant. Weeds were controlled by tillage in the monocrop. No control of weeds was performed in the no-till area. This experimental design is described in detail by Almagro et al. (2023) and Sánchez-Navarro et al. (2022).

Case study 2

(CS2; Figure 3.2) centered on a commercial mandarin orchard (*C. reticulata* var. Clemenvilla) covering 2.3 ha, with 970 trees cultivated with a spacing of 6 m between rows and 4 m between trees within the same row, planted in 2000. The farm is in the Region of Murcia, southeastern Spain (37°57′31″N, 0°56′17″W). The climate is semiarid Mediterranean with a mean annual

Figure 3.2 Mandarin monocrop (left) and mandarin diversified with fava bean (right) in CS2. Photo courtesy of Raul Zornoza.

temperature of 18°C and a mean annual rainfall of 280 mm. The mean potential evapotranspiration reaches 1,300 mm/yr (Agricultural Information System of Murcia, http://siam.imida.es). The soil is classified as Calcaric Regosol (IUSS, 2014) and has a silt loam texture. Two different treatments were established in a split-plot design with three replicates in March 2018. Plots of 12 m × 28 m were established, with the long side of each following the direction of the maximum slope, including three rows of six trees. The average plot slope was 12%. Treatments were (a) mandarin monoculture with no alley cropping, and (b) mandarin diversified with fava bean grown from September to December for human consumption and barley/vetch grown from February to June for feed, as alley cropping. Fava bean seeds were manually sown, and drip irrigation was used in three rows in each alley in early September, with a spacing of 100 cm between rows and 40 cm between plants (2.5 plants/m^2). Secondary crop residues were incorporated into the soil after harvest of pods. Barley/vetch seeds (1: 3 ratio) were manually sown at 150 kg/ha covering the entire alley surface in early February. Pruning residues were chopped in situ and left as mulch on the soil surface in the monoculture. Pruning residues were removed from the system and not applied as mulch in the diversification system because the alley was covered with the barley/vetch crop.

A drip irrigation system was installed in all trees rows, with one line per tree row and three pressure-compensated emitters (4 L/h) per tree. Irrigation was scheduled weekly and applied at night or in the early morning. The frequency of irrigation varied according to the evaporative demand: one or two times per week in winter, two to seven times per week in spring and autumn, and 7–14 times per week in summer. The electrical conductivity (EC) of the irrigation water varied between 1.0 and 1.5 dS/m. Three drip irrigation lines were established for irrigation of alley crops in the diversification system with pressure-compensated emitters (4 L/h) every 40 cm. Tillage was performed (0–30 cm) immediately before sowing the alley crops to prepare the soil and after the harvest of aerial parts for animal feed to incorporate crop residues into the soil. Water needs were estimated as the product of precipitation, reference crop evapotranspiration (ET0) and the crop coefficients (between 0.25 and 0.55) proposed by the Agricultural Information System of Murcia (http://siam.imida.es) for this area. Weed control was carried out in monocrop by tillage (chisel plowing four or five times per year at 20-cm depth), while weeds were not controlled in the diversification system. Fertilizers were applied as fertigation by use of the commercial products Neptuno PK 28, Neptuno Triton, and Neptuno Pandora (Medifer, Constantino Gutiérrez, SA), as a mixture of soluble nitrogen, phosphorus, potassium, calcium, magnesium, and chelated micronutrients. No irrigation was performed in the alleys of monocrop. This experimental design is described in detail by Sánchez-Navarro et al. (2023).

Case study 3

(CS3; Figure 3.3) involved an irrigated organic melon crop (*Cucumis melo* L.) covering 1,720 m^2, cultivated with a spacing of 200 cm between rows and 120 cm between plants. This crop was in the Tomás Ferro Experimental Farm of the Universidad Politécnica de Cartagena, southeastern Spain (37°41′N, 0°57′E). The climate is semiarid Mediterranean with a total annual precipitation of 275 mm and a mean annual temperature of 18 °C. The annual potential evapotranspiration surpasses 900 mm (Agricultural Information System of Murcia, http://siam.imida.es). The soil is classified as Haplic Calcisol (loamic, hypercalcic; IUSS, 2014) and has a clay loam texture. Three different treatments were established in a completely randomized experimental design with three replicates beginning in May 2018. Plots of 12 m × 10 m (120 m^2) were established. Treatments were (a) melon monocrop with intensive tillage (chisel plowing at the beginning of each crop at 30- to 40-cm depth) and

Figure 3.3 Melon monocrop (left) and melon-cowpea mixed intercropping (right) from CS3. Photo courtesy of Virginia Sanchez-Navarro.

elimination of crop residues, and (b) mixed intercropping system (melon and cowpea plants alternating in the same row), with reduced tillage (chisel plowing at the beginning of each crop at 15- to 20-cm depth) and addition of crop residues. Cowpea seeds in the mixed intercropping system were sown in all melon rows between two melon plants, with a spacing of 200 cm between rows and 120 cm between plants (0.4 plants/m). Crop residues were incorporated in the soil after harvest in the mixed intercropped system. All crops were drip irrigated and grown combined with organic management. The irrigation was scheduled according to climatic conditions, crop coefficient, and evapotranspiration rate. The average irrigation amount was 3,016 m^3/ha per crop cycle for all treatments. Thus, with the same quantity of water used to produce melon combined with monocropping, we grew two associated crops in the mixed intercropping system. Weed control was done by hand-hoeing. In both systems, compost (derived from sheep manure) was added annually at the beginning of each cycle (April), as a traditional practice in the region, with a dose of 14,000 kg/ha. Fertilizers were applied as fertigation by use of organic fertilizer NORGAN (plan-based fertilizer with 45% humic and fulvic acids, 3.2% N, 7% K_2O, Fyneco SL, Spain) in melon monocrop, while the mixed intercropping system received 30% less fertilizer than the melon monocrop to test the reduced need for fertilizers owing to the presence of the legume. This reduction rate in fertilizer application was based on the nitrogen contribution of legume cultivation to soil. This experimental design is described in detailed by Marcos-Pérez et al. (2023).

Soil Sampling

Two soil samplings were carried out at 0- to 10-cm soil depth: after the first crop cycle (8 April 2019, 21 February 2019, and 10 September 2018 for CS1, CS2, and CS3, respectively) and after two crop cycles (24 March 2021, 17 February 2021, and 11 September 2020 for CS1, CS2, and CS3, respectively). Three composite samples (derived from five samples) were randomly taken in each plot in CS1 and CS2 (nine composite soil samples per treatment and case study), while two composite samples (derived from five samples) were randomly taken in each plot in CS3 (six composite soil samples per treatment). In CS1 and CS2, soil samples were collected in the alleys between the tree rows, 2 m from the tree trunks. In CS3, soil samples were collected between two plants in the crop line. Each soil sample at depths of 0–10 cm was divided into three subsamples. The first subsample was air dried for at least 7 days, sieved <2 mm, and kept at room temperature for most physicochemical analyses. The second subsample was sieved <2 mm and kept at 4 °C for the determination of NH_4^+ and NO_3– content, that was performed within

1 week after soil sampling. The third subsample was sieved <2 mm and kept at −20 °C for the biological analyses.

Soil Physicochemical Analyses

Soil Structure

Bulk density (BD) was measured using the cylinder method (cores with 5-cm diameter and 5-cm length). Soil aggregation was determined using the wet sieving method of air-dried soil (Elliott, 1986). The following size distribution of aggregates was carried out: aggregate size distribution floating (AgFLO); aggregate size distribution 250–2,000 µm (Ag250–2,000); aggregate size distribution 100–250 µm (Ag100–250); aggregate size distribution 53–100 µm (Ag53–100); and aggregate size distribution <53 µm (Ag<53).

Water Retention

Soil water content at wilting point −50 kPa (SWW) and soil water content at field capacity −1,500 kPa (SWFC) were measured using disturbed samples sieved at 2 mm (Keller et al., 2007). Actual field soil moisture content (Fma) was measured gravimetrically by drying a soil sample under controlled conditions (Reynolds, 1970).

Nutrient Availability

Soil EC and pH were calculated using deionized water (1 : 5 and 1 : 2.5, respectively). Available phosphorus was analyzed using Burriel-Hernando solution (0.2 g of $CaCO_3$, 0.17 g of $MgCO_3$, 5 mL of glacial acetic acid, and 0.2 mL of H_2SO_4 in 2 L of deionized water) at a 1 : 25 soil : extractant ratio (Díez, 1982). Cation exchange capacity (CEC) was determined using $BaCl_2$ as the exchangeable salt (ISO, 1995). Exchangeable sodium, calcium, potassium, and magnesium were measured in the $BaCl_2$ extract from CEC. Available boron was extracted with deionized water (1 : 5 w/v) at 50 °C (Tang et al., 1997). Bioavailable micronutrients (iron, manganese, copper, and zinc) were extracted using the chelating agent DTPA (1.2 w/v; Kabata-Pendias, 2000; Reichman, 2002). Phosphorus, calcium, magnesium, sodium, potassium, boron, zinc, copper, manganese, and iron concentrations were determined using ICP-MS (Agilent 7900). NH_4^+ was extracted with 2 M KCl at a 1 : 10 soil : extractant ratio (Keeney & Nelson, 2015) and colorimetrically measured (Kandeler & Gerber, 1988). NO_3- was extracted with deionized water at a 1:10 soil : extractant ratio and measured by ion chromatography (Metrohm 861).

Carbon Sequestration

Total organic carbon (TOC) and total nitrogen (Nt) were analyzed using an elemental CHN analyzer (CHN 628, Leco). $CaCO_3$ was calculated from the inorganic carbon content, which was analyzed using an elemental CHN analyzer (CHN 628, Leco). Particulate organic carbon (POC) was determined according to the method of Cambardella and Elliott (1992).

DNA Extraction and Bioinformatics

DNA was used as an indicator of the soil microbial content. DNA extraction from soil was carried out using 0.5 g of soil and the DNeasy PowerSoil Kit (QIAGEN) according to the manufacturer's instructions. Purity and concentration values were obtained using a NanoDrop™ 2000/2000c Spectrophotometer (Thermo Fisher Scientific, Waltham, MA, USA) and a Qubit® 2.0 Fluorometer (Invitrogen, Thermo Fisher Scientific, USA) combined with the Qubit dsDNA HS Assay Kit (Thermo Fisher Scientific), respectively. Amplification of bacterial 16S hypervariable regions was carried out with an Ion 16S Metagenomics Kit (Thermo Fisher Scientific). The library was prepared using an Ion Xpress Plus gDNA Fragment Library Preparation Kit (Thermo Fisher Scientific) combined with an Ion Xpress™ Barcode Adapters kit (Thermo Fisher Scientific) to enable pooling several samples for sequencing reactions. An Agilent 2100 Bioanalyzer instrument was used to evaluate concentration, purity, and size distributions of the barcoded libraries for further dilutions with a suitable Agilent High Sensitivity DNA Kit. Prepared and diluted library amplicons were processed for template preparation by using Ion Sphere Particles (ISPs) via an Ion OneTouch 2 System with a suitable Ion PGM™ Hi-Q View OT2 Kit (Thermo Fisher Scientific), followed by enrichment of ISPs using Ion OneTouch™ ES. Sequencing was carried out using an Ion PGM System, Ion PGM Torrent Server, and a suitable Ion PGM Hi-Q™ View Sequencing kit (Thermo Fisher Scientific) in competence with sequencing chips, Ion 316™ Chip v2 kit. All purification processes between incubation and the amplification reactions during library preparation were processed using DynaMagTM 2 magnetic racks (Thermo Fisher Scientific) and an AMPure XP Purification Kit (Beckman Coulter). Purification of ISPs after the enrichment was done with a DynaMagTM 2 magnetic rack and Dynabeads™ MyOne™ Streptavidin C1 Beads (Özbolat et al., 2023).

Bacterial 16S raw reads were trimmed for barcodes and primers interfaced with the BaseCaller application. Sequences were denoised via ACACIA (Caporaso et al., 2010; Bragg et al., 2012), and low-quality sequences were excluded using the Quantitative Insights into Microbial Ecology (QIIME) pipeline (Caporaso et al., 2010). Bacterial 16S reads with a $Q < 25$ (quality threshold) were then removed, and the remaining sequences were filtered for chimeras using VSEARCH (Rognes et al., 2016), with the ribosomal database project (RDP database). The usable reads were then consigned to operational taxonomic units (OTUs) based on 97% similarity with the SILVA 128 reference database. Low confidence OTUs were removed. The total number of reads analyzed, ranging from 56,149 to 147,290, were grouped into 21,311 OTUs. Subsequently, the number of sequences was rarefied at 56,149 for each sample. Rarefaction curves showed that the sampling and sequencing were close to the plateau for the three cropping systems, with a coverage value of >0.90 (Özbolat et al., 2023).

Functional Genes

The copy number of the microbial functional genes *amoA* (ammonium-oxidizing, nitrifying bacteria), *nirK* (nitrite reductase, denitrifying bacteria), and *narK* (nitrate reductase, denitrifying bacteria) involved in the nitrogen cycle were quantified using the quantitative PCR (qPCR) analysis. Primers used and PCR cycling conditions are shown in Tables 3.2 and 3.3, respectively.

Crop Production

In CS1, almond crop yield (CROPYLD) was calculated by weighing all the almonds harvested directly from the trees in each plot on 29 July 2019 and 3 August 2021. Thyme was harvested by cutting the aerial part of all plants from the entire surface of each plot, which were at full blossom on 23 April 2021. Overall crop yield (OVERALLCROPYLD) was calculated by the addition of almond and thyme yields in each plot. This information is described in detail by Sánchez-Navarro et al. (2022).

In CS2, mandarin yield (CROPYLD) was calculated by weighing all the mandarins harvested directly from the trees in each plot on 24 January 2019 and 3 February 2021. Barley/vetch yield was determined by weighing the aerial biomass cut on 13 June 2019 and 22 June 2020 (no commercial production was obtained in 2018). Fava beans were harvested by collecting all the pods in each plot when the seeds were fresh, on 9 January 2019 and 12 February 2021, and weighed as crop yield. Overall crop yield (OVERALLCROPYLD) was calculated by the addition of mandarin and fava bean and vetch/barley yields in each plot, as described in detail by Sánchez-Navarro et al. (2023).

In CS3, melon crop yield (CROPYLD) was determined by weighing all the fruits per plot when they were ripe and ready for consumption on 6 August 2018 and 24 August 2020 for melons under the monocrop system and on 6 August 2018 and 12 August 2020 for melons under the diversification system. With regard to cowpea yield, all the pods in each plot were harvested when the seeds were dried at the end of the crop cycle on 7 August 2018 and 7 September 2020. Overall crop yield (OVERALLCROPYLD) was calculated by the addition of melon and cowpea yields in each plot, as described in detail by Marcos-Pérez et al. (2023).

Table 3.2 Primers Used in This Study.

Function	Target gene	Gene definition	Primer set	Primer sequence (5′–3′)	Annealing temperature (°C)	Amplicon size (bp)	Reference
Nitrification	*amoA*	Ammonia monooxygenase	amoA1F amoA2R	GGGGTTTCTACTGGTGGT CCCCTCKGSAAAGCCTTCTTC	57	491	Rotthauwe et al. (1997)
Denitrification	*nirK*	Nitrite reductase	nirK876F nirK1040R	ATYGGCGGVCAYGGCGA GCCTCGATCAGRTTRTGGTT	58	165	Henry et al. (2004)
Denitrification	*narG*	Nitrate reductase	narG1960m2F narG250m2R	TAYGTSGGGCAGGARAAACTG CGTAGAAGAAGCTGGTGCTGTT	56	110	López-Gutiérrez et al. (2004)

Table 3.3 PCR Cycling Conditions

amoA	nirK	narG
Initial denaturation: 95 °C, 10 min	Initial denaturation: 95 °C, 10 min	Initial denaturation: 95 °C, 10 min
39 × 95 °C, 15 s; 58 °C, 30 s; 72 °C, 45 s	6 × 95 °C, 15 s; 63 to 58 °C, 30 s (−1 °C by cycle); 72 °C, 30 s; 80 °C, 15 s	6 × 95 °C, 15 s; 65 to 60 °C, 30 s (−1 °C by cycle); 72 °C, 30 s; 80 °C, 15 s
1 × 95 °C, 15 s; 58 °C, 30 s; 72 °C, 45 s	40 × 95 °C, 15 s; 60 °C, 30 s; 72 °C, 30 s; 80 °C, 15 s	40× 95 °C, 15 s; 60 °C, 30 s; 72 °C, 30 s; 80 °C, 15 s
Final extension: 72 °C, 10 min	Final extension: 72 °C, 10 min	Final extension: 72 °C, 10 min

Note. Adapted from Álvaro-Fuentes et al. (2019).

Statistical Analysis

The average values of soil and crop properties measured at the beginning of the experiment (2018 for the melon crop and 2019 for the almond and mandarin crops) were subtracted from values at the end of the experiment (2020 for the melon crop and 2021 for the almond and mandarin crops) and divided by the initial values to obtain the increment data (time Δ data) according to the following equation (Eq. 1):

$$\Delta \text{data}(\%) = [(\text{Final value} - \text{Initial value}) / \text{Initial value}] \times 100 \qquad \text{(Eq. 1)}$$

Although the case studies were not equivalent to one another and represented different diversification systems and management practices, the relative increases or decreases for all properties could be used to compare all the properties since units could be homogenized to a percentage of variation in all crops and treatments. The Δ data were checked to ensure normal distribution using the Kolmogorov–Smirnov test and ln-transformed when necessary to ensure normal distribution. The Δ data of soil physicochemical and microbiological properties and crop production were submitted to two-way ANOVA (crop type and cropping system) to assess the differences between crop type and cropping system at the end of the experiment. A principal components analysis (PCA) was performed with the Δ data of soil physicochemical and microbiological parameters and crop production to study the structure of dependence and correlation established among the variables studied. In addition to the variation rates (%) of different properties, crop yield and overall crop yield data for the last crop cycle were checked to ensure normal distribution using the Kolmogorov–Smirnov test. The crop yield and overall crop yield data were submitted to one-way ANOVA for the factor cropping system. The crop yield and overall crop yield data were checked to ensure normal distribution using the Kolmogorov–Smirnov test and ln-transformed when necessary to ensure normal distribution. Statistical analyses were performed with the software IBM SPSS for Windows, Version 26. The absolute values of soil parameters are shown in Tables 3.4, 3.5, 3.6, and 3.7.

Results

Soil Physicochemical and Biological Properties

Properties related to soil structure (ΔBD, ΔAg100–250, ΔAg53–100), moisture (ΔSWW, ΔSWFC, ΔFMa), and soil fertility (ΔpH, ΔEC, $\Delta CaCO_3$, ΔCEC, ΔCa, ΔMg, ΔK, ΔP, ΔB, ΔCu, ΔMn, ΔFe, and ΔZn) were significantly influenced by crop type (Tables 3.8, 3.9, 3.10). In CS1 (almond), there was a higher increase in soil available phosphorus, compared with the other crops. In CS2 (mandarin), the increase of soil water content with time was higher, with higher increases in ΔSWW compared with the other crops. CS3 (melon) had the highest increase in soil properties linked to the structure (ΔBD), moisture (ΔSWFC), and nutrients (ΔpH, ΔEC, $\Delta CaCO_3$, ΔCu, ΔFe, and ΔMn), compared with the other crops. Regarding the effect of crop diversification and soil conservation practices, in CS1 (almond), ΔAg53–100, ΔFMa, ΔTOC, ΔNH_4, ΔTOC, ΔK, and ΔZn resulted in higher increase under diversification than under the monocrop system; however, ΔSWFC, ΔB, and ΔFe showed higher increases under monocropping. In CS2 (mandarin), ΔBD, ΔFMa, ΔSWFC, and ΔpH showed higher increases under diversification compared with the monocrop system; however, ΔTOC, ΔNH_4, ΔK, and ΔB only increased under monocropping compared with the diversified system. In CS3 (melon), the diversification system led to the increase of ΔAg53–100 and ΔTOC, while the monocrop system was related to increases of ΔFMa and nutrients (ΔK, ΔB, ΔMn, ΔFe, and ΔZn). These results suggest that crop type resulted in an increase of soil moisture content in CS2 and CS3, as well as an increase of soil nutrient content in CS1. In turn, crop diversification linked to soil conservation practices led to improvement of the soil structure in all crops, while the monocrop system was related to a higher soil nutrient content in CS2.

The variation in the number of bacterial OTUs was significantly influenced by crop type, with the highest increases in soils from CS1, with no significant effect of crop diversification. Functional genes were not significantly affected by crop type or crop diversification. However, the cropping system significantly affected the ΔDNA (Table 3.11), with increases in DNA content with crop diversification in CS1 and CS2. DNA content tended to decrease with time in CS3, although this decrease was more intense in the monoculture compared with the diversified

Table 3.4 Absolute Values of the Studied Soil Physicochemical Properties in Almond, Mandarin, and Melon in Monocultures and in Diversification Systems at the Beginning of the Experiment and After 2 Years.

Crop	Cropping system	Soil sampling	BD	SWW	SWFC	FMa	AgFLO	Ag250–2,000	Ag100–250	Ag53–100	Ag<53
			g/cm^3	cm^3/cm^3			%				
Almond	MN	I	1.17 ± 0.01	0.13 ± 0.00	0.24 ± 0.00	0.10 ± 0.00	0.23 ± 0.09	2.08 ± 0.38	17.04 ± 1.24	17.69 ± 0.44	63.18 ± 1.64
		F	1.11 ± 0.06	0.14 ± 0.01	0.25 ± 0.01	0.16 ± 0.02	0.08 ± 0.01	2.48 ± 1.26	21.54 ± 0.87	17.72 ± 0.45	58.26 ± 0.85
	D	I	1.27 ± 0.03	0.15 ± 0.00	0.23 ± 0.01	0.10 ± 0.01	0.11 ± 0.03	1.54 ± 0.43	14.13 ± 0.49	40.64 ± 0.31	46.83 ± 0.56
		F	1.18 ± 0.07	0.14 ± 0.01	0.25 ± 0.01	0.20 ± 0.01	0.08 ± 0.02	4.91 ± 1.70	20.75 ± 1.49	16.95 ± 0.95	57.39 ± 1.12
Mandarin	MN	I	1.24 ± 0.03	0.15 ± 0.01	0.28 ± 0.00	0.08 ± 0.00	1.29 ± 0.17	14.78 ± 2.17	18.88 ± 1.52	15.30 ± 0.71	51.04 ± 1.59
		F	1.19 ± 0.03	0.12 ± 0.00	0.27 ± 0.00	0.13 ± 0.00	0.50 ± 0.06	11.07 ± 1.25	21.60 ± 2.13	16.55 ± 0.65	50.78 ± 3.23
	D	I	1.16 ± 0.04	0.13 ± 0.01	0.25 ± 0.00	0.08 ± 0.01	0.51 ± 0.12	12.79 ± 1.54	15.82 ± 2.11	22.11 ± 3.23	49.28 ± 2.32
		F	1.52 ± 0.02	0.12 ± 0.00	0.26 ± 0.00	0.14 ± 0.00	0.16 ± 0.03	5.41 ± 1.05	14.46 ± 1.29	15.65 ± 0.54	64.48 ± 2.59
Melon	MN	I	1.41 ± 0.13	0.14 ± 0.00	0.25 ± 0.00	0.07 ± 0.00	0.10 ± 0.02	7.12 ± 1.78	30.48 ± 3.14	24.89 ± 0.74	37.51 ± 3.29
		F	1.31 ± 0.03	0.14 ± 0.00	0.26 ± 0.00	0.10 ± 0.01	0.04 ± 0.00	2.56 ± 1.38	31.28 ± 2.33	21.37 ± 0.56	40.12 ± 2.08
	D	I	1.56 ± 0.05	0.16 ± 0.00	0.27 ± 0.00	0.11 ± 0.02	0.04 ± 0.01	6.21 ± 1.90	37.53 ± 1.93	26.05 ± 1.25	34.89 ± 2.88
		F	1.19 ± 0.02	0.14 ± 0.00	0.28 ± 0.00	0.06 ± 0.00	0.11 ± 0.07	1.37 ± 0.23	33.34 ± 2.79	26.40 ± 2.36	38.89 ± 4.32

Note. Values are mean ± standard error. MN, monocrop; D, diversification system; I, initial (at the beginning of the experiment); F, final (after 2 years); BD, bulk density; SWW, soil water content at wilting point; SWFC, soil water content at field capacity; FMa, actual field soil moisture content; AgFLO, aggregate size distribution floating; Ag250–2000, aggregate size distribution 250–2000 μm; Ag100–250, aggregate size distribution 100–250 μm; Ag53–100, aggregate size distribution 53–100 μm; Ag<53, aggregate size distribution <53 μm.

Table 3.5 Absolute Values of the Studied Soil Physicochemical Properties in Almond, Mandarin, and Melon in Monocultures and Diversification Systems at the Beginning of the Experiment and After 2 Years.

Crop	Diversification type	Soil sampling	TOC (g/kg)	POC (g/kg)	$CaCO_3$ (%)	Nt (g/kg)	NH_4 (mg/kg)	pH	EC (dS/m)
Almond	MN	I	4.51 ± 0.17	1.17 ± 0.10	58.41 ± 0.44	0.73 ± 0.02	2.68 ± 0.15	8.38 ± 0.03	0.20 ± 0.01
		F	4.59 ± 0.34	1.15 ± 0.10	58.15 ± 0.37	0.92 ± 0.03	1.54 ± 0.22	8.54 ± 0.02	0.20 ± 0.02
	D	I	3.74 ± 0.26	0.86 ± 0.11	59.54 ± 0.43	0.54 ± 0.03	1.55 ± 0.28	8.37 ± 0.04	0.26 ± 0.04
		F	4.62 ± 0.30	1.20 ± 0.28	59.39 ± 0.57	0.59 ± 0.08	1.78 ± 0.23	8.61 ± 0.05	0.24 ± 0.03
Mandarin	MN	I	10.52 ± 0.98	5.08 ± 0.23	54.31 ± 0.26	1.29 ± 0.08	4.45 ± 0.36	7.60 ± 0.03	1.36 ± 0.14
		F	10.38 ± 0.40	4.28 ± 0.21	53.66 ± 0.24	1.26 ± 0.03	5.82 ± 1.14	7.77 ± 0.03	1.05 ± 0.12
	D	I	8.89 ± 0.18	3.95 ± 0.41	51.80 ± 0.82	1.15 ± 0.05	4.96 ± 0.58	7.53 ± 0.07	1.17 ± 0.28
		F	8.50 ± 0.47	2.68 ± 0.22	51.95 ± 0.60	1.10 ± 0.03	3.50 ± 0.29	8.24 ± 0.06	0.61 ± 0.19
Melon	MN	I	9.85 ± 0.38	2.13 ± 0.10	26.18 ± 0.37	1.13 ± 0.05	0.92 ± 0.14	8.53 ± 0.02	0.41 ± 0.13
		F	9.86 ± 0.40	1.78 ± 0.13	30.50 ± 1.70	1.08 ± 0.03	0.41 ± 0.21	7.98 ± 0.03	0.50 ± 0.01
	D	I	11.47 ± 0.27	1.92 ± 0.08	21.74 ± 0.81	1.35 ± 0.04	3.00 ± 1.01	8.39 ± 0.03	0.31 ± 0.02
		F	11.72 ± 0.45	2.44 ± 0.10	24.20 ± 0.80	1.35 ± 0.01	0.77 ± 0.22	7.73 ± 0.05	0.55 ± 0.03

Note. Values are mean ± standard error. MN, monocrop; D, diversification system; I, initial (at the beginning of the experiment); F, final (after 2 years); TOC, total organic carbon; POC, particulate organic carbon; Nt, total nitrogen; EC, electrical conductivity.

Table 3.6 Absolute Values of the Studied Soil Physicochemical Properties in Almond, Mandarin, and Melon in Monocultures and Diversification Systems at the Beginning of the Experiment and After 2 Years.

Crop	Diversification type	Soil sampling	CEC	Ca	Mg	K	P	B	Cu	Mn	Fe	Zn
			cmol/kg				mg/kg					
Almond	MN	I	11.36 ± 0.41	8.85 ± 0.35	1.97 ± 0.11	0.55 ± 0.02	0.42 ± 0.16	0.38 ± 0.05	0.63 ± 0.05	9.85 ± 1.23	1.46 ± 0.16	0.16 ± 0.02
		F	13.84 ± 0.35	11.25 ± 0.39	1.96 ± 0.08	0.55 ± 0.03	3.98 ± 1.60	0.31 ± 0.02	0.78 ± 0.04	2.69 ± 0.18	1.55 ± 0.07	0.24 ± 0.01
	D	I	10.26 ± 0.52	7.84 ± 0.43	1.92 ± 0.09	0.45 ± 0.03	0.34 ± 0.06	0.44 ± 0.06	0.91 ± 0.09	12.71 ± 0.77	2.10 ± 0.21	0.20 ± 0.02
		F	13.56 ± 0.43	10.26 ± 0.47	2.48 ± 0.14	0.57 ± 0.05	1.72 ± 0.53	0.38 ± 0.04	0.77 ± 0.05	2.36 ± 0.04	1.56 ± 0.11	0.29 ± 0.08
Mandarin	MN	I	15.35 ± 0.58	11.77 ± 0.55	2.07 ± 0.38	0.87 ± 0.04	1.26 ± 0.37	1.14 ± 0.14	1.03 ± 0.06	19.44 ± 1.55	4.87 ± 0.38	0.37 ± 0.04
		F	17.08 ± 0.24	14.36 ± 0.17	1.39 ± 0.04	1.05 ± 0.02	1.71 ± 0.46	1.84 ± 0.11	0.70 ± 0.02	3.46 ± 0.10	2.24 ± 0.07	0.49 ± 0.03
	D	I	14.70 ± 0.61	10.27 ± 0.16	2.93 ± 0.26	0.68 ± 0.04	2.59 ± 0.53	0.80 ± 0.10	0.90 ± 0.06	21.86 ± 2.32	7.30 ± 2.10	0.34 ± 0.06
		F	16.17 ± 0.20	12.26 ± 0.28	2.67 ± 0.10	0.62 ± 0.02	2.30 ± 0.50	0.82 ± 0.05	0.65 ± 0.02	3.20 ± 0.16	2.29 ± 0.11	0.34 ± 0.02
Melon	MN	I	22.30 ± 0.31	11.96 ± 0.12	8.50 ± 0.35	0.92 ± 0.08	22.19 ± 1.14	0.66 ± 0.01	0.79 0.14±	3.75 ± 0.43	1.72 ± 0.32	1.61 ± 0.23
		F	17.76 ± 0.09	12.04 ± 0.09	4.20 ± 0.03	0.66 ± 0.04	16.40 ± 1.36	0.42 ± 0.02	2.14 ± 0.04	15.15 ± 0.21	5.44 ± 0.07	4.12 ± 0.07
	D	I	25.24 ± 0.90	12.28 ± 0.10	10.98 ± 0.97	0.98 ± 0.06	59.48 ± 3.77	0.79 ± 0.07	1.43 ± 0.08	7.18 ± 0.36	2.70 ± 0.20	5.20 ± 0.22
		F	19.65 ± 0.36	12.72 ± 0.19	4.98 ± 0.14	0.96 ± 0.04	53.99 ± 1.26	0.67 ± 0.04	2.38 ± 0.51	15.51 ± 3.20	4.79 ± 1.04	6.87 ± 1.60

Note. Values are mean ± standard error. MN, monocrop; D, diversification system; I, initial (at the beginning of the experiment); F, final (after 2 years); CEC, cation exchange capacity.

Table 3.7 Absolute Values of the Studied Soil Microbiological Properties in Almond, Mandarin, and Melon in Monocultures and in Diversification Systems at the Beginning of the Experiment and After 2 Years.

Crop	Diversification type	Soil sampling	OTU	amoA	nirK	narG	DNA
			No. of reads	copies/g soil			ng DNA/g soil
Almond	MN	I	853 ± 199	3,908,692 ± 1,579,216	3,556,910 ± 1,437,087	3,361,475 ± 1,358,126	1,420 ± 151
		F	1,177 ± 53	67,353 ± 134,366	521,184 ± 95,390	223,979 ± 33,071	1,410 ± 129
	D	I	853 ± 322	1,101,056 ± 270,068	1,001,961 ± 245,761	946,908 ± 232,258	1,181 ± 127
		F	1,210 ± 188	1,462,499 ± 260,792	497,407 ± 104,846	450,871 ± 164,759	1,541 ± 266
Mandarin	MN	I	804 ± 192	13,495,937 ± 4,338,035	11,606,505 ± 4,338,035	12,281,303 ± 3,947,612	3,322 ± 277
		F	275 ± 11	16,137,389 ± 5,187,084	13,362,896 ± 5,187,084	13,359,896 ± 4,294,307	3,106 ± 333
	D	I	1,270 ± 432	1,158,216 ± 624,345	996,066 ± 624,345	1,053,976 ± 568,154	644 ± 365
		F	343 ± 86	1,384,904 ± 746,543	1,384,904 ± 746,543	1,146,541 ± 618,052	2,752 ± 296
Melon	MN	I	618 ± 13	14,681,835 ± 3,018,446	12,626,378 ± 2,595,863	13,360,470 ± 2,746,789	3,364 ± 217
		F	306 ± 26	10,435,878 ± 1,130,908	12,866,279 ± 2,645,185	13,534,156 ± 2,782,494	1,960 ± 282
	D	I	774 ± 15	8,626,313 ± 2,205,470	7,418,630 ± 1,896,704	7,849,945 ± 2,006,977	3,497 ± 561
		F	315 ± 19	10,774,141 ± 3,699,655	8,353,404 ± 1,631,225	7,390,159 ± 1,535,329	2,449 ± 812

Note. Values are mean ± standard error. MN, monocrop; D, diversification system; I, initial (at the beginning of the experiment); F, final (after 2 years); OUT, operational taxonomic units; *amoA*, ammonia monooxygenase; *nirK*, nitrite reductase; *narG*, nitrate reductase.

Table 3 8 Variation Rates (%) of the Studied Soil Physicochemical Properties in Almond, Mandarin, and Melon in Monocultures and After 2 Years of Diversification Compared with Initial Values at the Beginning of the Experiment.

Crop	Cropping system	ΔBD	ΔSWW	ΔSWFC	ΔFMa	ΔAgFLO	ΔAg250–2,000	ΔAg100–250	ΔAg53–100	ΔAg<53
Almond	MN	−5.23 ± 5.38	5.80 ± 1.88	4.25 ± 1.41	68.93 ± 16.88	10.89 ± 45.08	−4.20 ± 35.87	31.05 ± 9.79	0.65 ± 3.77	−7.07 ± 3.66
	D	−7.11 ± 4.89	−4.72 ± 5.01	5.21 ± 1.85	119.39 ± 20.20	108.44 ± 83.98	709.01 ± 368.18	47.66 ± 11.18	−58.18 ± 2.58	31.64 ± 3.50
Mandarin	MN	−3.48 ± 2.42	−14.96 ± 3.94	−3.98 ± 0.98	62.80 ± 7.73	−56.56 ± 6.26	−2.85 ± 23.14	14.38 ± 5.06	10.24 ± 7.29	−0.29 ± 6.13
	D	32.78 ± 6.00	−6.75 ± 4.20	3.17 ± 1.50	93.24 ± 18.07	−20.65 ± 47.22	−58.02 ± 5.67	2.50 ± 12.97	−17.00 ± 11.00	33.16 ± 8.24
Melon	MN	−3.54 ± 10.29	−1.65 ± 2.86	5.67 ± 0.53	41.85 ± 6.05	−46.54 ± 16.67	−69.53 ± 9.24	7.33 ± 15.30	5.11 ± 4.66	10.89 ± 11.62
	D	−23.60 ± 2.65	−7.30 ± 1.07	4.66 ± 2.21	−27.78 ± 18.60	152.20 ± 96.05	5.79 ± 89.03	−9.79 ± 10.53	25.18 ± 12.35	15.49 ± 17.19
F value										
Crop type (C)		12.08***	4.48*	7.56**	19.69***	2.25 ns	2.65 ns	7.91**	11.79***	0.16***
Cropping system (CS)		1.12 ns	1.73 ns	3.3 ns	0.08 ns	11.99*	2.45 ns	0.06 ns	10.37**	9.91**
C x CS		10.56**	2.42 ns	4.70*	10.09**	4.71 ns	2.24 ns	1.74 ns	10.07**	1.67 ns

Note. Values are mean ± standard error. MN, monocrop; D, diversification system; BD, bulk density; SWW, soil water content at wilting point; SWFC, soil water content at field capacity; FMa, actual field soil moisture content; AgFLO, aggregate size distribution floating; Ag250–2000, aggregate size distribution 250–2,000 µm; Ag100–250, aggregate size distribution 100–250 µm; Ag53–100, aggregate size distribution 53–100 µm; Ag<53, aggregate size distribution <53 µm.

* $P < .05$.

** $P < .01$.

*** $P < .001$.

ns, not significant ($P > .05$).

Table 3.9 Variation Rates (%) of the Studied Soil Physicochemical Properties in Almond, Mandarin, and Melon in Monocultures and After 2 Years of Diversification Compared with Initial Values at the Beginning of the Experiment.

Crop	Diversification type	ΔTOC	ΔPOC	$\Delta CaCO_3$	ΔNt	ΔNH_4	ΔpH	ΔEC
Almond	MN	1.61 ± 6.44	5.41 ± 14.32	−0.44 ± 0.48	27.61 ± 4.98	−42.20 ± 7.99	1.91 ± 0.32	−1.71 ± 6.69
	D	26.19 ± 8.16	52.44 ± 31.78	−0.25 ± 0.68	11.52 ± 14.54	114.61 ± 70.78	2.81 ± 0.25	−2.27 ± 13.11
Mandarin	MN	4.41 ± 8.66	−14.37 ± 5.38	−1.19 ± 0.49	0.55 ± 5.82	39.73 ± 28.96	2.19 ± 0.58	−22.02 ± 6.25
	D	−4.27 ± 5.08	−23.32 ± 12.75	0.39 ± 1.31	−2.38 ± 5.69	−19.30 ± 12.22	9.45 ± 0.63	−48.59 ± 7.36
Melon	MN	0.31 ± 3.70	−15.81 ± 7.90	16.46 ± 6.12	−3.63 ± 5.00	−53.89 ± 23.95	−6.49 ± 0.41	47.98 ± 23.29
	D	2.20 ± 3.19	27.71 ± 6.89	11.44 ± 1.82	0.31 ± 2.63	−31.39 ± 47.51	−7.91 ± 0.74	82.26 ± 17.22
F value								
Crop type (C)		2.70***	3.50*	22.88***	2.77 ns	8.36 ns	291***	24.58***
Cropping system (CS)		0.94 ns	4.55 ns	0.02 ns	1.01 ns	0.16 ns	22.46***	0.48 ns
C × CS		3.64*	1.69 ns	0.14 ns	1.12 ns	24.81*	31.24***yyy	2.12 ns

Note. Values are mean ± standard error. MN, monocrop; D, diversification system; TOC, total organic carbon; POC, particulate organic carbon; Nt, total nitrogen; EC, electrical conductivity.

* $P < .05$.

*** $P < .001$.

ns, not significant ($P > .05$).

Table 3.10 Variation Rates (%) of the Studied Soil Physicochemical Properties in Almond, Mandarin, and Melon in Monocultures and After 2 Years of Diversification Compared with Initial Values at the Beginning of the Experiment.

Crop	Diversification type	ΔCEC	ΔCa	ΔMg	ΔK	ΔP	ΔB	ΔCu	ΔMn	ΔFe	ΔZn
Almond	MN	22.46 ± 2.72	27.67 ± 2.77	1.26 ± 4.29	0.76 ± 3.77	1562 ± 652	−12.44 ± 6.58	34.26 ± 15.05	−69.85 ± 3.75	15.58 ± 11.24	67.66 ± 20.27
	D	34.34 ± 6.55	33.22 ± 7.50	33.50 ± 13.75	26.75 ± 10.12	403 ± 140	−5.15 ± 15.73	−9.38 ± 9.86	−80.76 ± 1.38	−20.80 ± 8.25	97.72 ± 90.21
Mandarin	MN	12.11 ± 3.07	24.30 ± 6.64	−23.56 ± 7.49	22.16 ± 4.96	164 ± 95	70.60 ± 12.63	−31.26 ± 3.03	−81.33 ± 1.47	−52.62 ± 2.28	35.12 ± 7.96
	D	11.22 ± 3.98	19.54 ± 2.66	−3.88 ± 8.31	−5.62 ± 6.34	168 ± 169	9.75 ± 7.9	−26.23 ± 3.90	−84.43 ± 1.24	−50.20 ± 9.99	24.73 ± 19.89
Melon	MN	−20.28 ± 1.29	0.71 ± 1.23	−50.25 ± 2.10	−25.40 ± 9.09	−26.60 ± 3.11	−36.46 ± 3.53	205.65 ± 53.64	325.03 ± 47.65	275.87 ± 82.64	177.44 ± 40.57
	D	−21.76 ± 3.07	3.56 ± 0.73	−52.95 ± 5.15	−1.44 ± 3.76	−7.69 ± 6.36	−12.74 ± 8.52	64.27 ± 35.11	111.01 ± 41.70	80.75 ± 42.15	33.54 ± 31.39
F value											
Crop type (C)		73.46***	16.82***	38.04***	8.91**	6.01*	23.94***	27.17***	87.17***	39.97***	8.42**
Cropping system (CS)		0.48 ns	0.04 ns	5.17*	1.46 ns	0.03 ns	2.20 ns	7.71*	12.96**	15.60**	4.90*
C × CS		1.44 ns	0.50 ns	2.19 ns	10.59**	0.63 ns	10.03**	0.73 ns	10.57**	6.27*	4.90*

Note. Values are mean ± standard error. MN, monocrop; D, diversification system; CEC, cation exchange capacity.
* $P < .05$.
** $P < .01$.
*** $P < .001$.
ns: not significant ($P > .05$).

Table 3.11 Variation Rates (%) of the Studied Soil Microbiological Properties in Almond, Mandarin, and Melon in Monocultures and After 2 Years of Diversification Compared with Initial Values at the Beginning of the Experiment.

Crop	Diversification type	ΔOTU	Δ*amoA*	Δ*nirK*	Δ*narG*	ΔDNA
Almond	MN	57.91 ± 22.53	−56.37 ± 22.38	−65.40 ± 18.67	−86.36 ± 5.19	8.02 ± 21.55
	D	79.86 ± 28.11	78.12 ± 56.41	−39.38 ± 15.96	−41.96 ± 18.92	40.60 ± 29.93
Mandarin	MN	−53.57 ± 13.06	63.11 ± 47.23	136.44 ± 128.05	88.95 ± 115.42	−3.81 ± 13.80
	D	−48.89 ± 28.67	164.14 ± 131.55	113.98 ± 98.42	135.69 ± 123.68	765 ± 206
Melon	MN	−50.35 ± 4.71	−12.36 ± 20.71	18.42 ± 31.81	13.46 ± 25.86	−38.67 ± 12.47
	D	−59.25 ± 2.62	86.97 ± 81.65	54.67 ± 49.88	30.27 ± 46.62	−19.69 ± 28.99
F value						
Crop type (C)		25.64***	1.12 ns	4.19 ns	1.14 ns	4.31 ns
Cropping system (CS)		0.13 ns	3.70 ns	1.11 ns	0.45 ns	28.90*
C × CS		0.30 ns	0.04 ns	2.14 ns	0.45 ns	8.55*

Note. Values are mean ± standard error. MN, monocrop; D, diversification system; OUT, operational taxonomic unit; *amoA*, ammonia monooxygenase; *nirK*, nitrite reductase; *narG*, nitrate reductase.

* $P < .05$.

*** $P < .001$.

ns, not significant ($P > .05$).

system. In this respect, the combination of diversified crops and soil conservation practices increased soil microbial content in CS1 and CS2.

The absolute values of these properties can be found in reports by Almagro et al. (2023), Bonetti et al. (2023), Marcos-Pérez et al. (2023), Martínez-Mena et al. (2021), and Sánchez-Navarro et al. (2022, 2023).

Crop Production

In CS1 (almond), ΔCROPYLD was not influenced by cropping system. However, ΔOVERALLCROPYLD was affected by the introduction of thyme as the alley crop, with the highest increases under diversification compared with the monocrop system. In CS2 (mandarin) and CS3 (melon), ΔCROPYLD and ΔOVERALLCROPYLD were not significantly influenced by the cropping system, although there was an increasing trend with crop diversification compared with monoculture (Table 3.12).

Interrelationship Between Soil Properties and Crop Production

The PCA results showed that 100% of the total variation could be explained by the first two PCs (Figure 3.4). PC1, which explained 58% of the variation, separated CS1 and CS2 (positive scores) from CS3 (negative scores). CS1 was related to the highest increases in Δ*nirK*, ΔFMa, ΔCa, ΔK, ΔNt, ΔAg53–100, Δ*narG*, ΔAg100–250, ΔPOC, Δ*amoA*, $\Delta CaCO_3$, ΔEC, and ΔP, and to the highest decreases in ΔpH, ΔAg<53, ΔZn, ΔFe, ΔCu, ΔAg250–2,000, ΔOTUB, and ΔMn (Table 3.13). In turn, PC1 separated the diversified system and the monocrop system in CS3. Diversification was related to higher increases in Δ*nirK*, ΔFMa, ΔAg100–250, ΔPOC, Δ*amoA*, $\Delta CaCO_3$, ΔEC, and ΔP, and to higher decreases in ΔpH, ΔAg<53, ΔZn, ΔFe, ΔCu, ΔAg250–2,000, ΔOTUs, and ΔMn. PC2, which explained 42% of the variation, separated the diversification system (positive scores) and the monocrop system (negative scores) in CS1 and CS3 (Table 3.13). Diversification was related to higher increases in ΔMg, ΔTOC, ΔDNA, ΔCEC, ΔB, ΔCROPYLD, ΔOVERALLCROPYLD, ΔSWFC, and ΔNH_4, and to higher decreases in ΔSWW, ΔBD, and ΔAgFLO. In turn, PC2 separated the monocrop system (positive scores) and the diversification system (negative scores) in CS2. It is important to highlight that increases in crop yield were in the same PC2, and so significantly correlated to increases in TOC, DNA, CEC, B, SWDC, Mg, and NH_4^+ and decreases in BD. These findings confirm that improvement in soil properties is related to increases in crop production.

Table 3.12 Variation Rates (%) and Data of Crop Production (kg/ha) in Almond, Mandarin, and Melon in Monoculture and After 3 Years of Diversification Compared with Initial Values at the Beginning of the Experiment.

		ΔCROPYLD (main crop)	ΔOVERALLCROPYLD (main + secondary crop)	CROPYLD last cycle (main crop)	CROPYLD last cycle (secondary crop)	OVERALLCROPYLD (main + secondary crop)
Crop	**Diversification type**	**%**	**%**		**kg/ha**	
Almond	MN	−30.01 ± 12.69	−30.01 ± 12.69	82 ± 11	—	82 ± 11
	D	−23.60 ± 30.25	1824 ± 374	118 ± 34	1,795 ± 364	1,913 ± 348
F value						
Cropping system		1.07 ns	42.56**	1.00 ns	—	27.63**
Crop	**Diversification type**	**ΔCROPYLD (main crop)**	**ΔOVERALLCROPYLD (main + secondary crop)**	**CROPYLD last cycle (main crop)**	**CROPYLD last cycle (secondary crop)**	**–**
Mandarin	MN	−41.63 ± 8.13	−41.63 ± 8.13	11,978 ± 876	—	11,978 ± 87
	D	−23.44 ± 19.71	−31.24 ± 15.47	11,423 ± 3,297	2,691 ± 270	14,114 ± 3,363
F value						
Cropping system		0.73 ns	0.35 ns	0.02 ns	—	0.37 ns
Crop	**Diversification type**	**ΔCROPYLD (main crop)**	**ΔOVERALLCROPYLD (main + secondary crop)**	**CROPYLD last cycle (main crop)**	**CROPYLD last cycle (secondary crop)**	**–**
Melon	MN	−43.30 ± 0.99	−43.30 ± 0.99	8,561 ± 247	—	8,561 ± 247
	D	−57.88 ± 7.09	−57.40 ± 7.24	10,843 ± 1,415	165 ± 52	11,009 ± 1,456
F value						
Cropping system		4.15 ns	3.72 ns	2.52 ns	—	2.74 ns

Note. Values are mean ± standard error. MN, monocrop; D, diversification system; CROPYLD, crop yield; OVERALLCROP YIELD, sum of the main and secondary crops yield. Main crops: almond (CS1), mandarin (CS2), and melon (CS3). Secondary crops: thyme (CS1), fava bean and vetch/barley (CS2), and cowpea (CS3).

** $P < .01$.

ns, not significant ($P > .05$).

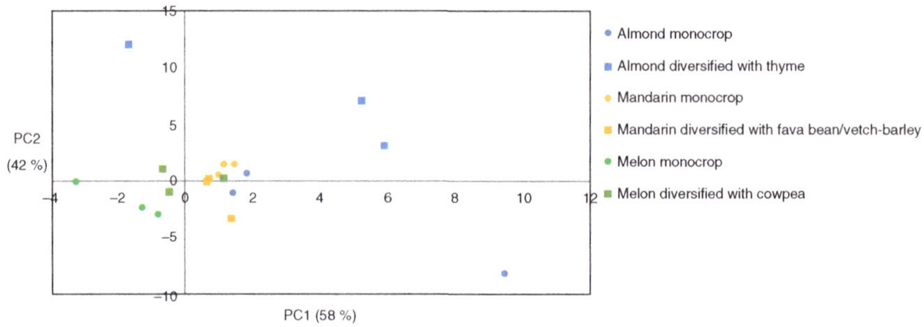

Figure 3.4 PCA factor scores of variations in soil physicochemical and microbiological properties and crop production in monocultures and diversified crops of study. Color represents crop type: blue = almond crop (CS1); yellow = mandarin crop (CS2); green = melon crop (CS3). Figure type represents cropping system: square = diversification system; circle = monocrop system.

Discussion

The establishment of the different crop types (almond, mandarin, and melon) under diversification systems with soil conservation practices was related to improved soil health, which was mainly due to the increase in soil physicochemical properties associated with the soil structure and to the water content. Nutrient availability and carbon sequestration were increased with the adoption of diversified practices in almond and melon crops and with monocrop in the mandarin crop. In turn, the diversification system had a positive impact on soil microbial content in the woody crops. Finally, almond and melon crop yields were increased by the establishment of diversified crop systems.

Soil Fertility and Carbon Sequestration

The almond crop demonstrated improvements in soil organic carbon and nutrient availability, the mandarin crop showed improvements in water retention, and melon crop had improvements in soil structure, water retention, and nutrient content. Thus, the melon crop was more closely associated with overall improvement in soil health, which can be attributed to the root system, which had shallow and deep roots, and organic fertilization (Garcia et al., 2013). In this regard, the beneficial effects of different kinds of crops on soil physical health such as aggregation or soil organic carbon content are attributed to the different root systems and the use of organic fertilizers (Alhameid et al., 2017).

Crop diversification associated with soil conservation management showed a significant improvement of the soil structure in all crops (aggregate stability in

Table 3.13 Matrix of PCA Obtained with Variation Rates (%) of Soil Physicochemical Biological Parameters at 0- to 10-cm Soil Depth and Crop Production.

Variance explained	PC1 (58%)	PC2 (42%)
Δ*nirK*	**1.000**	—
ΔpH	**−1.000**	—
ΔFMa	**0.995**	—
ΔCa	**0.995**	—
ΔAg<53	**−0.991**	—
ΔK	**0.986**	—
ΔZn	**−0.982**	—
ΔNt	**0.977**	—
ΔFe	**−0.973**	—
ΔAg53–100	**0.955**	—
Δ*narG*	**0.954**	—
ΔAg100–250	**0.926**	—
ΔCu	**−0.919**	—
ΔPOC	**0.884**	—
Δ*amoA*	**0.863**	—
ΔAg250–2,000	**−0.854**	—
$\Delta CaCO_3$	**0.847**	—
ΔOTU	**−0.840**	—
ΔMn	**−0.818**	—
ΔEC	**0.818**	—
ΔP	**0.805**	—
ΔMg	—	**0.999**
ΔSWW	—	**−0.999**
ΔBD	—	**−0.997**
ΔAgFLO	—	**−0.996**
ΔTOC	—	**0.989**
ΔDNA	—	**0.980**
ΔCEC	—	**0.962**
ΔB	—	**0.926**
ΔOVERALLCROPYLD	—	**0.917**
ΔCROPYLD	—	**0.913**
ΔSWFC	—	**0.902**
ΔNH_4	—	**0.806**

Note. See previous tables for definitions of abbreviations.

CS1 and CS3 and BD in CS2), soil water retention in mandarin and melon crops, and nutrient availability in the almond crop. Regarding carbon sequestration and storage, we observed a higher organic carbon content in the diversified system for almond and melon crops after two crop cycles. These results show that the presence of different crops (variety of biomass crop and root exudates) at the same time led to a higher increase in carbon-rich root exudates, moisture, and nutrients (Farmaha et al., 2022; McDaniel et al., 2014, Tiemann et al., 2015). Moreover, the choice of the crops involved in the different diversification systems combined with soil conservation practices are key factors in improving soil fertility not only by the nutrients from crop residues and root exudates, but also processes such as soil organic matter mineralization and solubilization from minerals related to higher microbial content (Yan et al., 2023).

Among monocrops, the mandarin monocrop system resulted in better results than the diversified systems in terms of carbon sequestration in soil and nutrient availability. The mandarin monocrop system resulted in higher increases of TOC content and nutrients. This outcome is not in line with previous results in literature, and with results provided by almond and melon crops in this study, where crop diversification tended to improve soil health. This opposite outcome could be explained by the need for more time being needed to achieve the improvement of soil nutrients and carbon sequestration in this Mediterranean area with high organic matter mineralization rates and leaching under irrigation. In turn, an important determinant that might be responsible for these results is a lower complementarity between fava bean, barley, and vetch and the mandarin trees involved in the diversification system. There may have been competition for nutrients or the presence of different root system depths or microclimate variations (Raseduzzaman and Jensen, 2017). Furthermore, it should be noted that the interaction between management and environment in a field trial of a mandarin crop requires a medium- or long-term trial to understand the effect of the cropping system on soil properties (Hoffmann et al., 2020). In addition, the improvement of the soil nutrients content can be caused by a wide range of compounds secreted by the plant roots, with the quantity and quality of these compounds depending on plant genotype, plant age, and different biotic and abiotic factors (Jones et al., 2004). Alternatively, a priming effect could have taken place as a result of the introduction of legumes as secondary crops and the elimination of pruning remains. The cultivation of legumes represents an extra contribution of nitrogen to the soil, since they are able to fix the atmospheric nitrogen (lower C/N ratio; Jensen et al., 2020). The elimination of pruning remains represents a loss of carbon to the soil (higher C/N ratio; Shahbaz et al., 2017). Thus, both practices could involve lower TOC content in the soil of a diversified mandarin crop.

Biological Properties

Crop diversification combined with soil conservation practices increased the DNA content of the soil, which could indicate greater soil microbial biomass. The adoption of reduced tillage in the almond crop, the adoption of reduced tillage and incorporation of crop residues in the mandarin crop, or the reduced tillage depth, organic fertilization, and crop residues in the melon crop combined with the crop diversification could promote greater benefits such as higher water storage capacity (Benjamin et al., 2003), nutrient cycling (Garcia et al., 2008; Reddy et al., 2009), and an increased population of beneficial microorganisms, mainly owing to the higher amount of crop residues incorporated into the soil (Wang et al., 2007; Simmons & Coleman, 2008). In terms of soil biological health, the highest soil DNA concentrations, which were found under diversification systems, can be attributed to the net balance between factors such as timing, quantity and quality of root metabolites; decomposition of aboveground crop residues; reduced tillage; and organic fertilization (Follett, 2001; Roger-Estrade et al., 2010). Under reduced tillage, above- and belowground biomass production increases the amount of organic matter in soil, and thus, the soil aggregation is improved (Anders et al., 2012).

Crop Production

In this study, crop diversification not only enhanced soil fertility and microbiology, but also crop productivity. This fact was observed in almond and melon crops, in which the diversification of the main crops with secondary crops led to higher yields. In this respect, in the melon crop, although the crop productivity did not show significant differences by cropping system, the main and overall crop yield were higher under the diversified system. This leads to significant savings in water and fertilizers because two crops were cultivated with the same amount of water and a lower fertilizer supply compared with the melon monocrop. This study validated the improvement of the soil health after the establishment of diversification systems in almond and melon crops, with a positive relation between the increases of crop yield (CROPYLD and OVERALCROPYLD) and the increases of soil physicochemical (TOC, CEC, B, Mg, NH_4^+, and SWFC) and microbiological (DNA content) properties. Diversified crops benefit the soil microorganisms mainly because of higher levels of microbial-derived products (metabolites and enzymes) that can enhance microbial functions such as nutrient cycling (Kumar Singh et al., 2019). In this respect, Luo et al. (2018) and Liu et al. (2022) observed an increase of nutrient cycling and microorganisms' activity because of soil conservation agriculture being established in the agroecosystem. Acosta-Martínez et al. (2007) observed a positive effect on soil health parameters such as the microbial population composition combined with soil conservation practices. Sainju et al. (2021) reported that soil conservation practices such as continuous cropping with increased incorporation of crop residues led to a higher nutrient availability, while no-tillage led to decreased mineralization of crop residues, and thus the establishment of these practices resulted in the increase of the soil nutrient content, increasing crop yield.

It is important to highlight that the multivariate analysis demonstrated in all crops that increased crop yield was associated with increases in soil organic carbon content, microbial biomass, and nutrient availability and decreases in BD, confirming that practices that contribute to enhance soil health are related to higher production under Mediterranean conditions, both under rainfed and irrigated systems, independently of the crop type or cropping system. This confirmation aligns with previous results that suggest that investing in sustaining soil health can have important benefits such as high yields (Sainju et al., 2021; Bonetti et al., 2023).

Conclusions

Our results suggest that various main crops (almond, mandarin, and melon) had a different effect on soil physical, chemical, and microbiological properties. As a general pattern, the establishment of crop diversification combined with soil conservation management practices in almond and melon crops seems to represent a good alternative for soil health in the short term compared with monocrop systems, related to higher soil organic carbon levels, aggregate stability, water storage, nutrient cycling, microbial biomass, and overall crop yield. However, for the mandarin crop, the monocrop system was linked to increased soil organic carbon, nutrient content, and microbial richness in the short term. Finally, our results confirm that crop yield was positively related to soil organic carbon content, microbial biomass, and nutrient availability in all case studies.

Acknowledgments

This work was supported by the European Commission Horizon 2020 project Diverfarming (grant agreement 728 003).

References

Abid, M., & Lal, R. (2008). Tillage and drainage impact on soil quality. Aggregate stability, carbon and nitrogen pools. *Soil and Tillage Research*, *100*(1–2), 89–98. https://doi.org/10.1016/j.still.2008.04.012

Acosta-Martínez, V., Mikha, M. M., & Vigil, M. F. (2007). Microbial communities and enzyme activities in soils under alternative crop rotations compared to wheat-fallow for the Central Great Plains. *Applied Soil Ecology*, *37*(1–2), 41–52. https://doi.org/10.1016/j.apsoil.2007.03.009

Adekalu, K. O., Okunade, D. A., & Osunbitan, J. A. (2006). Compaction and mulching effects on soil loss and runoff from two southwestern Nigeria agricultural soils. *Geoderma*, *137*(1–2), 226–230. https://doi.org/10.1016/j.geoderma.2006.08.012

Alam, M. K., Bell, R. W., & Biswas, W. K. (2019). Decreasing the carbon footprint of an intensive rice-based cropping system using conservation agriculture on the Eastern Gangetic Plains. *Journal of Cleaner Production*, *218*, 259–272. https://doi.org/10.1016/j.jclepro.2019.01.328

Alhameid, A., Ibrahim, M., Kumar, S., Sexton, P., & Schumacher, T. E. (2017). Soil organic carbon changes impacted by crop rotational diversity under no-till farming in South Dakota, USA. *Soil Science Society of America Journal*, *81*(4), 868–877. https://doi.org/10.2136/sssaj2016.04.0121

Almagro, M., Díaz-Pereira, E., Boix-Fayos, C., Zornoza, R., Sánchez-Navarro, V., Re, P., Fernández, C., & Martínez-Mena, M. (2023). The combination of crop diversification and no tillage enhances key soil quality parameters related to soil functioning without compromising crop yields in a low-input rainfed almond orchard under semiarid Mediterranean conditions. *Agriculture, Ecosystems and Environment*, *345*, Article 108320. https://doi.org/10.1016/j.agee.2022.108320

Anders, M. M., Brye, K. R., Olk, D. C., & Schmid, B. T. (2012). Rice rotation and tillage effects on soil aggregation and aggregate carbon and nitrogen dynamics. *Soil Science Society of America Journal*, *76*(3), 994–1004. https://doi.org/10.2136/sssaj2010.0436

Andersson, G. K. S., Ekroos, J., Stjernman, M., Rundlöf, M., & Smith, H. G. (2014). Effects of farming intensity, crop rotation and landscape heterogeneity on field bean pollination. *Agriculture, Ecosystems and Environment*, *184*, 145–148. https://doi.org/10.1016/j.agee.2013.12.002

Bardgett, R. D., & van der Putten, W. H. (2014). Belowground biodiversity and ecosystem functioning. *Nature*, *515*(7528), 505–511. https://doi.org/10.1038/nature13855

Benjamin, J. G., Nielsen, D. C., & Vigil, M. F. (2003). Quantifying effects of soil conditions on plant growth and crop production. *Geoderma*, *116*(1–2), 137–148. https://doi.org/10.1016/S0016-7061(03)00098-3

Bertrand, C., Eckerter, P. W., Ammann, L., Entling, M. H., Gobet, E., Herzog, F., Mestre, L., Tinner, W., & Albrecht, M. (2019). Seasonal shifts and complementary use of pollen sources by two bees, a lacewing and a ladybeetle species in European agricultural landscapes. *Journal of Applied Ecology*, *56*, Article 11. https://doi.org/10.1111/1365-2664.13483

Blanco-Canqui, H., Shaver, T. M., Lindquist, J. L., Shapiro, C. A., Elmore, R. W., Francis, C. A., & Hergert, G. W. (2015). Cover crops and ecosystem services: Insights from studies in temperate soils. *Agronomy Journal*, *107*(6), 2449–2474. https://doi.org/10.2134/agronj15.0086

Bonanomi, G., De Filippis, F., Cesarano, G., La Storia, A., Ercolini, D., & Scala, F. (2016). Organic farming induces changes in soil microbiota that affect agro-ecosystem functions. *Soil Biology and Biochemistry*, *103*, 327–336. https://doi.org/10.1016/j.soilbio.2016.09.005

Bonetti, J. d. A., Nunes, M. R., Fink, J. R., Tretto, T., & Tormena, C. A. (2023). Agricultural practices to improve near-surface soil health and crop yield in subtropical soils. *Soil and Tillage Research*, *234*, Article 105835. https://doi.org/10.1016/j.still.2023.105835

Bragg, L., Stone, G., Imelfort, M., Hugenholtz, P., & Tyson, G. W. (2012). Fast, accurate error-correction of amplicon pyrosequences using *Acacia*. *Nature Methods*, *9*(5), 425–426. https://doi.org/10.1038/nmeth.1990

Cambardella, C. A., & Elliott, E. T. (1992). Particulate soil organic-matter changes across a grassland cultivation sequence. *Soil Science Society of America Journal*, *56*(3), 777–783. https://doi.org/10.2136/SSSAJ1992.03615995005600030017X

Campbell, A. J., Wilby, A., Sutton, P., & Wäckers, F. (2017). Getting more power from your flowers: Multi-functional flower strips enhance pollinators and pest control agents in apple orchards. *Insects*, *8*, Article 101. https://doi.org/10.3390/insects8030101

Caporaso, J. G., Kuczynski, J., Stombaugh, J., Bittinger, K., Bushman, F. D., Costello, E. K., Fierer, N., Peña, A. G., Goodrich, J. K., Gordon, J. I., Huttley, G. A., Kelley, S. T., Knights, D., Koenig, J. E., Ley, R. E., Lozupone, C. A., McDonald, D., Muegge, B. D., Pirrung, M., ... Knight, R. (2010). QIIME allows analysis of high-throughput community sequencing data. *Nature Methods*, *7*(5), 335–336. https://doi.org/10.1038/nmeth.f.303

Cerda, R., Allinne, C., Gary, C., Tixier, P., Harvey, C. A., Krolczyk, L., Mathiot, C., Clément, E., Aubertot, J. N., & Avelino, J. (2017). Effects of shade, altitude and management on multiple ecosystem services in coffee agroecosystems. *European Journal of Agronomy*, *82*, 308–319. https://doi.org/10.1016/j.eja.2016.09.019

Chabert, A., & Sarthou, J. P. (2020). Conservation agriculture as a promising trade-off between conventional and organic agriculture in bundling ecosystem services. *Agriculture, Ecosystems and Environment*, *292*, Article 106815. https://doi.org/10.1016/j.agee.2019.106815

Congreves, K. A., Hayes, A., Verhallen, E. A., & Van Eerd, L. L. (2015). Long-term impact of tillage and crop rotation on soil health at four temperate agroecosystems. *Soil and Tillage Research*, *152*, 17–28. https://doi.org/10.1016/j.still.2015.03.012

Congreves, K. A., Voroney, R. P., & Van Eerd, L. L. (2014). Amending soil with used cooking oil to reduce nitrogen losses after cole crop harvest: A ^{15}N study.

Nutrient Cycling in Agroecosystems, *100*(3), 257–271. https://doi.org/10.1007/s10705-014-9643-8

Degrune, F., Theodorakopoulos, N., Colinet, G., Hiel, M. P., Bodson, B., Taminiau, B., Daube, G., Vandenbol, M., & Hartmann, M. (2017). Temporal dynamics of soil microbial communities below the seedbed under two contrasting tillage regimes. *Frontiers in Microbiology*, *8*, Article 1127. https://doi.org/10.3389/fmicb.2017.01127

Diekötter, T., Wamser, S., Wolters, V., & Birkhofer, K. (2010). Landscape and management effects on structure and function of soil arthropod communities in winter wheat. *Agriculture, Ecosystems and Environment*, *137*(1–2), 108–112. https://doi.org/10.1016/j.agee.2010.01.008

Díez, J. A. (1982). Consideraciones sobre la utilización de la técnica extractiva de Burriel-Hernando para la evaluación de fósforo asimilable en suelos. *Anales de Edafología y Agrobiología*, *41*, 1345–1353.

Elliott, E. T. (1986). Aggregate structure and carbon, nitrogen, and phosphorus in native and cultivated soils. *Soil Science Society of America Journal*, *50*(3), 627–633. https://doi.org/10.2136/sssaj1986.03615995005000030017x

Eurostat. (2015). Statistics explained: Your guide to European statistics. http://ec.europa.eu/eurostat/statisticsexplained/

Farmaha, B. S., Sekaran, U., & Franzluebbers, A. J. (2022). Cover cropping and conservation tillage improve soil health in the southeastern United States. *Agronomy Journal*, *114*(1), 296–316. https://doi.org/10.1002/agj2.20865

Follett, R. F. (2001). Soil management concepts and carbon sequestration in cropland soils. *Soil and Tillage Research*, *61*(1–2), 77–92.

Food and Agriculture Organization. 2011. *The state of the world's land and water resources for Food and Agriculture Organization*.

Forrest, J. R. K., Thorp, R. W., Kremen, C., & Williams, N. M. (2015). Contrasting patterns in species and functional-trait diversity of bees in an agricultural landscape. *Journal of Applied Ecology*, *52*(3), 706–715. https://doi.org/10.1111/1365-2664.12433

Garcia, R. A., Crusciol, C. A. C., Calonego, J. C., & Rosolem, C. A. (2008). Potassium cycling in a corn-brachiaria cropping system. *European Journal of Agronomy*, *28*(4), 579–585. https://doi.org/10.1016/j.eja.2008.01.002

Garcia, R. A., Li, Y., & Rosolem, C. A. (2013). Soil organic matter and physical attributes affected by crop rotation under no-till. *Soil Science Society of America Journal*, *77*(5), 1724–1731. https://doi.org/10.2136/sssaj2012.0310

Haaland, C., Naisbit, R. E., & Bersier, L. F. (2011). Sown wildflower strips for insect conservation: A review. *Insect Conservation and Diversity*, *4*, 60–80. https://doi.org/10.1111/j.1752-4598.2010.00098.x

Haruna, S., & Nkongolo, N. (2019). Tillage, cover crop and crop rotation effects on selected soil chemical properties. *Sustainability*, *11*(10). Article 2770. https://doi.org/10.3390/su11102770

Henry, S., Baudoin, E., López-Gutiérrez, J. C., Martin-Laurent, F., Brauman, A., & Philippot, L. (2004). Quantification of denitrifying bacteria in soils by *nirK* gene targeted real-time PCR. *Journal of Microbiological Methods*, *59*(3), 327–335. https://doi.org/10.1016/j.mimet.2004.07.002

Hoffmann, M. P., Swanepoel, C. M., Nelson, W. C. D., Beukes, D. J., van der Laan, M., Hargreaves, J. N. G., & Rötter, R. P. (2020). Simulating medium-term effects of cropping system diversification on soil fertility and crop productivity in southern Africa. *European Journal of Agronomy*, *119*, Article 126089. https://doi.org/10.1016/j.eja.2020.126089

Holzschuh, A., Steffan-Dewenter, I., & Tscharntke, T. (2008). Agricultural landscapes with organic crops support higher pollinator diversity. *Oikos*, *117*(3), 354–361. https://doi.org/10.1111/j.2007.0030-1299.16303.x

International Union of Soil Sciences. (2014). World reference base for soil resources 2014: International soil classification system for naming soils and creating legends for soil maps. World Soil Resources Reports.

Isaacs, K. B., Snapp, S. S., Chung, K., & Waldman, K. B. (2016). Assessing the value of diverse cropping systems under a new agricultural policy environment in Rwanda. *Food Security*, *8*, 491–506. https://doi.org/10.1007/s12571-016-0582-x

ISO. (1995). ISO 13536:1995. Soil quality—Determination of the potential cation exchange capacity and exchangeable cations using barium chloride solution buffered at pH = 8,1.

Jensen, E. S., Carlsson, G., & Hauggaard-Nielsen, H. (2020). Intercropping of grain legumes and cereals improves the use of soil N resources and reduces the requirement for synthetic fertilizer N: A global-scale analysis. *Agronomy for Sustainable Development*. *40*, Article 5. https://doi.org/10.1007/s13593-020-0607-x

Jones, D. L., Hodge, A., & Kuzyakov, Y. (2004). Plant and mycorrhizal regulation of rhizodeposition. *New Phytologist*, *163*(3), 459–480. https://doi.org/10.1111/j.1469-8137.2004.01130.x

Kabata-Pendias, A. (2000). *Trace elements in soils and plants*. CRC Press. https://doi.org/10.1201/9781420039900

Kandeler, E., & Gerber, H. (1988). Short-term assay of soil urease activity using colorimetric determination of ammonium. *Biology and Fertility of Soils*, *6*(1), 68–72. https://doi.org/10.1007/BF00257924/METRICS

Keeney, D. R., & Nelson, D. W. (2015). Nitrogen—Inorganic forms. In A. L. Page (Ed.), *Methods of soil analysis: Part 2. Chemical and microbiological properties* (pp. 643–698). American Society of Agronomy, Inc., Soil Science Society of America, Inc. https://doi.org/10.2134/AGRONMONOGR9.2.2ED.C33

Keller, T., Arvidsson, J., & Dexter, A. R. (2007). Soil structures produced by tillage as affected by soil water content and the physical quality of soil. *Soil and Tillage Research*, *92*, 45–52. https://doi.org/10.1016/j.still.2006.01.001

Kumar Singh, A., Cabral, C., Kumar, R., Ganguly, R., Kumar Rana, H., Gupta, A., Rosaria Lauro, M., Carbone, C., Reis, F., & Pandey, A. K. (2019). Beneficial effects of dietary polyphenols on gut microbiota and strategies to improve delivery efficiency. *Nutrients*, *11*(9), Article 2216. https://doi.org/10.3390/nu11092216

Kumari, V. V., Balloli, S. S., Kumar, M., Ramana, D. B. V., Prabhakar, M., Osman, M., Indoria, A. K., Manjunath, M., Maruthi, V., Chary, G. R., Sarath Chandran, M. A., Gopinath, K. A., Venkatesh, G., Rao, M. S., Singh, V. K., & Timsina, J. (2024). Diversified cropping systems for reducing soil erosion and nutrient loss and for increasing crop productivity and profitability in rainfed environments. *Agricultural Systems*, *217*, Article 103919. https://doi.org/10.1016/j.agsy.2024.103919

Kuntz, M., Berner, A., Gattinger, A., Scholberg, J. M., Mäder, P., & Pfiffner, L. (2013). Influence of reduced tillage on earthworm and microbial communities under organic arable farming. *Pedobiologia*, *56*(4–6), 251–260. https://doi.org/10.1016/j.pedobi.2013.08.005

Lal, R. (2004). Soil carbon sequestration to mitigate climate change. *Geoderma*, *123*(1–2), 1–22. https://doi.org/10.1016/j.geoderma.2004.01.032

Li, J., Huang, L., Zhang, J., Coulter, J. A., Li, L., & Gan, Y. (2019). Diversifying crop rotation improves system robustness. *Agronomy for Sustainable Development*, *39*, Article 38. https://doi.org/10.1007/s13593-019-0584-0

Liu, B., Xia, H., Jiang, C., Riaz, M., Yang, L., Chen, Y., Fan, X., & Xia, X. (2022). 14 year applications of chemical fertilizers and crop straw effects on soil labile organic carbon fractions, enzyme activities and microbial community in rice-wheat rotation of middle China. *Science of the Total Environment*, *841*, Article 156608. https://doi.org/10.1016/j.scitotenv.2022.156608

López-Gutiérrez, J. C., Henry, S., Hallet, S., Martin-Laurent, F., Catroux, G., & Philippot, L. (2004). Quantification of a novel group of nitrate-reducing bacteria in the environment by real-time PCR. *Journal of Microbiological Methods*, *57*(3), 399–407. https://doi.org/10.1016/j.mimet.2004.02.009

Luo, G., Li, L., Friman, V. P., Guo, J., Guo, S., Shen, Q., & Ling, N. (2018). Organic amendments increase crop yields by improving microbe-mediated soil functioning of agroecosystems: A meta-analysis. *Soil Biology and Biochemistry*, *124*, 105–115. https://doi.org/10.1016/j.soilbio.2018.06.002

Marcos-Pérez, M., Sánchez-Navarro, V., Martinez-Martinez, S., Martínez-Mena, M., García, E., & Zornoza, R. (2023). Intercropping organic melon and cowpea combined with return of crop residues increases yields and soil fertility. *Agronomy for Sustainable Development*, *43*(4), Article 53. https://doi.org/10.1007/s13593-023-00902-y

Maron, J. L., Marler, M., Klironomos, J. N., & Cleveland, C. C. (2011). Soil fungal pathogens and the relationship between plant diversity and productivity. *Ecology Letters*, *14*(1), 36–41. https://doi.org/10.1111/j.1461-0248.2010.01547.x

Martínez-Mena, M., Boix-Fayos, C., Carrillo-López, E., Díaz-Pereira, E., Zornoza, R., Sánchez-Navarro, V., Acosta, J. A., Martínez-Martínez, S., & Almagro, M. (2021). Short-term impact of crop diversification on soil carbon fluxes and balance in rainfed and irrigated woody cropping systems under semiarid Mediterranean conditions. *Plant and Soil*, *467*(1–2), 499–514. https://doi.org/10.1007/s11104-021-05101-w

McDaniel, M. D., Tiemann, L. K., & Grandy, A. S. (2014). Does agricultural crop diversity enhance soil microbial biomass and organic matter dynamics? *A meta-analysis. Ecological Applications*, *24*(3), 560–570. https://doi.org/10.1890/13-0616.1

Ministerio de Agricultura Alimentación y Medio Ambiente. (2013). Encuesta sobre superficies y rendimientos de cultivos. Resultados nacionales y autonómicos, (in Spanish).

Ministerio de Agricultura Alimentación y Medio Ambiente. (2023). Ministerio de Agricultura, Pesca y Alimentación. Información Subsectorial (in Spanish).

Myaka, F. M., Sakala, W. D., Adu-Gyamfi, J. J., Kamalongo, D., Ngwira, A., Odgaard, R., & Nielsen, N. E. (2006). Yields and accumulations of N and P in farmer-managed intercrops of maize-pigeonpea in semi-arid Africa. *Plant and Soil*, *285*(1–2), 207–220. https://doi.org/10.1007/s11104-006-9006-6

Naresh, R. K., Dwivedi, A., Gupta, R. K., Rathore, R. S., Dhaliwal, S. S., Singh, S. P., Kumar, P., Kumar, R., Singh, V., Singh, V., & Singh, O. (2016). Influence of conservation agriculture practices on physical, chemical and biological properties of soil and soil organic carbon dynamics in the subtropical climatic conditions. *Journal of Pure and Applied Microbiology*, *10*(2), 1061–1080.

OECD/FAO. (2018). *OECD-FAO agricultural outlook 2018–2027*. OECD Publishing.

Özbolat, O., Sánchez-Navarro, V., Zornoza, R., Egea-Cortines, M., Cuartero, J., Ros, M., Pascual, J. A., Boix-Fayos, C., Almagro, M., de Vente, J., Díaz-Pereira, E., & Martínez-Mena, M. (2023). Long-term adoption of reduced tillage and green manure improves soil physicochemical properties and increases the abundance of beneficial bacteria in a Mediterranean rainfed almond orchard. *Geoderma*, *429*, Article 116218. https://doi.org/10.1016/j.geoderma.2022.116218

Palomo-Campesino, S., García-Llorente, M., Hevia, V., Boeraeve, F., Dendoncker, N., & González, J. A. (2022). Do agroecological practices enhance the supply of ecosystem services? A comparison between agroecological and conventional horticultural farms. *Ecosystem Services*, *57*, Article 101474. https://doi.org/10.1016/j.ecoser.2022.101474

Preissel, S., Reckling, M., Schläfke, N., & Zander, P. (2015). Magnitude and farm-economic value of grain legume pre-crop benefits in Europe: A review. *Field Crops Research*, *175*, 64–79. https://doi.org/10.1016/j.fcr.2015.01.012

Raseduzzaman, M., & Jensen, E. S. (2017). Does intercropping enhance yield stability in arable crop production? A meta-analysis. *European Journal of Agronomy*, *91*, 25–33. https://doi.org/10.1016/j.eja.2017.09.009

Reddy, S. S., Nyakatawa, E. Z., Reddy, K. C., Raper, R. L., Reeves, D. W., & Lemunyon, J. L. (2009). Long-term effects of poultry litter and conservation tillage on crop yields and soil phosphorus in cotton–cotton–corn rotation. *Field Crops Research*, *114*(2), 311–319. https://doi.org/10.1016/j.fcr.2009.09.001

Reichman, S. M. (2002). The responses of plants to metal toxicity: A review focusing on copper, manganese and zinc. In S. M. Reichman (Ed.), *Symptoms and visual evidence of toxicity* (pp. 22–26). Australian Minerals & Energy Environment Foundation.

Reynolds, S. G. (1970). The gravimetric method of soil moisture determination. Part I: A study of equipment, and methodological problems. *Journal of Hydrology*, *11*(3), 258–273. https://doi.org/10.1016/0022-1694(70)90066-1

Roger-Estrade, J., Anger, C., Bertrand, M., & Richard, G. (2010). Tillage and soil ecology: Partners for sustainable agriculture. *Soil and Tillage Research*, *111*(1), 33–40. https://doi.org/10.1016/j.still.2010.08.010

Rognes, T., Flouri, T., Nichols, B., Quince, C., & Mahé, F. (2016). VSEARCH: A versatile open source tool for metagenomics. *PeerJ*, *4*, Article e2584. https://doi.org/10.7717/peerj.2584

Rotthauwe, J. H., Witzel, K. P., & Liesack, W. (1997). The ammonia monooxygenase structural gene *amoA* as a functional marker: Molecular fine-scale analysis of natural ammonia-oxidizing populations. *Applied and Environmental Microbiology*, *63*(12), 4704–4712. https://doi.org/10.1128/aem.63.12.4704-4712.1997

Rovira, A. D. (1969). Plant root exudates. *Botanical Review*, *5*, 35–37.

Sáez, F. (1995). Essential oil variability of *Thymus hyemalis* growing wild in southeastern Spain. *Biochemical Systematics and Ecology*, *23*(4), 431–438. https://doi.org/10.1016/0305-1978(95)00022-M

Sainju, U. M., Liptzin, D., Dangi, S., & Ghimire, R. (2021). Soil health indicators and crop yield in response to long-term cropping sequence and nitrogen fertilization. *Applied Soil Ecology*, *168*, Article 104182. https://doi.org/10.1016/j.apsoil.2021.104182

Sánchez-Navarro, V., Martínez-Martínez, S., Acosta, J. A., Almagro, M., Martínez-Mena, M., Boix-Fayos, C., Díaz-Pereira, E., Temnani, A., Berrios, P., Pérez-Pastor, A., & Zornoza, R. (2023). Soil greenhouse gas emissions and crop production with implementation of alley cropping in a Mediterranean citrus orchard. *European Journal of Agronomy*, *142*, Article 126684. https://doi.org/10.1016/j.eja.2022.126684

Sánchez-Navarro, V., Shahrokh, V., Martínez-Martínez, S., Acosta, J. A., Almagro, M., Martínez-Mena, M., Boix-Fayos, C., Díaz-Pereira, E., & Zornoza, R. (2022). Perennial alley cropping contributes to decrease soil CO_2 and N_2O emissions and increase soil carbon sequestration in a Mediterranean almond orchard. *Science of the Total Environment*, *845*, Article 157225. https://doi.org/10.1016/j.scitotenv.2022.157225

Shahbaz, M., Kuzyakov, Y., Sanaullah, M., Heitkamp, F., Zelenev, V., Kumar, A., & Blagodatskaya, E. (2017). Microbial decomposition of soil organic matter is mediated by quality and quantity of crop residues: Mechanisms and thresholds. *Biology and Fertility of Soils*, *53*, 287–301. https://doi.org/10.1007/s00374-016-1174-9

Sharma, A., Jat, M. L., Saharawat, Y., Singh, V., & Singh, R. (2012). Conservation agriculture for improving productivity and resource-use efficiency: Prospects and research needs in Indian context. *Indian Journal of Agronomy*, *57*(3s), 131–140.

Simmons, B. L., & Coleman, D. C. (2008). Microbial community response to transition from conventional to conservation tillage in cotton fields. *Applied Soil Ecology*, *40*(3), 518–528. https://doi.org/10.1016/j.apsoil.2008.08.003

Sünnemann, M., Alt, C., Kostin, J. E., Lochner, A., Reitz, T., Siebert, J., Schädler, M., & Eisenhauer, N. (2021). Low-intensity land-use enhances soil microbial activity, biomass and fungal-to-bacterial ratio in current and future climates. *Journal of Applied Ecology*, *58*(11), 2614–2625. https://doi.org/10.1111/1365-2664.14004

Tang, C., McLay, C. D. A., & Barton, L. (1997). A comparison of proton excretion of twelve pasture legumes grown in nutrient solution. *Australian Journal of Experimental Agriculture*, *37*(5), 563–570. https://doi.org/10.1071/EA96151

Tiemann, L. K., Grandy, A. S., Atkinson, E. E., Marin-Spiotta, E., & McDaniel, M. D. (2015). Crop rotational diversity enhances belowground communities and functions in an agroecosystem. *Ecology Letters*, *18*(8), 761–771. https://doi.org/10.1111/ele.12453

Tuomivirta, T., Canfora, L., & Thiele-Bruhn, S. (2019). In J. Álvaro-Fuentes, D. Lóczy, S. Thiele-Bruhn, & R. Zornoza (Eds.), *Handbook of plant and soil analysis for agricultural systems.*

Van Kessel, C., & Hartley, C. (2000). Agricultural management of grain legumes: Has it led to an increase in nitrogen fixation? *Field Crops Research*, *65*, 165–181.

Wang, Q. R., Li, Y. C., & Klassen, W. (2007). Changes of soil microbial biomass carbon and nitrogen with cover crops and irrigation in a tomato field. *Journal of Plant Nutrition*, *30*(4), 623–639. https://doi.org/10.1080/01904160701209410

Whitby, M. D., Kieran, T. J., Glenn, T. C., & Allen, C. (2020). Agricultural pests consumed by common bat species in the United States corn belt: The importance of DNA primer choice. *Agriculture, Ecosystems and Environment*, *303*, Article 107105. https://doi.org/10.1016/j.agee.2020.107105

Williams, H., Colombi, T., & Keller, T. (2020). The influence of soil management on soil health: An on-farm study in southern Sweden. *Geoderma*, *360*, Article 114010. https://doi.org/10.1016/j.geoderma.2019.114010

Yan, Z., Zhou, J., Liu, C., Jia, R., Mganga, K. Z., Yang, L., Yang, Y., Peixoto, L., Zang, H., & Zeng, Z. (2023). Legume-based crop diversification reinforces soil health and carbon storage driven by microbial biomass and aggregates. *Soil and Tillage Research*, *234*, Article 105848. https://doi.org/10.1016/j.still.2023.105848

4

Urban Soil Health: A Case Study from Northwestern Spain

Remigio Paradelo Núñez

Chapter Overview

Soil health assessment is an essential part of policies to protect soil and the environment. Maintaining and enhancing soil health can be particularly challenging in urban environments, where soils are exposed to many degradation processes and can be heavily disturbed by anthropic activities and contamination. However, information about the status of soil health in urban soils remains very scarce, particularly in Spanish cities. This chapter presents a summary of soil health indicators in the city of Santiago de Compostela in northwestern Spain. Physical, chemical, and biological components of soil health were studied in soils developed over different parent materials and under different land uses. Soils are rich in organic carbon, have medium to low fertility owing to acid pH and low nitrogen availability, and are contaminated by low to moderate levels of pollution. Soils have good structural conditions, although compaction issues are present. Spatial heterogeneity in soil properties, and consequently soil health, is high and in many cases unexplained. Lessons learned in this case study regarding the protection of soil health in urban soils are discussed. Given the positive influence of organic matter in all aspects of soil health, implementing measures for protecting, increasing, and/or maintaining the organic matter content of urban soils is the most important recommendation.

Soil Health Series: Volume 5 Soil Health and Sustainability in Spain and Portugal, First Edition.
Edited by Iñigo Virto and Rodrigo Antón.

Published 2025 by Soil Science Society of America, Inc. and John Wiley & Sons, Inc.

Introduction

Inappropriate or unsustainable use of soil and the increasing demand for land for urban development and infrastructure are negatively affecting soil health worldwide. This in turn disrupts the capacity of soils to carry out the vital services that they perform (Veerman et al., 2020). Soil degradation processes can lead to a collapse of landscapes and ecosystems, making societies more vulnerable to extreme weather events, food security risks, and even political instability (Lal et al., 2021). Protecting soil from degradation can be particularly challenging in urban environments. Between 1975 and 2015, the population has grown in about 92% of worldwide urban centers, with 46% of the cities experiencing a doubling of their population. The intensification of urbanization poses a major challenge for the environment, particularly for soils, which suffer the effects of urban development and associated anthropic activities, including severe disruption, contamination, and other degradation processes (Béchet et al., 2019).

The characteristics of urban soils are strongly influenced by the activities associated with urbanization, including alterations caused by mixing, filling, transportation, and other perturbations from construction-related activities (Levin et al., 2017). These activities often result in low-fertility soils with high amounts of artefacts and coarse fractions, poor structure, low water holding capacity, low permeability because of compaction and/or sealing, and high potential erosion risk (Prokof'eva et al., 2020). Soil organic carbon (SOC) content is low in the most degraded soils, which has negative impacts on other aspects of soil health, such as soil structure or biodiversity. In addition, owing to the presence of several sources of pollution near or within the cities, a wide variety of contaminants can accumulate in urban soils, including metals and metalloids, polycyclic aromatic hydrocarbons, and emerging pollutants such as pesticides, antibiotics, or microplastics (Li et al., 2018). All these properties lead to low levels of microbial activity and diversity (Schmidt et al., 2017), which negatively affects soil health. Consequently, urban soils usually present low functionality, especially with regard to regulation of the water cycle, plant production, or pollutant buffering (Morel et al., 2015).

Despite their important role in sustainability and human health in an increasingly urbanized world, soils in urban areas are still imperfectly studied and their health has not received enough attention to date (Lehmann et al., 2020), as soil monitoring has traditionally focused on agricultural and forest soils.

In this sense, information about the status of soil health in urban soils remains very scarce, and some works have only very recently started considering the health of urban soils beyond pollution (Salomon et al., 2022). At first glance, soil health in urban areas would be expected to be low because of the stressors that act on urban soils (low fertility, enrichment in pollutants, poor physical conditions, low organic matter content, low biological activity). Thus, adopting measures to restore and/or protect soil health will be necessary. However, closer investigation reveals that urban soils are also inherently diverse and spatially heterogeneous, in parallel with the complex mechanisms that govern urban soil properties and distribution. Consequently, comprehensive studies about all the components of urban soil health are necessary.

In this chapter, findings from a 5-year project focused on the urban soils of the city of Santiago de Compostela are summarized. The project encompassed study of several aspects of soil properties and functions, including soil types, general chemical properties, presence of pollutants, and microbiology. Results are presented following the five aspects of soil health considered in the new European framework for soil health assessment (Veerman et al., 2020).

Study Area

The city of Santiago de Compostela, located in the northwestern corner of the Iberian Peninsula, is the capital of the autonomous region of Galicia and an important political, administrative, religious, and tourist center, with an area of 222 km^2 and 99,000 inhabitants (Instituto Nacional de Estadística, 2023). The climate is warm and wet, and according to the Köppen–Geiger Climate Classification, the city lies within the temperate oceanic climate (Cfb) zone (Kottek et al., 2006). The mean annual air temperature is 13.0 °C, with August being the warmest month (mean air temperature 19 °C) and January being the coldest (mean air temperature 8 °C). The average annual precipitation is 1,787 mm. The relatively low values for potential evapotranspiration (>300 mm in summer and 50–100 mm in winter, with a total of 750–800 mm, calculated by the Penman method) result in a positive water balance (600–800 mm; Martínez Cortizas & Pérez Alberti, 1999). The city is in the contact zone between a granitic area and the Santiago Unit of the metamorphic massif known as the Ordes Complex. It therefore contains important geological diversity, with four main lithological units (Díaz-García, 1990): granitic rocks, schists, orthogneisses, and amphibolites (Figure 4.1). The city center has 25 green areas with surfaces that range from 1 to 40 ha, including lawn areas, forest areas, and public and private urban gardens that sometimes coexist at very short distances from one another.

Surface soils (0- to 20-cm depth) at 64 points were studied in green areas of the city of Santiago de Compostela, in addition to 13 soil profiles (as shown in Figure 4.1), and analyzed according to the methods shown in Table 4.1. Samples in urban grasslands (55%), urban forests (22%), and urban agricultural areas (22%) were included in the sampling network. Regarding lithology, 44% of the sampling

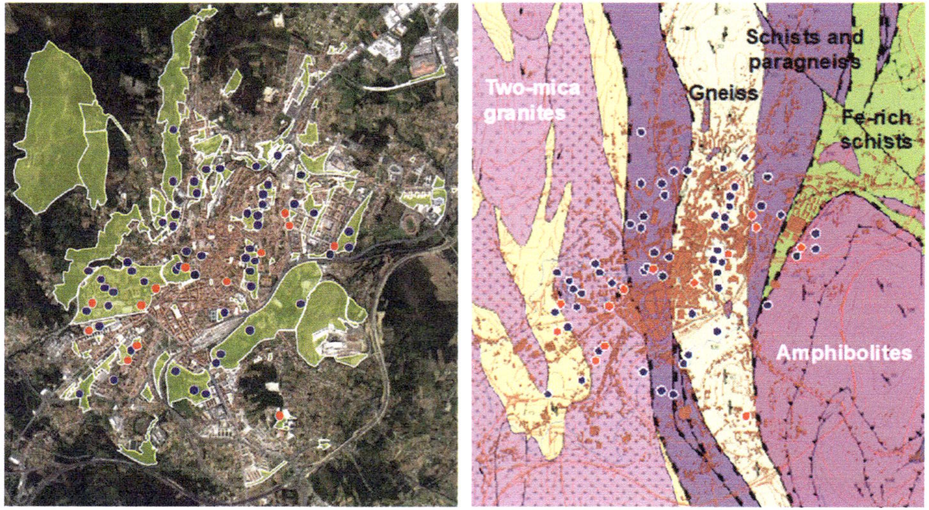

Figure 4.1 Location of Study Area, Sampling Points, and Lithology (Profile Observations in Red, and Surface Soil Sampling in Blue). Photo by María Almagro Bonmati.

Table 4.1 Summary of Soil Properties Measured and Methods Used.

Property	Analytical method	Reference
Physical properties		
Texture	Robinson pipette	Guitián and Carballas (1976)
Aggregate stability	Wet sieving	Kemper and Rosenau (1986)
Infiltration	Single ring infiltrometer	USDA (1999)
Bulk density	Steel cylinder	
Chemical properties		
pH	Water suspension (1 : 2.5)	Guitián and Carballas (1976)
Electrical conductivity	Filtered water extract (1 : 2)	Pansu and Gautheyrou (2006)
Exchangeable bases	Extraction with 1 N ammonium acetate pH 7	Pansu and Gautheyrou (2006)
Organic C	Walkley-Black wet dichromate digestion	Pansu and Gautheyrou (2006)
Total N	Dumas dry combustion	Bremner (1996)
Available N	Extraction with 2 N KCl	Keeney and Nelson (1982)
Available P	Extraction with 0.5 N sodium bicarbonate	Olsen and Sommers (1982)
Total trace elements	X-ray fluorescence	–
Polycyclic Aromatic Hydrocarbons	Miniaturized ultrasound-assisted extraction	Celeiro et al. (2018)
Biological properties	–	–
Basal respiration	Incubation and NaOH trap	Anderson (1982)

sites were on schist, 23% on granite or granite mixed with schist, 23% on gneiss, and 8% on amphibolites. According to the World Reference Base classification (IUSS-WRB, 2014), the soils of the green areas could be classified as Leptic Umbrisols, Haplic Umbrisols, Skeletic Regosols, and Urbic Technosols (Paradelo et al., 2022). These soils were generally acidic, coarse textured (dominant texture was sandy loam), and rich in organic matter, with differences in soil composition and chemical properties due to the diversity of parent materials and land uses (Paradelo et al., 2021). Soils developed over amphibolites presented heavier textures and greater amounts of iron compounds than soils developed over schist, gneiss, or granite. Among land uses, soils of urban gardens presented higher nutrient content, pH, and salinity than soils of urban forests or urban grasslands (Paradelo et al., 2021).

Results and Discussion

In Europe, restoration and preservation of healthy soils are key elements of the European Union Green Deal, while soil health is a central element of the EU soil strategy for 2030 (European Commission, 2021). The objective is to develop solutions for restoring soil health and functions, so that at least 75% of all soils in each EU member state are healthy by 2030 (Veerman et al., 2020), including explicitly urban soils. In addition to indicators related to vegetation cover, the following aspects must be considered when assessing soil health: soil nutrients and pH, soil pollutants, soil structure, SOC, and soil biodiversity.

Soil Nutrients and pH

Soil fertility is an important aspect of soil health in relation to plant production. Plants obtain essential nutrients and micro-nutrients from soils, while soil pH affects chemical and biological processes, including nutrient availability and the balance and functions of soil microbial communities.

Soil fertility is typically low in urban soils owing to a high content of rock fragments and gravel, dominant coarse textures, low organic matter content, and so forth (Levin et al., 2017). However, this situation can be different in urban agricultural soils, depending on specific soil properties and agricultural practices, which are highly variable (Burghardt et al., 2018; McDougal et al., 2019).

In the city of Santiago de Compostela, the soils are mostly acidic (pH in water with a mean value of 5.7). Typically, soil pH is higher in most cities (Morel et al., 2015) owing to the widespread presence of carbonate-bearing materials in urban soils. However, Santiago de Compostela has a low abundance of these materials, in addition to having climatic conditions that are likely to leach such materials and acidify the soils. The content of available nutrients, especially nitrogen, was low compared with agricultural soils (Paradelo et al., 2021). Although total nitrogen was abundant, ranging from 0.6 to 6.0 g/kg, its availability was very low, except in the urban garden soils. Available phosphorus concentrations were generally high but very variable, with an average concentration of 51 mg/kg. The highest levels were found in areas close to the city center, which suggests that high phosphorus concentrations might be connected to greater levels of anthropogenic pollution. Potassium was also abundant in the soils because of its richness in the parent materials.

Overall, fertility can be considered medium to low, owing to acid pH and low nitrogen availability; however, there is a high variability, and notable differences associated with land use and lithology of the parent material. Compared with other land uses, soils of urban allotment gardens presented higher nutrient content and pH. In terms of lithology, soils developed on amphibolites presented low phosphorus availability and high phosphorus retention capacity, owing to the presence of poorly crystalline compounds and andic properties (Paradelo et al., 2021).

Soil Pollutants, Salts, and Excess Nutrients

Soils are unhealthy when excess nutrients, salts, and/or pollutants are present in higher concentrations than allowed by health regulations or plant requirements, negatively affecting plant production, filtering functions, and so forth. High pollutant concentrations in the soil increase the risk of transference to other environmental compartments and the food chain. Enrichment in pollutants relative to non-urban areas is a specific feature of soils in most cities, so significant levels of pollutants can be expected in urban soils (Li et al., 2018). In addition, cultivating these soils can potentially increase human exposure to contaminants and represent a risk for food security (Weber et al., 2019). Thus, the health of urban soils used for food production is particularly important, given that their importance has increased in the last decades (Lal, 2020).

Salinity issues are not a problem in Santiago de Compostela owing to climatic conditions that ensure leaching of salts from the soil. In the study, electrical conductivity values were low in all soils (Table 4.2), with a maximum of 28.9 mS/m. Land use had a significant effect, however, with higher values in soils used for urban agriculture, likely related to fertilization (Paradelo et al., 2021).

In the case of overfertilized urban gardens, excess phosphorus could be a problem, but trace metals and arsenic should be the main issue with regard to pollution here. The levels of trace elements found in the urban soils of Santiago de Compostela followed the sequence: zinc > lead > chromium > copper > arsenic > nickel, with concentrations that were generally higher than in the non-urban soils of the region and regional backgrounds (Macías & Calvo, 2009), but not high enough to classify the soils as contaminated according to Spanish regulations. Copper, lead, and zinc were the elements that most frequently surpassed typical values in non-urban soils, followed by arsenic, with chromium and nickel concentrations being closer to background values for soils over similar parent materials. Among the elements studied, copper, lead, and zinc concentrations were correlated, with the highest concentrations occurring in soils of the green areas in the city center, likely in connection to fuel combustion emissions. Chromium and nickel concentrations were related to lithology of the parent material, with the highest concentrations in soils developed over amphibolite. Finally, arsenic concentrations were notably high in two precise locations without a clear connection to a known source of pollution. Overall, copper, lead, and zinc concentrations in the soils were mostly determined by anthropogenic pollution, whereas nickel and chromium concentrations were controlled by lithology (Herbón et al., 2021).

Table 4.2 Summary of Soil Properties (0–20 cm).

Property	Range	Median	Mean	Standard deviation
Physical properties				
Bulk density (Mg/m^3)	0.34–1.41	0.94	0.95	0.21
Aggregate stability (%)	49–94	88	86	8
Permeability (mm/h)	2–187	45	56	39
RUSLE K factor	0.03–0.15	0.06	0.07	0.03
Chemical properties				
pH	4.7–7.4	5.7	5.7	0.5
Electrical conductivity (mS/m)	5.4–28.9	11.5	12.8	5.4
Organic C (g/kg)	6–110	33	35	16
Total N (g/kg)	0.6–6.0	2.5	2.6	1.0
Available N (mg/kg)	3.0–26	8	9	5
Available P (mg/kg)	3–152	38	51	40
Exchangeable K ($cmol_c/kg$)	0.1–1.3	0.6	0.6	0.3
Total Pb (mg/kg)	20–566	71	97	91
Total Cu (mg/kg)	17–188	46	58	37
Total Zn (mg/kg)	55–484	96	116	67
Total Ni (mg/kg)	91–11	25	29	16
Total Cr (mg/kg)	17–277	58	74	56
Total As (mg/kg)	13–205	33	37	29
Total polycyclic aromatic hydrocarbons (μg/kg)	4–4728	210	508	894
Biological properties				
Respiration (mg CO_2-C·kg^{-1}·d^{-1})	0.5–26.5	6.5	7.5	4.8

Note. Full datasets can be found in Herbón et al. (2021), Paradelo et al. (2021), Gómez-Brandón et al. (2022), and Paradelo et al. (2023, 2025).

Regarding organic pollutants, the soils presented low polycyclic aromatic hydrocarbon pollution compared with other cities (Li et al., 2018), in line with the low population density and low degree of industrialization of Santiago de Compostela. Among the compounds considered, benzo(a)pyrene presented the highest risk, as its concentrations surpassed the precautionary limits established by Spanish law in 38% of the soils. In any case, values were far from reaching levels that would classify the soils as polluted. Spatial distribution was very heterogeneous and without a clear pattern based on factors such as location in the city, soil properties or vegetation at a site, or correlations to soil physicochemical properties. Overall, polycyclic aromatic hydrocarbon concentrations were closely related to those of lead, but not to other inorganic pollutants or edaphic properties, suggesting that their main origin must be a common source that is exogeneous and anthropic, likely a combination of combustion and traffic (Paradelo et al., 2023).

Overall, most urban soils in Santiago de Compostela presented low to moderate levels of pollution. None of the soils analyzed would be considered legally polluted according to Spanish laws, although in some cases they surpassed the trace element levels that would require further investigation of risk assessment (Herbón et al., 2021; Paradelo et al., 2023).

Soil Structure

General soil physical conditions, including aggregate stability, bulk density, the absence of soil sealing, and erosion risk, must be considered in a discussion of soil structure. Good soil physical condition allows for healthy root growth, reaching all parts of the soil, and allowing infiltration of rainwater to prevent runoff and soil loss. Soil physical condition is often problematic in urban soils, with poor structure, low water holding capacity owing to a high content of coarse elements, and low permeability because of compaction and/or sealing, all this resulting in high potential erosion risk (Prokof'eva et al., 2020).

In Santiago de Compostela, the soils presented heterogeneous physical properties (Table 4.2), as is common in urban soils. As a result of high organic matter and iron content, the soils presented good structural conditions, with low bulk densities (maximum of 1.4 Mg/m^3) and high aggregate stability (mean 88%). Compaction issues were widespread, particularly in the subsoil, but they depended on land use: soils under urban agriculture use generally presented higher permeability and lower compaction levels than urban grasslands and forests (Paradelo et al., 2025). The high infiltration values observed despite compaction were likely caused by an abundance of coarse fragments creating channels for preferential flow. Overall, soils have a low potential erosion risk as shown by low erodibility (K factor in the RUSLE equation) values (0.03–0.15), but sound management is essential to keep actual erosion rates down, because high slopes, climatic factors, and human behavior could lead to increased erosion at some points (Paradelo et al., 2025).

Soil Organic Carbon

SOC is of paramount importance in soil functioning, and it is important for adsorbing nutrients, retaining water, and improving soil structure, the workability of soils, and plant productivity. In addition, an increase in SOC stock allows drawing down CO_2 from the atmosphere, thereby contributing to climate regulation. SOC content in urban soils varies greatly and depends on factors such as land use or agricultural practices, in particular those involving the use of organic amendments (Allory et al., 2019). In any case, they could be low in the most degraded soils, with negative impacts on other aspects of soil health, such as soil structure or biodiversity.

In Santiago de Compostela, urban soils were rich in organic matter, with a mean organic carbon content (0–20 cm) of 35 g/kg (Paradelo et al., 2021), similar to natural and agricultural soils in the region, for which the most common interval of variation ranges between 30 and 40 g/kg (Calvo de Anta et al., 2015). Galician soils are rich in organic carbon mainly because of climatic conditions. Specifically, high water availability throughout the year and warm temperatures result in high plant productivity, with large returns of plant inputs to soil, producing significant accumulation of organic matter in the soils. Therefore, this aspect is positive but maintaining measures and management practices to preserve and/or increase organic matter content requires ongoing attention.

Soil Biodiversity

The presence of functionally diverse bacterial, fungal, and animal communities is essential for soil functions and ecosystem services, including soil structure, litter decomposition, organic carbon storage, and nutrient cycling. Soil biodiversity is still poorly understood in urban soils (Guilland et al., 2018), but considering the other known properties, like low fertility and organic matter content, compaction or possible presence of pollutants, low microbial activity and diversity can be expected in many cases (Schmidt et al., 2017).

Studies on soil biology in Santiago de Compostela have shown a high variability in all the microbiological variables among soils, reflecting the typical spatial heterogeneity of urban soils (Gómez-Brandón et al., 2022). Respiration and enzymatic activities were highly correlated with soil organic matter contents, but not affected by land use within the city, whereas spatial heterogeneity was particularly influential for random microbial community composition (Probst et al., 2023), which cannot be related to factors such as land use, vegetation, or lithology of the soils. Since soil organic matter seems to be the main driving force of biological activity in these urban soils, measures for increasing and/or maintaining its levels will always be positive for this aspect of soil health.

Recommendations

In view of the results summarized here, several recommendations can be made to protect soil health in this and other cities. First, practices aimed at conserving and increasing soil organic matter levels must be favored, given the positive effect on all aspects of soil health: SOC storage, improved soil structure and general physical properties, increased soil biological activity and biodiversity, and reduced mobility of soil pollutants.

In addition, management practices oriented to reducing soil sealing and erosion may be necessary to improve the functions related to the water cycle. Impermeable areas should be limited by avoiding sealing of new surfaces or even reduced by desealing activities. Regarding nonsealed areas, excessive compaction by machinery during landscaping and other work should be limited, and compaction by walkers should be controlled to avoid the development of desire lines (paths for preferential transit), for example, by considering the main directions of walkers in the design of green areas. These goals can be especially relevant in cities with high rainfall, such as Santiago de Compostela, where soils are wet most of the year and thus susceptible to compaction. The measures mentioned earlier to conserve organic matter are also helpful in this sense, as organic matter reduces the soil susceptibility to compaction and increases porosity and infiltration.

Restoration of areas degraded by soil compaction, erosion, or pollution is another necessary action to protect urban soil health. In this sense, the use of waste produced in the city, such as urban compost, is an interesting idea for increasing the levels of organic matter and improving the properties of degraded soils. Construction and demolition waste is another abundant material in urban areas that can be used for soil restoration or in landscaping and for the development of new green areas where the original soil has been removed, provided the waste does not contain toxic compounds in excessive concentrations. This material has the additional advantage of increasing the soil pH due to significant calcium carbonate content, which can be useful in the context of acid soils. These proposed applications of waste would contribute to increase material circularity in urban areas, while also greening the city and improving the properties of degraded soils.

Conclusion

Several conclusions on urban soil health can be drawn from the studies on properties and functions of the soils of Santiago de Compostela. Regarding soil nutrients and pH, fertility in these soils is medium to low owing to acid pH and low nitrogen availability, although phosphorus and potassium availability is adequate. Soils are rich in organic carbon, which is also the main driver of biological activity. In relation to soil structure, the soils present good structural conditions, with low bulk densities and high aggregate stability, although compaction issues are also present. Regarding pollutants, most urban soils in Santiago de Compostela present low to moderate levels of pollution that are likely explained by a low population density and minimal industrial activities. In any case, the heterogeneity in soil properties, and consequently soil health, is very high in urban areas, so further research in urban soils and a wider database are necessary. Recommendations for guaranteeing urban soil health inevitably include soil organic matter conservation and increase. Given the positive influence of organic matter on all aspects of soil health, measures for protecting, increasing, and/or maintaining its content in soil should be adopted in order to guarantee urban soil functioning and health.

References

Allory, V., Cambou, A., Moulin, P., Schwartz, C., Cannavo, P., Vidal-Beaudet, L., & Barthès, B. G. (2019). Quantification of soil organic carbon stock in urban soils using visible and near infrared reflectance spectroscopy (VNIRS) in situ or in laboratory conditions. *Science of the Total Environment*, *686*, 764–773.

Anderson, J. P. E. (1982). Soil respiration. In A. L. Page, R. H. Miller, & D. R. Keeney (Eds.), *Methods of soil analysis, part 2* (pp. 831–871). American Society of Agronomy, Soil Science Society of America.

Béchet, B., Le Bissonnais, Y., & Ruas, A. (2019). *Sols artificialisés. Déterminants, impacts et leviers d'action*. Éditions Quae.

Bremner, J. M. (1996). Nitrogen total. In D. L. Sparks (Ed.), *Methods of soil analysis: Part 3: Chemical methods* (pp. 1085–1121). Soil Science Society of America.

Burghardt, W., Heintz, D., & Hocke, N. (2018). Soil fertility characteristics and organic carbon stock in soils of vegetable gardens compared with surrounding arable land at the center of the urban and industrial area of Ruhr, Germany. *Eurasian Soil Science*, *51*, 1067–1079.

Calvo de Anta, R., Luis-Calvo, E., Casas-Sabarís, F., Galiñanes-Costa, J. M., Matilla-Mosquera, M., Macías-Vazquez, F., Camps-Arbestain, M., & Vazquez-García, N. (2015). Soil organic carbon in northern Spain (Galicia, Asturias, Cantabria and País Vasco). *Spanish Journal of Soil Science*, *5*, 41–53.

Celeiro, M., Dagnac, T., & Llompart, M. (2018). Determination of priority and other hazardous substances in football fields of synthetic turf by gas chromatography-mass spectrometry: A health and environmental concern. *Chemosphere*, *195*, 201–211.

Díaz García, F. (1990). Geochemical characterization of the basic and acid rocks in the Santiago Unit (NW Spain, Ordenes Complex). *Geotectonic implications. Revista de la Sociedad Geológica de España*, *3*, 261–273. (in Spanish, with English abstract)

European Commission. (2021). *COMMUNICATION FROM THE COMMISSION TO THE EUROPEAN PARLIAMENT, THE COUNCIL, THE EUROPEAN ECONOMIC AND SOCIAL COMMITTEE AND THE COMMITTEE OF THE REGIONS EU Soil Strategy for 2030 Reaping the benefits of healthy soils for people, food, nature and climate. COM(2021)699 Final*. https://eur-lex.europa.eu/legal-content/EN/TXT/PDF/?uri=CELEX:52021DC0699

Gómez-Brandón, M., Herbón, C., Probst, M., Fornasier, F., Barral, M. T., & Paradelo, R. (2022). Influence of land use on the microbiological properties of urban soils. *Applied Soil Ecology*, *175*, Article 104452

Guilland, C., Maron, P. A., Damas, O., & Ranjard, L. (2018). Biodiversity of urban soils for sustainable cities. *Environmental Chemistry Letters*, *16*, 1267–1282.

Guitián, F., & Carballas, T. (1976). *Técnicas de análisis de suelos*. Pico Sacro.

Herbón, C., Barral, M. T., & Paradelo, R. (2021). Potentially toxic trace elements in the urban soils of Santiago de Compostela (Northwestern Spain). *Applied Sciences*, *11*(9), Article 4211

Instituto Nacional de Estadística. (2023). Población del Padrón Continuo por Unidad Poblacional a 1 de enero. https://www.ine.es/nomen2/inicio_r.do

IUSS Working Group-WRB. (2014). *World reference base for soil resources 2014. International soil classification system for naming soils and creating legends for soil maps* (World Soil Resources Reports No. 106). FAO.

Keeney, D. R., & Nelson, D. W. (1982). Nitrogen inorganic forms. In A. L. Page (Ed.), *Methods of soil analysis. Part 2. Chemical and microbiological properties* (pp. 643–698). American Society of Agronomy.

Kemper, W. D., & Rosenau, R. C. (1986). Aggregate stability and size distribution. In A. Klute (Ed.), *Methods of soil analysis, part I. Physical and mineralogical methods* (pp. 427–442). American Society of Agronomy.

Kottek, M., Grieser, J., Beck, C., Rudolf, B., & Rubel, F. (2006). World map of Köppen-Geiger climate classification updated. *Meteorologische Zeitschrift*, *15*, 259–263.

Lal, R. (2020). Home gardening and urban agriculture for advancing food and nutritional security in response to the COVID-19 pandemic. *Food Security*, *12*, 871–876.

Lal, R., Bouma, J., Brevik, E., Dawson, L., Field, D. J., Glaser, B., Hatano, R., Hartemink, A. E., Kosaki, T., Lascelles, B., Monger, C., Muggler, C., Ndzana, G. M., Norra, S., Pan, X., Paradelo, R., Reyes-Sánchez, L. B., Sandén, T., Singh, B. R., ... Zhang, J. (2021). Soils and sustainable development goals of the United Nations: An International Union of Soil Sciences perspective. *Geoderma Regional*, *25*, Article e00398

Lehmann, J., Bossio, D. A., Kögel-Knabner, I., & Rillig, M. C. (2020). The concept and future prospects of soil health. *Nature Reviews Earth & Environment*, *1*, 544–553.

Levin, M. J., Kim, K.-H., Morel, J. L., Burghardt, W., Charzynski, P., & Shaw, R. K. (2017). *Soils within cities*. Schweizerbart Science.

Li, G., Sun, G.-X., Ren, Y., Luo, X. S., & Zhu, Y. G. (2018). Urban soil and human health: A review. *European Journal of Soil Science*, *69*, 196–215.

Macías, F., & Calvo, R. (2009). *Niveles Genéricos de Referencia de Metales Pesados y Otros Elementos Traza en Suelos de Galicia*. Xunta de Galicia.

Martínez Cortizas, A. M., & Pérez Alberti, A. (1999). *Atlas climático de Galicia*. Xunta de Galicia.

McDougal, R., Kristiansen, P., & Rader, R. (2019). Small-scale urban agriculture results in high yields but requires judicious management of inputs to achieve sustainability. *Proceedings of the National Academy of Sciences*, *116*, 129–134.

Morel, J. L., Chenu, C., & Lorenz, K. (2015). Ecosystem services provided by soils of urban, industrial, traffic, mining, and military areas (SUITMAs). *Journal of Soils and Sediments*, *15*, 1659–1666.

Olsen, S. R., & Sommers, L. E. (1982). Phosphorus. In A. L. Page (Ed.), *Methods of soil analysis. Part 2. Chemical and microbiological properties*. American Society of Agronomy.

Pansu, M., & Gautheyrou, J. (2006). *Handbook of soil analysis*. Springer-Verlag.

Paradelo, R., Celeiro, M., Herbón, C., Barral, M. T., & García-Jares, C. (2023). Polycyclic aromatic hydrocarbon concentrations and spatial distribution in the soils of Santiago de Compostela (northwestern Spain). *Geoderma Regional*, *34*, Article e00703

Paradelo, R., Herbón, C., & Barral, M. T. (2021). Composition and chemical properties of the soils of the city of Santiago de Compostela, northwestern Spain. *Journal of Environmental Quality*, *50*, 7–21.

Paradelo, R., Herbón, C., & Barral, M. T. (2025). Physical properties of the urban soils of Santiago de Compostela (Spain). *Journal of Soils and Sediments* 25, 462–471. https://doi.org/10.1007/s11368-024-03833-7

Paradelo, R., Herbón, C., Barral, M. T., Pontevedra-Pombal, X., & García-Rodeja, E. (2022). Urban soils in Santiago de Compostela (Galicia, NW Spain). In M. Switoniak & P. Charzynski (Eds.), *Soil sequences atlas V* (pp. 87–105). Nicolaus Copernicus University Press.

Probst, M., Gómez-Brandón, M., Herbón, C., Barral, M. T., & Paradelo, R. (2023). Fungal-bacterial associations in urban allotment garden soils. *Applied Soil Ecology*, *188*, Article 104896

Prokof'eva, T. V., Umarova, A. B., Bykova, G. S., Suslenkova, M. M., Ezhelev, Z. S., Kokoreva, A. A., Gasina, A. I., & Martynenko, I. A. (2020). Morphological and physical properties in diagnostics of urban soils: case study from Moscow, Russia. *Soil Science Annual*, *71*(4), 309–320.

Salomon, M. J., Watts-Williams, S. J., McLaughlin, M. J., & Cavagnaro, T. R. (2022). Spatiotemporal dynamics of soil health in urban agriculture. *Science of the Total Environment*, *805*, Article 150224

Schmidt, D. J. E., Pouyat, R., Szlavecz, K., Setälä, H., Kotze, D. J., Yesilonis, I., Cilliers, S., Hornung, E., Dombos, M., & Yarwood, S. A. (2017). Urbanization erodes ectomycorrhizal fungal diversity and may cause microbial communities to converge. *Nature Ecology and Evolution*, *1*, Article 0123

USDA. (1999). *Soil quality test kit guide*. https://www.nrcs.usda.gov/sites/default/files/2022-10/Soil%20Quality%20Test%20Kit%20Guide.pdf

Veerman, C., Pinto Correia, T., Bastioli, C., Biro, B., Bouma, J., Cienciala, E., Emmett, B., Frison, E. A., Grand, A., Hristov, L., Kriaučiūnienė, Z., Pogrzeba, M., Soussana, J.-F., Vela, C., & Wittkowski, R. (2020). *Caring for soil is caring for life*. European Commission, Publications Office of the European Union.

Weber, A. M., Mawodza, T., Sarkar, B., & Menon, M. (2019). Assessment of potentially toxic trace element contamination in urban allotment soils and their uptake by onions: a preliminary case study from Sheffield, England. *Ecotoxicology and Environmental Safety*, *170*, 156–165.

5

Management Practices and Land Use to Enhance Soil Health of Highly Weathered Degraded Acid Soils in the Iberian Peninsula

Chiquinquirá Hontoria, Ignacio Mariscal-Sancho, Fernando Peregrina, and Rafael Espejo

Chapter Overview

Acid soils in the Iberian Peninsula are mainly found in the western third of the peninsula, on the siliceous domain of the Hesperian massif. These soils are poor, and while they can function under native trees or shrubby vegetation, they are very vulnerable when put under cultivation. In this chapter we focus specifically on Palexerults located on very old and stable platforms (*rañas*) in western Spain. The main limitations of these soils arise from their high acidity and aluminum toxicity, which are exacerbated by the loss of organic matter owing to cultivation and tillage, and from the formation of perched water tables near the surface. Therefore, increasing pH, enriching calcium, and reducing soil perturbation are highlighted as solutions to the problems. Through several studies, we show the effects of the land use (e.g., from native forest to tilled crops), the application of calcium amendments (mainly sugar foam and red gypsum), no-tillage, and combinations of these factors on chemical, physical, and biological properties of the soil, measured at different depths and times. We also show their relationship to biomass production and greenhouse gas emissions. The lessons learned here contribute to a better understanding of the processes that have taken place and provide management recommendations to enhance agricultural production and the recovery of degraded acid soils.

Soil Health Series: Volume 5 Soil Health and Sustainability in Spain and Portugal, First Edition.
Edited by Iñigo Virto and Rodrigo Antón.

Published 2025 by Soil Science Society of America, Inc. and John Wiley & Sons, Inc.

Introduction

In this chapter, we show the problems of a large area in the southwest of the Iberian Peninsula that is dominated by acid soils under a Mediterranean climate. The cultivation of this area over decades has led to a strong degradation of soil health and a loss of productivity. Different management practices have been proposed for the recovery of these soils, such as liming, non-disturbance of the soil, and pasture improvement with legume species, either alone or in combination. We present here the main results and lessons learned after long-term research.

Soil Problems in the Rañas of the Southwestern Iberian Peninsula

Soil health is defined as the continued capacity of the soil to function as a vital living ecosystem that supports plants, animals, and humans, with soil organic matter (SOM) being key to this capacity. The cultivation of natural ecosystems and their agricultural management in a traditional way breaks the pre-existing natural balance, leading to a significant loss of SOM and subsequent soil degradation. These outcomes lead to damage in different environmental compartments, with effects on human health and planetary health, in addition to reduced productivity and profitability owing to the increasing inputs needed to sustain productivity. In Mediterranean environments, this degradation process has been occurring practically since the Neolithic, but it did not worsen until the introduction of industrial agriculture and the replacement of blood traction by mechanical traction, which also reduced the input of manure. A good example of degradation of soil health due to a SOM decrease caused by cultivation can be found in the Ultisols of the *rañas* of Extremadura (Spain; Figure 5.1). This example can be

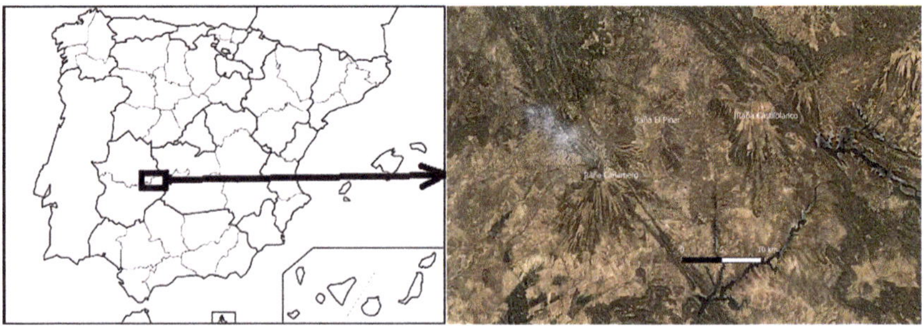

Figure 5.1 Location in the Iberian Peninsula and satellite image of the study area (Google Earth, 2024, lat. 39°27′N–39°13′N, long. 5°27′W-5°58′W, Spain).

extended to most of the *raña* formations of the Hesperian massif, which roughly coincides with the western half of the Iberian Peninsula. The *rañas* are continental detritic formations of the glacis-piedmont type, related to the quartzite elevations of the mountain ranges of the Hesperian massif of the Iberian Peninsula. Their soils present characteristics derived from the action of at least one subtropical climatic cycle, with limitations derived from a very acid pH, aluminum toxicity, and a tendency to waterlogging in the rainy seasons owing to its almost flat topography and the presence of argillic horizons at a shallow depth (Espejo, 1987; Mariscal-Sancho et al., 2010).

The natural vegetation in this area under Mediterranean climate is a cork oak forest (*Quercus suber* L.) with a dense replacement scrub (Figure 5.2). Under this forest the original soil is a plinthic Palehumult, an Ultisol with an umbric horizon that is very rich in SOM and an acidic Bt horizon, which in depth shows a plinthic segregation of iron oxides and colors (Soil Survey Staff, 2014). The deforestation and subsequent cultivation of these soils became widespread with the use of apatite as a fertilizer about 120 years ago. The main crops were rye (*Secale cereale* L.) alternating with fallow and continuously tilled olive groves (*Olea europaea* L.; Figure 5.2) and vineyards (*Vitis vinifera* L.). At the beginning, tillage led to a rapid SOM mineralization that released nutrients. However, it was no longer effective within a short time owing to the continuous export of these nutrients by crops and the release of Al^{3+} previously blocked by organic compounds, which increased the toxicity problems that these soils exhibit for Mediterranean crops. The strong loss of SOM, which was maximal in the tilled olive groves and vineyards, reduced soil health, and large amounts of CO_2 were emitted to the atmosphere (Mariscal-Sancho, 2008; Mariscal-Sancho et al., 2010). In addition to the SOM loss and consequent soil degradation, the formation of perched water tables above the Bt

Figure 5.2 Cork oak and olive groves are the most contrasting land uses in the study area. Photos by Rafael Espejo.

horizon during periods with abundant rainfall is incompatible with good seed germination and crop development. Consequently, after a few decades, the cultivated fields were abandoned, but already in such a state of degradation that complete natural self-recovery was hardly viable.

The recovery of these degraded soils requires the use of calcium amendments that reduce aluminum toxicity and the introduction of soil management methods that, if possible, exclude tillage to reverse the degradation processes by increasing the SOM content. Among these management practices would be the establishment of pastures and cultivation by direct sowing or no-tillage. Since these *raña* formations are in economically depressed areas, liming materials based on easily accessible and very economical industrial by-products would be recommended to promote the circular economy. To evaluate the impact of these practices, alone or in combination, on the recovery of the health of these soils, several experiments were carried out. The characteristics and results of these studies are explained in the following sections.

Characteristics of the Study Area

The study area is located in the southwestern part of the Iberian Peninsula, in the *Raña de Cañamero* (lat. 39°22′N–39°17′N, long. 5°21′W-5°16′W; Extremadura, Spain), which is a broad glacis-piedmont plateau that is very old and geomorphologically stable and covers an extensive flat surface (<1% slope) at an altitude between 580 and 620 masl (Figure 5.1). The climate is temperate with dry and warm summers and wet and cool winters (Csa according to the Köppen classification). The mean annual temperature is 15.3 °C, with January having the lowest mean temperature (6.2 °C) and August the highest (25.2 °C). Mean annual evapotranspiration (ETo Penman-Monteith) is 1,365 mm, with the lowest value in December (20 mm) and the highest in July (234 mm). Mean annual precipitation is 856 mm, with December being the wettest month (126 mm) and July the driest (6 mm; Hontoria et al., 2018). The wet season extends from November to April, concurrent with the lowest temperatures, and the dry season extends from May to October, concurrent with the highest temperatures. The great interannual weather variability leads to agricultural years with excess rainfall in autumn and early spring, while other years present deficits in winter to spring.

The soils in the cultivated and degraded areas are plinthic Palexerults (Soil Survey Staff, 2014), with an Ap horizon, an acidic pH decreasing with depth, an exchange complex dominated by Al^{3+}, and a low content of exchangeable basic cations. Kaolinite is the dominant mineral in the clay fraction. Nonporous quartzite rock fragments are frequent to abundant throughout the soil profile (Espejo, 1987). A representative profile of the study area is shown in Table 5.1.

Table 5.1 Soil Properties of a Representative Profile in the Study Area.

	SOC	pH		Al^{3+}	Exchangeable basic cations ($cmol_c$/kg)				RF	Texture %		
Horizon (cm)	g/kg	H_2O	KCl	$cmol_c$/kg	Ca^{2+}	Mg^{2+}	Na^+	K^+	%	Sand	Silt	Clay
Ap (0–25)	19.6	5.3	4.3	0.76	1.12	0.21	0.05	0.65	52	80.1	6.1	13.8
AB (25–35)	9.2	5.0	4.0	1.01	0.85	0.16	0.04	0.08	47	69.1	5.6	25.3
Bt (35–90)	1.4	4.9	3.7	1.75	0.82	0.11	0.03	0.07	56	59.6	4.7	35.4
Btv (90–200)	nd	4.6	3.4	2.23	1.12	0.10	0.05	0.06	75	56.0	5.1	38.9

Note. SOC, soil organic carbon; RF, rock fragments >2 mm. Al^{3+} is KCl-extractable.

Experimental Trials and Measurements

Experimental Trials

Soil and crop response to calcium amendments and its modulation by soil tillage system and land use were studied in two long-term field trials. These trials were preceded by a series of field, laboratory, and greenhouse experiments in which the effects on soil and biomass production of different calcium amendments were evaluated (Peregrina et al. 2006, 2007a, 2007b, 2008a, 2008b; Mariscal-Sancho et al. 2009a). The calcium amendments evaluated were gypsum, red gypsum (a by-product of the titanium production from ilmenite, which is very rich in gypsum), phosphogypsum (a by-product of obtaining phosphoric acid from phosphorite), limestone rock, and sugar factory foam (hereafter sugar foam), a by-product of sugar production from beet, which is rich in lime and other nutrients (Peregrina et al., 2006). Initially, the study of these by-products was of interest because, in addition to being calcium amendments, they could boost the circular economy.

Based on the results of these studies, the best calcium amendments were selected and evaluated in long-term field trials. These field trials were established in an area abandoned for 15 years after more than 50 years of cultivation. In 2002, a long-term trial (amendment × use) was established to evaluate the effect of applying or not applying a calcium amendment based on sugar foam on three land uses: tilled forage crop, natural pasture, and improved pasture with legume species and some grasses. The experimental design was a split-plot with four replicates and 16 plots of 6 m × 12 m each. The second long-term field trial (tillage × amendment) was established in 2005 to evaluate the response of the soil and a forage crop to applying or not applying a calcium amendment with and without tillage (four treatments). The experiment had a split-plot design with four replicates and plot dimensions of 4 m × 16 m. The calcium amendment consisted of a mixture of sugar foam and red gypsum that was incorporated into the first 7 cm of the soil at the beginning of the trial and deposited on the surface in subsequent applications in no-tillage plots. Red gypsum was applied because, owing to its greater solubility, it affects deeper horizons (Peregrina et al., 2006, 2008b). Three applications of calcium amendment were made during the trial. The first was supplemented with dolomitic coverter residue and $MgSO_4$ to replenish the magnesium lost from the exchange complex resulting from the application of gypsum (Peregrina et al., 2008b). A mixture of *Avena sativa* L. (57%), *Vicia sativa* L. (26%), and Triticale (17%) was sown in most years, but in the last ones, Triticale was removed. Moderate or low NPK fertilization was applied in autumn. The crop was harvested every July at a height of 20 cm above ground and residues were left on the soil surface (more information in Hontoria et al., 2018).

Measurements and Data Analysis

Many chemical, physical, and biological soil properties, as well as greenhouse gas (GHG) emissions, were measured in this research (Table 5.2). Soil sampling for chemical and physical properties was done routinely in autumn, every year or every 2 years, at a depth of up to 50 cm for chemical properties and up to 25 cm for organic carbon and aggregation variables. Other physical properties and biological properties were mostly evaluated in the surface layer (up to 10 cm) and the latter at different times of the year. GHG emissions were evaluated once or twice a month. Soil cover was measured randomly twice per plot by taking digital images of a 0.5×0.5 m^2 area and processing these images with SigmaScan Pro 5 software. Yield was measured in each season at harvest time by taking two random 1×1 m^2 subsamples from each plot. Data were usually analyzed with a general linear model (GLM) ANOVA, considering the design of each trial in each case. Variables were log or Box-Cox transformed when necessary. Differences between means were evaluated using Fischer's Least Significant Difference (LSD) test for $P < .05$. Relationships between variables were analyzed using simple and multiple regressions, as well as principal component analysis. The statistical packages Statgraphics Centurion and SPSS in different versions were used.

Effects of Land Use and Management on Chemical Properties

Effects of Land Use

The type of vegetation and land use have a great effect on the chemical properties of these soils (Table 5.3). Under climatic vegetation (cork oak forest), the Ah horizon (umbric) has a higher content of exchangeable basic cations (especially calcium

Table 5.2 Soil Measurements and Laboratory Methods.

Soil measurements	Method	
Soil chemical properties		
pH, electrical conductivity	1 : 2.5 soil/water	
Basic cations	Ammonium acetate pH 7 extraction + atomic absorption spectrophotometry	Gómez-Paccard et al., 2013
Aluminum cation (Al^{3+})	1 M KCl extraction + titration	Gómez-Paccard et al., 2013
Aluminum forms		Gómez-Paccard et al., 2013
Total organic carbon	Wet oxidation/dry combustion	
Particulate/mineral-associated organic carbon	Dispersion + sieving	Gómez-Paccard et al., 2013
Permanganate oxidizable carbon	Oxidation + colorimetry	Vázquez et al., 2017
Dissolved organic carbon	K_2SO_4 extraction + oxidation	Vázquez et al., 2019c
Water-soluble organic carbon	Water extraction (1 : 5) + TOC analyzer	García-Marco et al., 2016
Total nitrogen	Kjeldahl method/dry combustion	
Dissolved nitrogen	K_2SO_4 extraction	Vázquez et al., 2017
Ammonium and nitrates	KCl extraction + Electrode/Flow-injection analyzer	Vázquez et al., 2019c
Phosphorus	Extraction + colorimetry (molybdenum blue method)	Vázquez et al., 2020a
Soil physical properties		
Fine earth bulk density	Core method	Gómez-Paccard et al., 2015
Total porosity	Calculated from bulk density and particle density	Gómez-Paccard et al., 2015
Soil moisture	Drying at 105 °C to constant weight	
Soil water retention characteristics	Pressure plates, −33 kPa and −1,500 kPa	Gómez-Paccard et al., 2015
Saturated hydraulic conductivity	Constant head permeameter method	Gómez-Paccard et al., 2015

(*Continued*)

Table 5.2 (Continued)

Soil measurements	Method	
Water infiltration	Flooded/pounded method with a single ring	Gómez-Paccard et al., 2015
Level of the perched water layer	Piezometer	Gómez-Paccard et al., 2013
Water-stable aggregate percentage	Wet-sieving (0.250 mm)	Gómez-Paccard et al., 2013
Water-stable aggregate distribution	Wet-sieving (2, 0.250, and 0.053 mm)	Hontoria et al., 2018
Soil temperature	Digital sensor	García-Marco et al., 2016
Soil biological properties		
Microbial biomass carbon and nitrogen	Fumigation-extraction method	Vázquez et al., 2019c
Enzymatic activities (dehydrogenase, β-glucosidase, β-glucosaminidase, arylesterase, cellulase, protease, casein protease urease, arylsulfatase, acid phosphomonoesterase)	Different methods described in the references	Vázquez et al., 2017, 2019b, 2019c, 2020a; Giagnoni et al., 2020;
Basal respiration	Aerobic incubation + alkaline trap + titration	Vázquez et al., 2017
Carbon and nitrogen mineralization	Incubation	Vázquez et al., 2019a
Potentially mineralizable nitrogen	Anaerobic incubation + KCl extraction + colorimetry	Vázquez et al., 2019b
Denitrification capacity	Anaerobic incubation with KNO_3 and acetylene + gas chromatography	García-Marco et al., 2016
Glomalin	Sodium citrate extraction + Bradford essay	Gómez-Paccard et al., 2013
Root colonization	Stein + magnified intersections method	
Bacterial and archeal *16S*, *ITS*, *amoA*, *amoB*, *nirK*, *nirS*, *nosZ* (clade I) genes	Soil DNA extraction + qPCR with SYBR® Green Supermix	Vázquez et al., 2019a, 2019c; Vázquez et al., 2020c

Table 5.2 (Continued)

Soil measurements	Method	
Gas emission		
In situ respiration	Field chamber with EGM-4 infrared detector	Mariscal-Sancho et al., 2010
N_2O and CH_4 fluxes	Static chambers + gas chromatography	García-Marco et al., 2016

Table 5.3 Content of Exchangeable Basic Cations and Aluminum in the Surface Horizons Under the Different Uses and Vegetation.

	Soil depth	Ca^{2+}	Mg^{2+}	Na^{+}	K^{+}	Al^{3+}
Vegetation/Use	(cm)			$cmol_c$/kg		
Cork oak	0–5	4.13 d	1.21 c	0.14 b	0.43 b	1.11 a
75 % Shrub after abandonment	0–5	1.58 bc	0.25 b	0.08 ab	0.12 a	1.11 a
Pasture after abandonment	0–5	1.26 ab	0.18 a	0.05 a	0.10 a	1.22 a
Olive	0–5	0.96 a	0.15 a	0.06 ab	0.40 b	1.26 a
Cork oak	5–25	2.95 c	0.56 b	0.10 d	0.17 b	1.92 b
75 % Shrub after abandonment	5–25	0.85 ab	0.11 a	0.07 c	0.09 a	1.46 a
Pasture after abandonment	5–25	0.88 ab	0.10 a	0.04 ab	0.07 a	1.75 ab
Olive	5–25	0.75 a	0.08 a	0.03 b	0.09 a	1.66 ab

Note. Different letters indicate significant differences according to LSD ($P < .05$) within a column and a soil depth.

and magnesium) than in the Ap horizon (ochric) of soils cultivated with olive groves (Mariscal-Sancho et al., 2009b). These differences are due to (1) the higher SOM content in the cork oak forest, which provides a higher effective cation exchange capacity (ECEC) and allows the retention of the basic cations provided by the vegetation residues by first humidifying and then mineralizing them, and (2) the cation exportation through harvesting that occurs in the cultivated soils. The elimination of natural vegetation and soil tillage causes a reduction in the SOM of the Ap horizon, which induces a reduction in soil chemical fertility by loss of basic cations such as Ca^{2+} and Mg^{2+} and an increase in acidity (Mariscal-Sancho et al. 2009b).

Although the vegetation type and land use does not affect the exchangeable aluminum, the phytotoxicity of this aluminum is lower in cork oak soils since the higher exchangeable calcium content implies an increase in the Ca/Al ratio and a decrease in the aluminum saturation in the ECEC (Mariscal-Sancho et al., 2009b). The calcium-aluminum balance (CAB), obtained by the relation between calcium and aluminum activities in the soil solution (Noble et al., 1988), allows the evaluation of aluminum phytotoxicity to roots. In the soil solution of the Ap horizon of the tilled soils, CAB was of the order of 23 (Peregrina et al., 2006), well below the value of 31, above which no aluminum damage would appear in 90% of the roots (Menzies et al., 1994).

Evaluation of Different Calcium Amendments

Regarding the use of different calcium amendments to recover soils degraded by tillage, we found that sugar foam had the same capacity to increase the pH and reduce the exchangeable aluminum as limestone, and it also provided other nutrients such as magnesium in sufficient quantities to increase the content in the soil (Peregrina et al., 2006). This result implies an increase in exchangeable basic cations, in addition to the reduction of exchangeable aluminum, which allows an improvement in soil chemical fertility (Peregrina et al., 2006). This positive effect on the removal of exchangeable aluminum remains under field conditions for 4 years (González-Fernández et al., 2012), at which time a new dose of calcium amendment should be applied.

Gypsum also reduces aluminum phytotoxicity because it causes a reduction of exchangeable aluminum as it is displaced by calcium from gypsum dissolution (salt effect). In addition, it increases the Ca/Al ratio of the soil solution and the CAB value above 30 and favors the formation of Al-SO_4 pairs that are less phytotoxic than Al^{3+} in the soil solution (Peregrina et al., 2006). These positive effects are more limited in time than in the case of limestone since gypsum is much more soluble and leaches more easily. Thus, in the field, gypsum is leached from the Ap horizon after 2 years (Peregrina et al., 2008b). This leaching of gypsum causes the positive effects mentioned above to reach the AB and Bt subsurface horizons (Peregrina et al., 2008b), which allows the roots to explore these horizons and thus take advantage of a greater volume of available water, which is very important due to the xeric regime of these soils. However, the application of gypsum has the undesired effect of displacing magnesium from the exchange complex owing to the high calcium concentration, which causes the magnesium loss by leaching together with SO_4^{2}– (Peregrina et al., 2006). This effect has been observed in the field and can be compensated by the application of a magnesium source such as the by-product dolomitic converter residue (Peregrina et al., 2008b).

The gypsiferous by-products red gypsum and phosphogypsum have a similar capacity to gypsum to reduce exchangeable aluminum and increase the Ca/Al ratio,

CAB, and the formation of Al-SO_4 pairs. However, phosphogypsum is problematic because its composition contains appreciable amounts of fluoride (12.3 g/kg) in the form of HF (Peregrina et al., 2006). When dissolved in water (1 : 20) its pH is acidic (3.94) with fluorine levels on the order of 15 mg/L (1 : 20) (Mariscal-Sancho et al., 2009a). When the phosphogypsum dose is increased to 8.4 Mg/ha, the hydrofluoric acid causes a dissolution of soil silicates, increasing aluminum and silicon in the soil solution. This increase in aluminum causes greater activity of Al-F pairs in the soil solution, and greater assimilation of aluminum and fluorine by the roots, which reduces biomass production compared with that produced with a gypsum or red gypsum amendment (Mariscal-Sancho et al., 2009a).

Overall, considering all the results obtained and as a way to improve chemical fertility and reduce aluminum toxicity in cultivated soils, we recommend the application of a combination of (1) sugar foam as a liming product at a dose of 3.9 Mg/ha; (2) red gypsum at a dose of 7.5 Mg/ha to reduce aluminum toxicity in the subsurface horizons; and (3) dolomitic converter residue at a dose of 100 kg/ha to compensate for the losses of magnesium displaced by the calcium in the gypsum.

Effect of the Tillage System on the Calcium Amendment

Regarding the modulating effect of the tillage system on the amendment (tillage × amendment trial), we found that, under traditional tillage and no-tillage, the combination of sugar foam, red gypsum, and dolomitic converter residue eliminated the exchangeable aluminum in the upper 10 cm of the soil (Gómez-Paccard et al., 2013), while increasing calcium levels in the Ap and AB horizons (Gómez-Paccard et al., 2013), confirming the effect of red gypsum in the subsurface horizons (Figure 5.3). The effect of tillage can be seen in the 0- to 5-cm layer, where the exchangeable aluminum was reduced in the no-tillage compared with traditional tillage, an effect attributed to the formation of complexes with organic matter (Gómez-Paccard et al., 2013). This reduction in exchangeable aluminum was not found at greater depths because the accumulation of plant debris owing to no-tillage only occurs on the top 5 cm.

Soil tillage on a forage crop versus no soil disturbance under both natural and improved pasture with legume species (amendment × use trial) did not affect exchangeable aluminum and pH, but nevertheless exchangeable magnesium and potassium increased in both types of pasture (natural and improved) with respect to the forage crop in the top 5 cm of the soil (Mariscal-Sancho et al., 2024). This result would be caused by the higher SOM content under pastures and the fact that the amendment remains on the surface of the pasture because it is not incorporated by tillage.

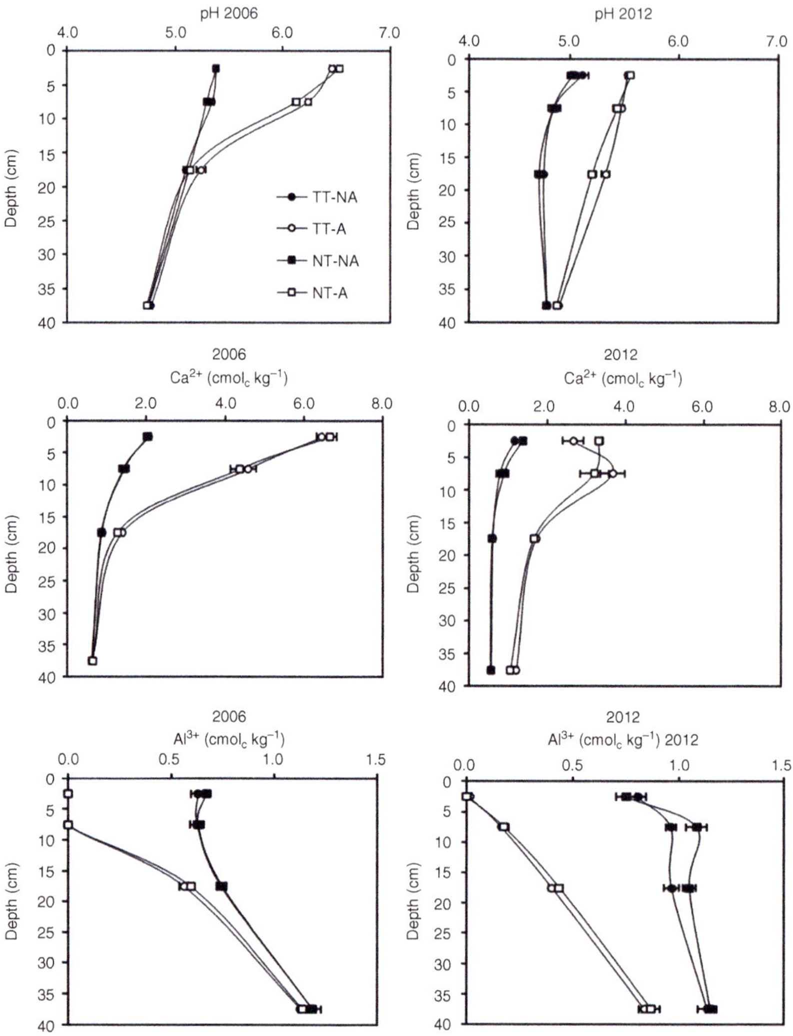

Figure 5.3 Vertical profiles of pH, calcium, and aluminum to a depth of 50 cm under the four tillage-amended combinations (TT, traditional tillage; NT, no-tillage; NA, non-amended; A, amended) after 1 and 7 years from the start of the field experiment. Adapted from Hontoria et al. (2018).

Effects of Land Use and Management on Soil Organic Carbon

In these soils, soil organic carbon (SOC) is strongly conditioned by land use and agricultural management. It responds mainly to the degree of soil disturbance, with a higher content at lower disturbances, and to a lesser extent, to the availability of nutrients favored by liming and fertilization. In terms of land use (Figure 5.4), SOC showed maximum values in the surface layer (top 10 cm) of the cork oak forest soil (60 g/kg), with lower values in the holm oak (*Quercus ilex* L.) forest (~40 g/kg). This is followed by soils that were once agricultural and are at different stages of abandonment, for example, with more than 75% scrub after decades of abandonment (~38 g/kg) or grassland after 15 years of abandonment with occasional cultivation for scrub control (33–22 g/kg). The lowest content was found in the olive grove that was continuously tilled in a traditional way (~20 g/kg), where SOC loss was at a maximum (Mariscal et al., 2007; Vázquez, 2019; Vázquez et al., 2020b). Cork oak contains three times more SOC in the top 10 cm than olive grove, and this ratio tended to be maintained in terms of carbon stock (12.8 vs. 36.1 Mg C/ha). At a depth of 50 cm, the cork oak stored ~130 Mg C/ha versus ~38 Mg C/ha for the olive grove (Mariscal-Sancho, 2008).

In the tillage × amendment trial (Gómez-Paccard et al., 2013; Hontoria et al., 2018; Vázquez, 2019), calcium amendment and especially no-tillage had a positive effect on SOC, with no clear interaction between the two factors. Both practices started to take effect in the short to medium term, although after several years of application only the topsoil (0–5 cm) was affected, revealing a strong stratification of SOC (Hontoria et al., 2018; Vázquez, 2019). The calcium amendment favored the accumulation of SOC and, above all, particulate organic carbon (POC)

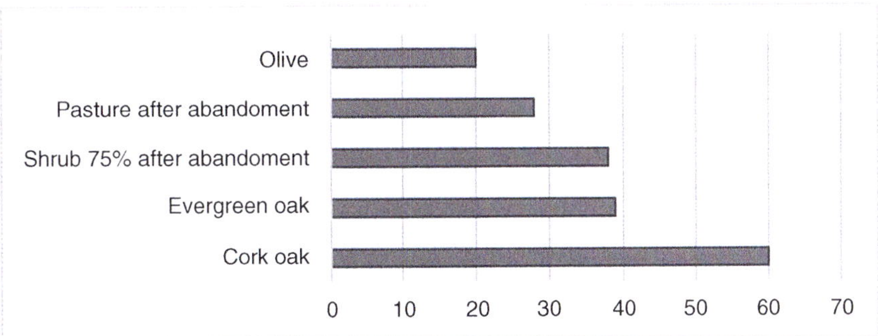

Figure 5.4 Concentration of soil organic carbon (g/kg) in the first 10 cm of soil under different land uses. Values are estimates based on different studies (Mariscal et al., 2007; Vázquez, 2019; Vázquez et al., 2020b).

in the topsoil thanks to the increase in biomass production, resulting from the mitigation of aluminum toxicity and the increase in nutrient availability, which led to an increase in the input of crop residues into the soil. Although the application of the calcium amendment led to an increase in certain enzymatic activities and microbiological activity (León et al., 2017; Vázquez et al., 2017; Vázquez et al., 2020a), resulting in part from an increased availability of labile carbon fractions (Vázquez et al., 2017), this increase in microbiological activity did not translate into a significant reduction in SOC (Caires et al., 2006; Paradelo et al., 2015).

No-tillage increased SOC in the topsoil, not only because of the reduction in mineralization losses (Six et al., 2002), but also because of a greater input of organic residues into the soil as plant biomass increased owing to more favorable growing conditions for the crop. Thus, no-tillage increased aggregate stability from the early years (Gómez-Paccard et al., 2013), as well as aggregate size (Hontoria et al., 2016). This improvement in soil structure together with the formation of biopores under no-tillage and the consequent improvements in soil-water-plant relationships led to an increase in crop production compared with traditional tillage, which was more pronounced under soil water deficit or water excess (Gómez-Paccard et al., 2015). The positive effect of no-tillage on SOC was initially weak, but the magnitude of the change increased over time (Figure 5.5) as meteorologically adverse years with a lack or excess of rainfall followed (years 5, 6, and 7 since start-up; Hontoria et al., 2018). As a result, year after year, there was a greater input of crop residues into the soil under no-tillage, which increased the SOC difference in favor of it, driven especially by the increase in POC (Hontoria et al., 2018). Thus, the SOC content in no-tillage versus traditional tillage increased by 5% at 3 years, 15% at 7 years, and almost 30% at 10 years from the start of the experiment (Hontoria et al., 2018; Vázquez et al., 2017).

No-tillage and calcium amendment had an additive effect, mainly during the first years of the trial, so that the combination of both obtained the highest SOC and POC values (Figure 5.5). After two applications of calcium amendment and 10 years from the start of the trial, no-tillage combined with calcium amendment increased SOC and POC values by 35% and 65%, respectively, with respect to the control in the top 5 cm (Vázquez et al., 2017; Vázquez, 2019). However, it is worth highlighting the beneficial role of no-tillage without calcium amendment, which also achieved relevant increases (30% and 48% of SOC and POC with respect to the control after 10 years; Vázquez, 2019) and for some samplings equaled or even exceeded the combination of no-tillage and calcium amendment (Vázquez et al., 2017). This ability of no-tillage to increase SOC even without calcium amendment may have contributed to reducing the influence of calcium amendment to the point of statistically disappearing.

In comparison with the SOC value (~28 g/kg) at the start of the trial (previously, cultivation was abandoned for 15 years), no-tillage with or without calcium

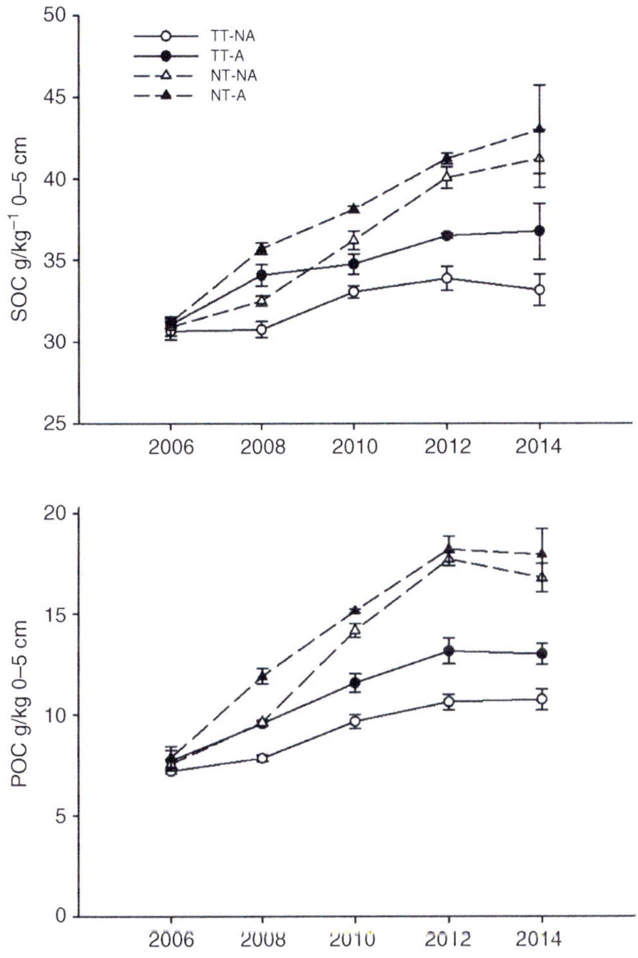

Figure 5.5 Evolution of soil organic carbon (SOC) and particulate organic carbon (POC) in the 0- to 5-cm layer after 10 years from the start of the field experiment. TT, traditional tillage; NT, no-tillage; A, application of calcium amendment; NA, no application of calcium amendment.

amendment managed to increase the SOC concentration by ~60% in the 0- to 5-cm layer after 10 years, which represents an increase of 5 Mg C/ha, equivalent to a sequestration of 19 Mg of CO_2 from the atmosphere. Traditional tillage increased somewhat more than 3 Mg C/ha and sequestered more than 12 Mg of CO_2 from the atmosphere. Even the control (tillage without calcium amendment) increased SOC thanks to the effect of fertilization on biomass production and probably also because of the fence that protected the field from grazing; the control accumulated less than 3 Mg C/ha (<11 Mg CO_2). With regard to the top 10 cm, values of ~38 g/kg of SOC were reached after 10 years of applying a fertilized forage crop under no-tillage (with and without calcium amendment), which is above the SOC reached after decades of abandonment and close to that shown in the top 10 cm of a holm oak forest (Figure 5.4). This result confirms the use of a non-tillage forage crop, reinforced with the use of calcium amendment and fertilization, as an adequate strategy for the recovery of these soils.

Instead of a forage crop, the recovery of degraded acid soils through SOC increase can be even better when pastures are used and additionally improved with legumes (amendment × use trial). Thus, after 11 years and for the 0- to 5-cm layer, improved pasture and natural pasture increased SOC by 33% and 24%, respectively, compared with a tilled forage crop (Mariscal-Sancho et al., 2024), despite the tilled forage crop having 60% more aerial biomass production than pastures. The positive effect on SOC reached up to 15-cm depth for the improved pasture. Apart from the above-mentioned negative effect that tillage of the forage crop exerted on SOC, the better performance of pasture, and especially of improved pasture, to accumulate SOC can be explained by (1) their higher root biomass production compared with the tilled forage crop (Bolinder et al., 2007); (2) the greater effectiveness of roots in stabilizing carbon in the soil (Jackson et al., 2017); (3) the protection against erosion and reduced temperature and moisture changes that occur the whole year in pastures versus only some months in the forage crop; and (4) the greater SOM accumulation that pastures with legumes favor (Conant et al., 2017), among others. POC followed a very similar pattern to SOC, but with an earlier response to different uses.

In the same trial, calcium amendment increased SOC when the top 15 cm was considered globally because of a greater production of aerial biomass, which was cumulatively 35% higher in the amended versus unamended plots. However, this difference in SOC was not found when the 0- to 5-cm and 5- to 15-cm layers were considered separately. For the top 5 cm, the amended plots tended to accumulate more SOC than the unamended plots after the first years ($P < .1$), but after 11 years this difference disappeared (Figure 5.6). As in the tillage × amendment trial, the undisturbed soil occurring in pastures triggered a set of soil health improvements that progressively reduced the influence of calcium amendment on SOC accumulation and after 11 years brought the SOC of unamended pastures closer to that of amended pastures, thus weakening the effect of calcium amendment. We also found no significant interaction between type of land use and calcium amendment after 11 years, but an additive effect was observed so that the highest SOC values were found in amended pastures improved with legumes. With respect to the SOC and POC values at the start of the trial, where the previous use was an unfertilized natural pasture resulting from abandonment, the improvement of the pasture composition and its fertilization allowed doubling the SOC

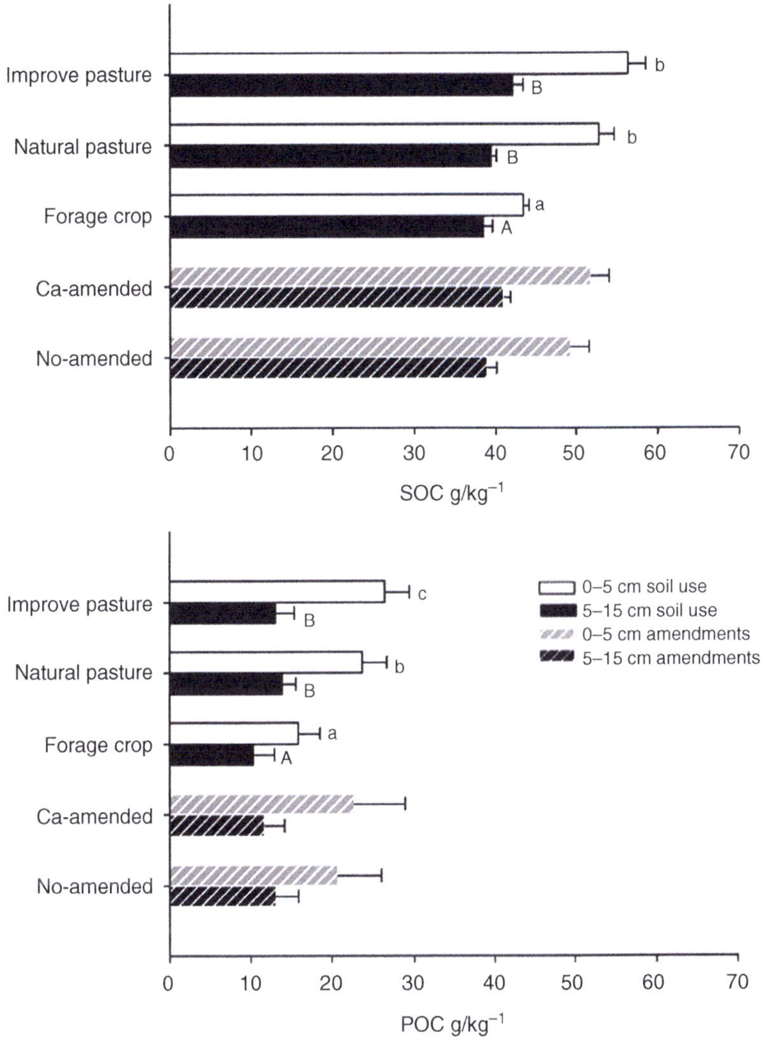

Figure 5.6 Soil organic carbon (SOC) and particulate organic carbon (POC) in different soil-uses (improved pasture, natural pasture, and tilled forage crop) with and without calcium amendment after 11 years from the start of the field experiment. Differences between means were significant only for soil use, and different lowercase and capital letters indicate significant differences between land uses for depths of 0–5 cm and 5–15 cm, respectively.

value and almost quadrupling the POC value in the top 5 cm after 11 years, with somewhat lower increased values for the 5- to 15-cm layer. The tilled forage crop also increased by 50% from the initial SOC thanks to fertilization that led to a remarkable increase in biomass production in this nutrient-poor soil. In terms of carbon stock, for the top 15 cm and after 11 years, improved pasture added more than 21 Mg C/ha; natural pasture, 18 Mg C/ha; and tilled forage crop, 16.8 Mg C/ha. Therefore, the use of pasture and the improvement of its composition is an appropriate strategy for the recovery of these degraded acid soils at greater depth, together with calcium amendment and fertilization.

Effects of Land Use and Management on Physical Properties

Physical properties were also closely related to the type of land use and were more affected by the tillage system than by whether or not a calcium amendment was applied. The effect of land use and management on topsoil physical properties is closely related to SOM content (Blanco-Canqui & Ruis, 2018) and the lack of soil disturbance influencing the amount and distribution of biopores (Kay & VandenBygaart, 2002). From least to most soil disturbance, forest soils (cork oak and evergreen oak) showed the lowest fine-earth soil bulk density values (0.7–0.8 g/cm), followed by scrub soil (0.8–0.9 g/cm), abandoned agricultural soils, either covered to a greater or lesser extent with scrub (0.9–1.0 g/cm), or with grassland occasionally tilled to grow rye (0.9–1.1 g/cm). Meanwhile, olive grove soil under traditional tillage, which was depleted in SOM, showed the highest values (1.1–1.3 g/cm; Mariscal et al., 2007; Mariscal-Sancho et al., 2009b; Vázquez et al., 2020b). The percentage of water-stable aggregates (WSAs) in the surface layer followed an inverse pattern to bulk density, with maximum values in forest soils and minimum values in olive grove soils and an intermediate position of abandoned or non-tilled agricultural soils. However, WSA showed greater sensitivity to land use than bulk density in a way that varied with the type of forest and with the years of crop cessation (Mariscal et al., 2007; Vázquez et al., 2020b). Thus, with a similar SOC and bulk density value, WSA at 6 years of tillage ceasing was higher than at 3 years (Mariscal et al., 2007). Despite the high values found in these soils, the level of WSA was thus confirmed as an early physical indicator of changes in land use.

The two long-term trials showed that no-tillage and pastures improved soil structure-related parameters, while calcium amendment had hardly any effect. In a forage crop (amendment × tillage trial), 4 years of no-tillage increased topsoil WSA by 10%, concurrent with an increase in SOC of 20% (Gómez-Paccard et al., 2013), while 6 years of no-tillage reduced bulk density by 13% compared with tillage, concurrent with an increase in SOC of just over 30% (Gómez-Paccard et al., 2015). The use of pasture compared with a tilled forage crop (amendment × use trial) increased WSA by 6% in the 5- to 15-cm layer (97% vs. 91%, respectively). Regarding aggregate size (Hontoria et al., 2012, 2016), the soil macroaggregates (>0.250 mm) dominated in the pasture soils, which presented the highest mean weight diameter of aggregates (MWD 1.2 mm), as well as in the no-till forage crop (MWD 0.9 mm), while microaggregates (0.053–0.250 mm) dominated in the tilled forage crop (MWD 0.8 mm). After 6 years, the use of pasture doubled the macroaggregate content of the topsoil compared with a tilled forage crop (60% vs. 30%), while not tilling a forage crop increased macroaggregates by about 40% compared with tilling, an effect also found in other soils of the Iberian Peninsula upon cessation of tillage (Plaza-Bonilla et al., 2013). The non-disturbance of the soil with the consequent increase in SOC, POC, earthworms, and glomalin, as well as the stimulation of mycorrhizal fungi and microbiological activity in general (Gómez-Paccard et al., 2013, 2015; Vázquez et al., 2020a), may explain the increase in water stability and size of the soil aggregates (Hontoria et al., 2016; Six et al., 2004). In contrast, the effect of the calcium amendment on the soil structure parameters was null or very weak in this sand-rich soil, where the effect from Ca^{2+} bridges alone may be reduced and be outweighed by that of organic matter and soil biota (Tester, 1990).

Effect of Management on Crop and Water-Soil Relations

In soils such as those of the Cañamero *raña*, crop problems in terms of the water-soil relations are frequent. In Mediterranean environments, periods of below-average rainfall are common, generating soil moisture deficits; however, periods with above-average rainfall are also frequent and can have adverse consequences in these soils. Thus, the fact that the *rañas* are almost horizontal platforms and their soils present thick Bt horizons favors the generation of shallow perched water tables in autumn, December–January, and March–April. In the case of both autumn and spring perched water tables, many problems are generated for crop germination and biomass development, respectively.

The real water storage capacity in these soils resides in the argillic horizons with the finest texture of the soil profile. As the mean pore diameter is larger in the overlying coarser-textured A horizons, the capillary rise of water is slowed and with it the loss of water by evaporation in early summer. Unfortunately, the predominance of aluminum in the Bt exchange complex, which also increases with depth (Espejo, 1987), inhibits root development in the Bt horizons, so enhancing the water storage capacity in the Ap and AB horizons is essential. This depends largely on the organic matter content, which can increase the water storage capacity. Moreover, under natural, non-degraded vegetation, the surface of these soils is covered by a thick layer of leaf litter, which slows water evaporation, such that the water content in the A horizon is higher than that of the Ap horizon of cultivated soils throughout the year (Mariscal-Sancho et al., 2010).

Similarly, under no-tillage, SOC increases, and a mulch is generated with crop residues in the shallowest part of the soil, which contributes to preserving soil moisture (Munawar et al., 1990) and leads to higher yields in Mediterranean environment (Plaza-Bonilla et al., 2017). In line with the above, in the tillage × amendment trial, SOC and aggregate size increased, and bulk density decreased in the top 5 cm of the soil under no-tillage, which, among other factors, led to an improvement in water properties (Gómez-Pacard et al., 2015). Thus, soil water content at both saturation and field capacity was significantly higher under no-tillage, with no differences in moisture between tillage systems at the permanent wilting point. In addition, surface infiltration and hydraulic saturated conductivity were more than two times higher in the no-tilled plots than in the tilled plots. Together with the increase in total porosity (lower bulk density), the conservation of root and earthworm biopores, which is favored by no-tillage, greatly facilitated the vertical displacement of water; in contrast, tillage altered the continuity of the pores (Lipiec et al., 2006).

This improvement is crucial when a shallow perched water table is generated, as untilled plots show a moisture content further away from saturation for longer than in tilled plots, resulting in better seed germination rates and better crop development and biomass production (Gómez-Paccard et al., 2015). Figure 5.7 illustrates the better soil-water relationships for crops in no-tillage relative to traditional tillage in both dry and rainy periods of two years; thus, untilled plots showed a higher soil water content during the period with higher temperatures and evapotranspiration (April–July and October), while tilled plots showed a soil water content much closer to saturation in some months (Gómez-Paccard et al., 2015).

Management Effects on Crop Productivity

The main cause of the low productivity of *raña* soils under cultivation is the aluminum toxicity, aggravated by the loss of SOM caused by tillage, followed by the very low content of exchangeable basic cations, aggravated by the extraction and subsequent export of the crops. An additional factor is the formation of perched water tables in rainy periods that lead to conditions close to saturation near the surface.

Studies carried out on the effect of different calcium amendments on biomass production (Peregrina et al., 2006, 2008b) indicate that, for the same equivalent dose in calcium to reach a pH of 6.2, gypsiferous amendments substantially

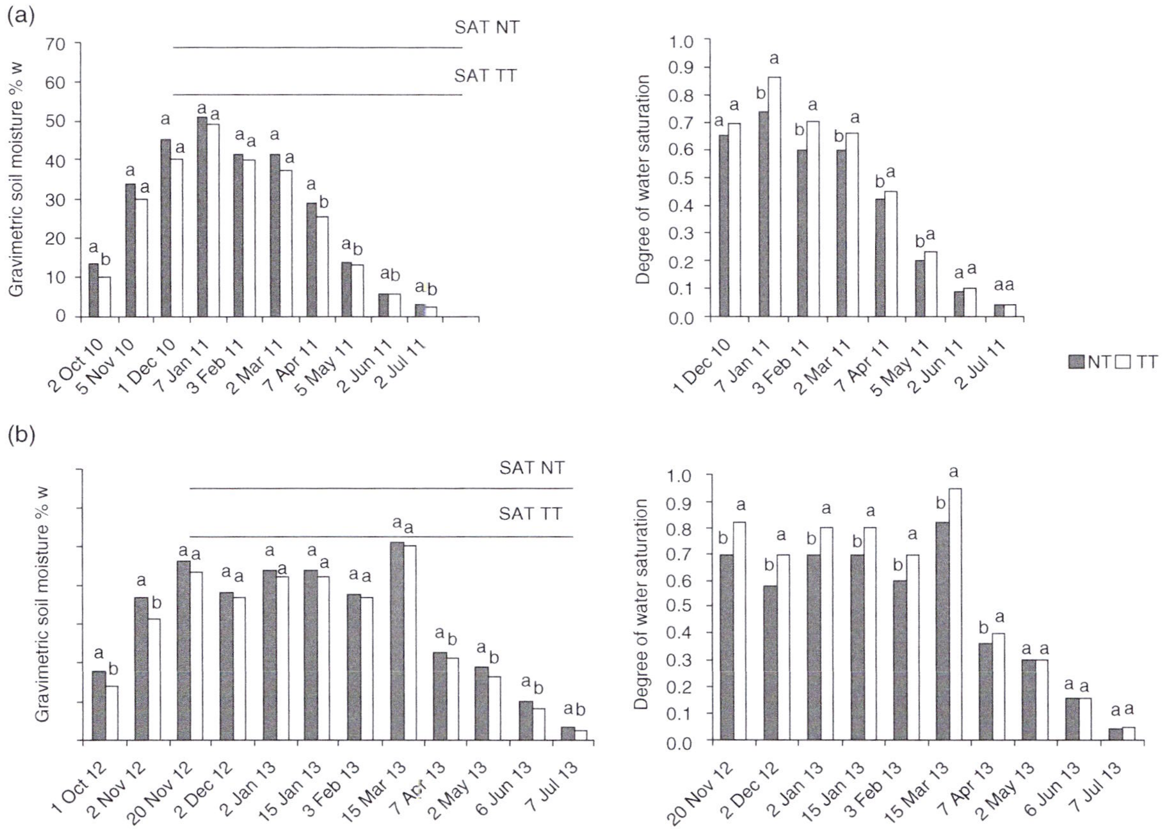

Figure 5.7 Gravimetric soil moisture and degree of water saturation as affected by tillage at 0- to 5-cm depth in (a) 2010–2011 and (b) 2012–2013 campaigns. NT, no-tillage; TT, traditional tillage. Different letters indicate significant differences between treatments for each factor ($P < .05$). Adapted from Gómez-Paccard et al. (2015).

increase biomass production in wheat with respect to non-amendments, at the same level or slightly lower than that obtained with pure limestone. Red gypsum showed better results than phosphogypsum owing to the higher zinc content of red gypsum, and/or the possible negative effect of the formation of Al-F ion pairs in phosphogypsum (Mariscal-Sancho et al., 2009a). The combination of gypsiferous amendments with the dolomitic coverter residue and the corresponding magnesium supply, led to a higher wheat yield than if it had not been combined. Finally, sugar foam increased wheat production to a greater extent than limestone thanks to its greater variety of nutrients.

In the amendment × tillage trial (Hontoria et al., 2018), the calcium amendment was effective in eliminating aluminum toxicity at least in the Ap horizon from the first agricultural year after its application, extending this effect throughout six of the nine years. The effect of no-tillage on productivity began to be noticed from the third year, concurrent with improvements at the physical level that take longer to manifest (Blanco-Canqui & Ruis, 2018) and that, as already seen, have important consequences on soil-water relations. Figure 5.8 shows information on annual precipitation over the period 2005–2014, as well as on average biomass production with or without calcium amendment application and under no-tillage or traditional tillage. The application of calcium amendment increased biomass production in seven out of nine years; in an agricultural year characterized by excess precipitation, crop production was very low regardless of whether calcium amendment was added owing to the formation of a perched water table that strongly reduced nascence. Production in only one year (seventh year) was lower in the amended plots because the higher nascence, and consequently higher plant density and higher water extraction in the amended plots, coped worse with an extraordinarily dry year.

Regarding the tillage system, in dry agricultural years biomass production was higher in no-tillage, regardless of the calcium amendment (Figure 5.8), because of its higher available soil water content as seen previously (Figure 5.7). This supports the findings of Soane et al. (2012) who indicated that no-tillage yields tend to match and even exceed tilled yields as mean precipitation decreases from northern to southern Europe. In rainy years (fifth and sixth year), biomass production was higher (even more than double) in no-tillage plots than in tilled plots (Figure 5.8) owing to the lower degree of saturation and better aeration conditions in the surface layer (Figure 5.7). These conditions favored germination and vegetative development, particularly of the legume species, which are more sensitive to soil water saturation (Muslera & Ratera, 1991).

Finally, the costs of applying the calcium amendments (mainly sugar foam) were evaluated. The increases in forage crop yields with the amendment allowed the investment to be amortized in two campaigns. Thus, the economic profitability of this application was verified by considering the duration of its effects (González-Fernández et al., 2012).

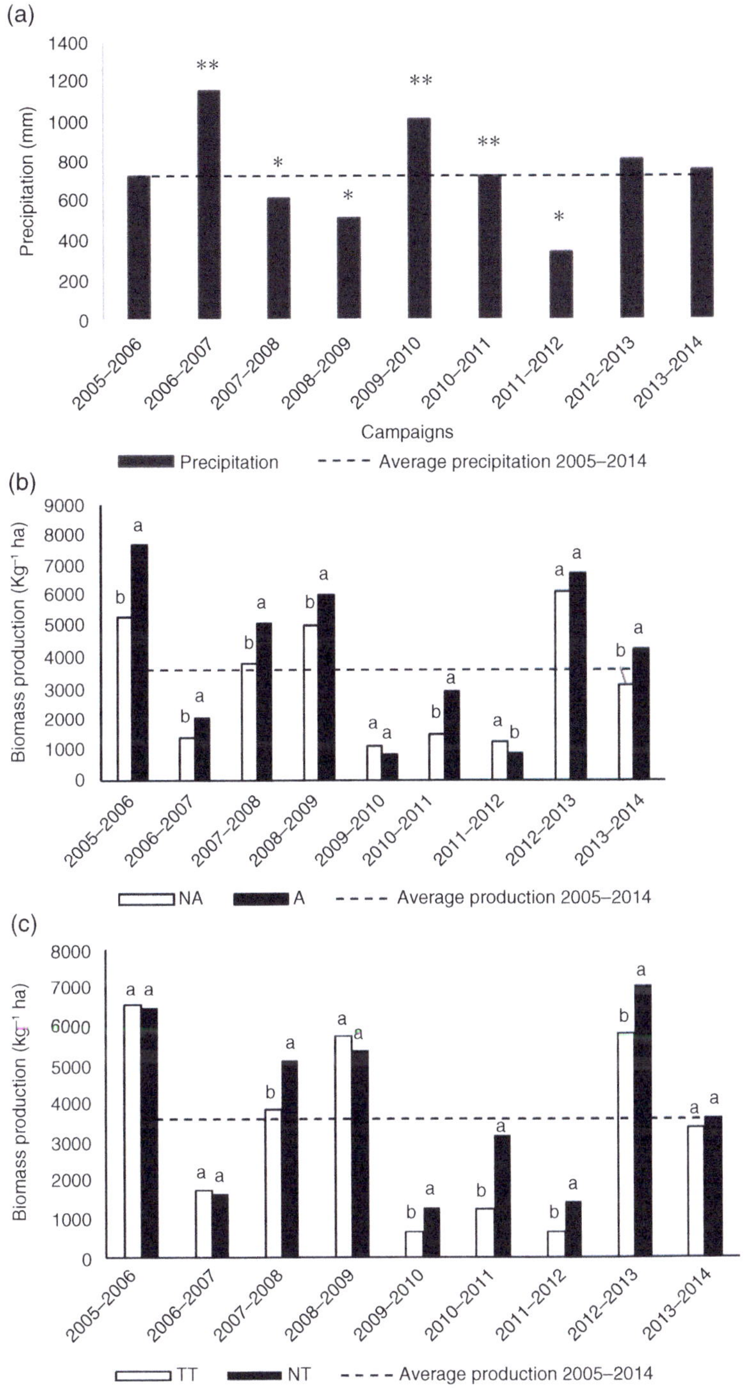

Figure 5.8 Forage biomass production in relation to precipitation: (a) annual precipitation (H, humid year; D, dry year; H*, water excess in autumn/winter); biomass production as affected by (b) calcium amendment and (c) tillage system. A, amended; NA, not amended; TT, traditional tillage; NT, not tillage. Different letters indicate significant differences between treatments for each factor ($P<.05$). Adapted from Hontoria et al. (2018).

Effects of Management on Microbiological Properties

The study of microbiological variables provided a valuable comprehensive assessment of the effects of different soil managements on carbon and nitrogen mineralization. First, the acidity of the starting soil often led to negative effects on microbial biomass carbon (MBC) and various enzymatic activities that affected SOM decomposition and nutrient recycling (Acosta-Martínez & Tabatabai, 2000; León et al., 2017). Second, it is important to highlight that the wide fluctuations of microbiological activities throughout the year, which are caused by the meteorological contrasts typical of the study area, complicated the interpretation of the results of the different soil managements (Vázquez et al., 2017). However, no-tillage was found to stimulate microbiological activity to a greater extent than traditional tillage, thanks in part to the moderation of the thermal and humidity oscillation caused by no-tillage. Thus, under no-tillage the temperature at 5-cm depth was lower in the hottest months (July and August) compared with traditional tillage, while water content in no-tillage was higher in the driest months for more than half of the study year (Vázquez et al., 2017). In fact, the greatest differences between the two tillage systems occurred in the hottest month during which no-tillage provided better temperature and humidity conditions for enzyme activity compared with traditional tillage (León et al., 2017).

In general, no-tillage increased carbon and nitrogen mineralization owing to an overall increase in organic carbon and nitrogen reserves, microbial biomass, and enzyme activity (León et al., 2017). Carbon mineralization was more sensitive to microbial biomass size than to microbial structure or activity. In this sense, MBC was confirmed as the main factor responsible for the observed variability in nitrogen and carbon mineralization of the studied soils according to the structural equation model in Figure 5.9, which shows that MBC was the variable that modulated the impact of tillage system on the initial rate of potential carbon mineralization and net nitrogen mineralization in the study area (Vázquez et al., 2019a).

In addition, the increase in soil pH produced by the calcium amendments modified the structure of the microbial communities, such that liming increased the population of bacteria by providing them with a pH close to their optimum. This increase in the bacterial population may explain the decrease in the MBC : MBN ratio recorded in the amended plots, as bacteria have less C : N than fungi (Vázquez et al., 2019a). Thus, according to Figure 5.9, sugar foam amendments decreased the MBC in contrast to NT.

The positive effect of calcium amendment on populations was also observed for ammonia-oxidizing bacteria (AOB) and ammonia-oxidizing archaea (AOA; Teutscherova et al. 2017; Vázquez et al., 2019c). The increase in AOB and AOA could be explained, as with total bacteria, by the reduction of soil acidity,

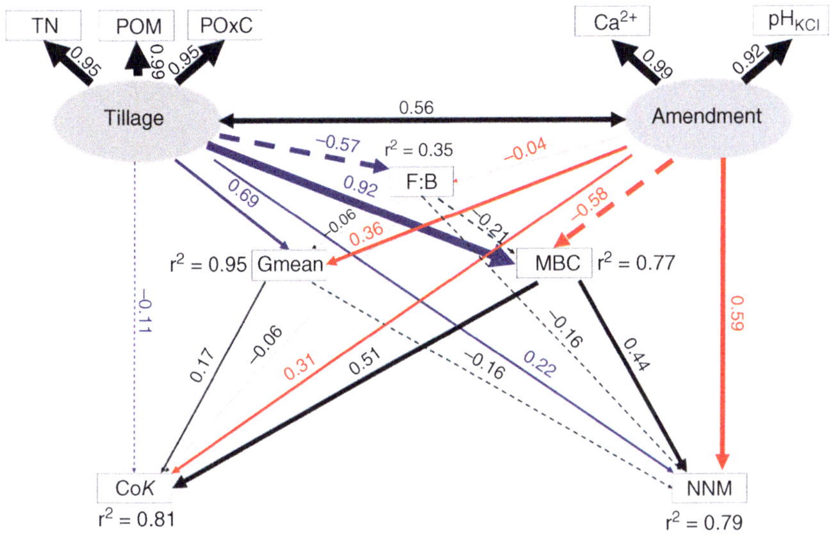

Figure 5.9 Structural equation model of the hypothesized relationships between tillage system and calcium amendment application and soil carbon and nitrogen mineralization and net nitrification. Endogenous variables are represented as squares and latent variables as ellipses. Blue arrows indicate direct effects of tillage, red arrows indicate direct effects of calcium amendment, and black arrows indicate indirect effects of both latent variables. Solid lines represent positive coefficients and dashed lines negative coefficients. The r^2 value represents the proportion of the total variance explained by the dependent variables. TN, total nitrogen; POM, particulate organic matter; PoxC, permanganate oxidizable carbon; Gmean, geometric mean of enzyme activities; F : B, fungal-to-bacterial ratio; MBC, microbial biomass carbon; CoK, initial rate of potential carbon mineralization; NNM, net nitrogen mineralization (adapted from Vázquez et al. 2019a).

alleviation of aluminum toxicity, and formation of aluminum-organic matter compounds (Gómez-Paccard et al., 2013). Gross nitrification rates appeared to be driven more by AOA ($R = 0.525$, $P = .001$) than by AOB ($R = 0.358$, $P = .044$), which is not surprising, given the low fertility and extreme environmental conditions of the study area, where archaea may present adaptive advantages over bacteria (Prosser & Nicol, 2012).

Our results confirmed the usefulness of calcium amendment not only to improve soil acidity and other chemical properties, as explained above, but also to increase microbial activity and nitrogen mineralization (which increases nitrogen availability). Therefore, one concern about the application of calcium amendments could be related to the possible depletion of soil carbon as a result of decreased soil acidity and the consequent stimulation of microbial activity (Paradelo et al., 2015). However, our amendments did not impair soil carbon accumulation for several reasons. First, the main amendment selected (sugar foam) included low organic matter content (86.7 g/kg). In addition, increased calcium has been shown to increase the stabilization of SOC through its ability to bridge negatively charged surfaces from SOC (Paradelo et al., 2015 Rowley et al., 2018).

Effects of Management and Land Use on Enzyme Activity and Arbuscular Mycorrhizal Fungi

The enzymatic activity of β-glucosidase often represents a bottleneck in the carbon cycle and reflects the ability of soil microorganisms to break down low molecular weight derivatives of cellulose, releasing glucose. It is therefore an indicator of mineralization processes affecting SOM (García et al., 2003), and it has been proposed as an indicator of soil quality (Knight & Dick, 2004). The activity of β-glucosidase was sensitive to the degree of degradation of the soils (Mariscal-Sancho et al., 2010). This enzyme, essential for the carbon cycle, in addition to depending on the content of labile fractions of SOM, was negatively affected by the abundance of *Cistus ladanifer* L. This shrub is abundant in the study area because it is a pyrophytic species (thermal shocks favor the germination of its seeds), and it tends to colonize areas after fires that ravage the study area with a certain cyclicity. Mariscal-Sancho et al. (2010) hypothesized that the allelopathic character of this plant could explain the decrease in β-glucosidase activity. The higher activity of β-glucosidase and also of β-glucosaminidase in untilled compared with tilled plots is in agreement with the higher MBC content León et al., (2017), which underscores a higher capacity to metabolize carbohydrates in the soil and accelerate the carbon cycle under the no-tillage condition (Acosta-Martínez et al., 2008).

The application of the calcium amendment had a strong enzyme-specific effect (Vázquez et al., 2019b). While the activity of β-glucosidase was not affected, the activity of casein protease (an enzyme that degrades high molecular weight protein compounds) increased and the activity of N-α-benzoyl-L-argininamide protease (BAA; targeting low molecular weight compounds to produce NH_4^+) decreased after calcium amendment application. These results suggest that degradation of large protein compounds is triggered. However, the resulting small molecular weight nitrogen compounds are relatively mobile and could therefore be transported to deeper soil layers. This possibility would explain the observed low BAA activity and the low release of NH_4^+, which is the expected product of arginine (small molecular weight N-compound) catalysis. This leads to a change in the way nitrogen is available in the soil after calcium amendment, which may have further implications for plant nutrition. In the long term, the application of calcium amendment and the consequent change in soil pH may lead not only to substantial changes in the nitrogen use efficiency of the soil biota, but also to a considerable alteration of the nitrogen acquisition preferences of plants. That is, part of the aforementioned enzymatic results can be attributed to changes in crops, which can promote important alterations in soil proteolytic activity (Vranova et al., 2013) by inducing changes in the rhizosphere (Baraniya et al., 2016) or through modifications in the quality of their residues. In our case, the harvested crop biomass showed different C : N ratios in the plots with calcium amendments compared to the plots without amendments. This change in the quality of plant residues left by the crop seems to play an additional role in the enzymatic activities. However, the mechanism by which calcium amendments affect the relationship between casein protease and BAA activity is still unclear.

No-tillage and calcium amendment enhanced the colonization of arbuscular mycorrhizal fungi (AMF) on oat and vetch roots of the crops (respectively 80% and 60% higher compared with unamended tilled plots). The limited phosphorus availability in the study soils, together with the increased nitrogen demand coming from the yield increase provided by no-tillage plus calcium amendment led to the strengthening of the mutualism between plants and AMF, where phosphorus acquisition through AMF was stimulated (Vázquez et al., 2020a). Therefore, the combination of no-tillage with calcium amendment increases nutrient use efficiency by stimulating crop-AMF interactions.

However, the abundance of AMF decreased with tillage mainly because of the mechanical disturbance from tillage (Gómez-Paccard et al., 2013), which caused direct disruption of roots and hyphae. Thus, the relative importance of total fungi in relation to bacteria increased in no-tillage, increasing the fungal-bacterial ratio (León et al., 2017). Glomalin-related soil proteins (GRSPs) are largely derived from AMF (Rillig, 2004), and indeed, the decrease in GRSP content in the tilled plots could be attributed to damage caused by tillage on AMF, as suggested by Wright et al. (1999) and Kabir (2005). Furthermore, an inverse relationship was found between AlK (aluminum extracted with KCl) and GRSPs in the 0- to 5-cm layer. Our results support the previous findings of Aguilera et al. (2011), who suggested that, in the absence of calcium amendment, GRSPs complex and immobilize part of the AlK, thus reducing aluminum toxicity. In such a way, AMF showed one more facet of their importance as plant growth-promoting microorganisms.

Basal respiration (BR) is defined as the rate of CO_2 emissions by respiration in soil originating from the mineralization of organic matter, and it is commonly interpreted as a key indicator to assess changes in soil health (Creamer et al., 2014). In Mediterranean areas, moisture becomes the main driver of soil respiration when below 10% gravimetric moisture, while temperature tends to have a negative effect on BR (Almagro et al., 2009). The increase in BR caused by no-tillage can be explained by the accumulation of SOC and MBC and increased enzymatic activities of untilled plots, especially in the top 5 cm of soil. This increase in BR in no-tillage, again, is linked to the improved soil moisture and temperature conditions in the months of water deficit in untilled plots compared with tilled ones.

In short, far from having a negative effect on soil biological parameters, the surface application of the industrial by-products studied (sugar foam and red gypsum) in no-tillage was confirmed to have increased the activity of certain enzymes related to carbon and nitrogen cycling and stimulated nitrogen mineralization and nitrification (León et al., 2017; Vázquez et al. 2019a).

Impact on GHG Emissions

The emission or consumption of GHGs is mainly the result of different and simultaneous microbial processes. CO_2 is produced by autotrophic and heterotrophic respiration, N_2O production is mainly the result of nitrification and denitrification, and CH_4 is emitted by methanogenic microorganisms and oxidized by methanotrophic microorganisms (García-Marco et al., 2016). One of the most important dimensions of the balance between GHG emissions and sequestration is closely related to the high capacity of the studied soils to sequester huge amounts of CO_2 in the form of SOC. This aspect has already been addressed previously, so in this section we focus on N_2O and CH_4 emissions.

Microbial processes affecting GHG emissions are highly dependent on soil physicochemical properties that regulate soil aeration and water-filled pore space after water application by precipitation or irrigation. Here we highlight that tillage decreased total porosity in the profile of the studied soils and reduced water infiltration, which is related to increase N_2O emissions. Therefore, tilled plots emitted 3.2 times the N_2O of untilled plots and had decreased CH_4 sequestration (tillage sequestered 0.4 times that sequestered by no-tillage; Table 5.4; García-Marco et al., 2016).

The application of calcium amendments had effects on several mechanisms regulating soil GHG production and consumption (Barton et al., 2013b; Čuhel et al., 2010), such as microbial activity and diversity, improved nutrient availability, and increased biomass production, as described in previous sections. In addition, calcium amendment increased crop growth, leading to improved nitrogen use efficiency and decreased N_2O emissions (Fageria, 2014). Thus, in these acid soils, calcium amendment mitigated N_2O emissions (at 10% confidence level; Table 5.4), but only when combined with tillage, in line with the results of Barton et al. (2013a). In this context, calcium amendment decreased N_2O emissions, probably by promoting higher N_2O reductase activity (N_2O-R) and thus leading to a complete reduction of N_2O to N_2 (Šimek & Cooper, 2002). This was possible as the N_2O-R activity is very sensitive to pH changes compared with other denitrifying reductase enzymes (Šimek & Cooper, 2002). The better soil incorporation of calcium amendment in tilled plots than in untilled ones probably accelerated the pH effect in the tilled plots, leading to a significant reduction of N_2O emission from September 2012 to June 2013.

Table 5.4 Cumulative Emissions of N_2O and CH_4 Fluxes Over the Periods of Basal Dressing Fertilization (September 1 to January 31) and the Top-Dressing Fertilization (February 1 to May 31).

	Cumulative fluxes					
	N_2O (mg N/m^2)			CH_4 (mg C/m^2)		
Treatment	**Basal dressing**	**Top dressing**	**Total**	**Basal dressing**	**Top dressing**	**Total**
*Tillage**						
Traditional	40.1	37.4	77.3	−3.58	0.825	−2.73
No-tillage	1.78	22.8	24.4	−3.94	−2.96	−6.98
*Calcium amendment**						
Amended	8.40	28.7	36.8	−4.40	−5.01	−9.52
Not-amended	33.5	31.5	64.8	−3.12	2.88	−0.190
Effects (P-values)						
T	.002	ns	.013	ns	ns	ns
A	.016	ns	.078	ns	.040	ns
T × A	.007	ns	.023	ns	.043	ns
*Significant T × A***						
TT amended	12.7 ± 6.3 b		43.3 ± 16.6 b		−6.60 ± 2.09 b	
TT	67.6 ± 18.4 a		111 ± 9 a		8.25 ± 2.83 a	
NT amended	4.10 ± 3.60 ns		30.4 ± 16.6 ns		−3.43 ± 3.73 ns	
NT	−0.54 ± 3.55		18.3 ± 9.8		−2.48 ± 3.35	

Note. Adapted from García-Marco et al. (2016). T, tillage system; A, amendment; ns, not significant.
* Data are means of eight replicates.
** Data are means of four replicates ± S.E. (standard error). Different lowercase letters in the interaction indicate significant differences between amendment treatments within a tillage by applying the Bonferroni significance test at $P < .05$.

In Mediterranean regions, soil rewetting with the first autumn rains after the dry summer period has been associated with significant pulses of N_2O emissions (García-Marco et al., 2016). In our Mediterranean acid soils, calcium amendment reduced cumulative N_2O emission by more than 70% compared with unamended soil after soil rewetting to 50% field capacity (Vázquez et al., 2020c). In addition to the effect of calcium amendment on the structure of AOB and denitrifying communities mentioned above, soil rewetting increased *nirK* transcription. This increase in *nirK* transcription again suggested that the mitigation of N_2O emissions observed in amended soils was caused by changes in the denitrification process, possibly with a greater or more efficient reduction of N_2O to N_2 (Vázquez et al., 2020c).

The calcium amendment also had positive effects in terms of CH_4 emission. The increase in pH could enhance CH_4 oxidation owing to the decreased toxicity of aluminum (Tamai et al., 2007; García-Marco et al., 2016), which could lead to changes in the size or structure of the microbial population responsible for CH_4 oxidation and thus decreased CH_4 emissions (Bååth & Arnebrant, 1994; Gulledge et al., 2004).

Conclusion

Our long-term research shows that the recovery of acid Ultisols degraded by tillage in the southwest of the Iberian Peninsula is not effective through mere tillage abandonment. It also requires specific agricultural management that not only improves soil health, but also provides sources of income for the population, enhances the circular economy, and contributes to reducing the environmental impact. To improve the chemical fertility of these degraded soils, the use of industrial by-products such as sugar foam and red gypsum at doses and in combinations tested in field conditions is able to reduce the aluminum toxicity and increase the exchangeable basic cations to promote crop productivity. In this way, the two objectives of improving economic profitability and recycling of raw materials, associated with the circular economy, are met.

If the loss of organic matter after cultivation was critical for the productivity decline, the recovery of these soils necessarily requires an increase in it. Therefore, practices such as no-tillage and its combination with calcium amendments, the use and improvement of pastures, and fertilization have proven to be key in the recovery process of these degraded acid soils by increasing SOC and thus leading to improvements in chemical, physical, and biological properties. However, these improvements were mainly limited to the topsoil except for the chemical properties that improved at greater depth thanks to the use of relatively soluble amendments.

No-tillage and calcium amendment increased SOC in the 0- to 5-cm layer, with no interaction between them and an additive effect that diminished over time. Over the years, the positive effect of the calcium amendment decreased as the positive effect of no-tillage increased owing to a better physical and chemical environment, which progressively led to higher biomass production and SOC accumulation. Key to this good performance of no-tillage, with and without calcium amendment, was the improvement of soil structure and increased porosity, which improved the soil-water relationship such that the no-tilled crop had more water in dry periods and more air in periods with excess rainfall when a perched water table formed near the surface. The use of pasture and the improvement of its composition achieved higher SOC values in the soil and affected a greater depth, so it can be a more effective alternative than the forage crop for soil recovery.

No-tillage also led to increased microbial activity and higher enzyme activity caused by sensitive amelioration of the extreme drought and high temperature conditions that typically occur for several months in the study area, in addition to increased SOC and decreased aluminum toxicity. The application of sugar foam and red gypsum as liming agents increased both gross mineralization and nitrification, while nitrogen immobilization was not affected. In addition, our results showed that, under these conditions, no-tillage can reduce the global warming potential of arable soils through reductions in N_2O emissions, increases in CH_4 oxidation, and improvements in crop yields that reduced available nitrogen.

This research provides scientific evidence supporting the benefits of agricultural practices, primarily those that do not disturb the soil, both economically through improvements in crop yields and environmentally through improved soil health, carbon sequestration, and reduced GHG emissions.

Acknowledgments

This research was supported by the Spanish Ministry of Science and Innovation (Projects AGL-4545-C03-02; AGF-0828-C03-02; CGL-2008-04361-C02-01; AGL-2012-39498), the Comunidad de Madrid (Spain), and European Structural and Investment Funds (Projects AGRISOST S2009/AGR-1630; AGRISOST S2013/ABI-2717). Most importantly, we would like to thank the many people who have participated over many years and made this research possible.

References

Acosta-Martínez, V., Acosta-Mercado, D., Sotomayor-Ramírez, D., & Cruz-Rodríguez, L. (2008). Microbial communities and enzymatic activities under different management in semiarid soils. *Applied Soil Ecology*, *38*(3), 249–260. https://doi.org/10.1016/j.apsoil.2007.10.012

Acosta-Martínez, V., & Tabatabai, M. A. (2000). Enzyme activities in a limed agricultural soil. *Biology and Fertility of Soils*, *31*, 85–91. https://doi.org/10.1007/s003740050628

Aguilera, P., Borie, F., Seguel, A., & Cornejo, P. (2011). Fluorescence detection of aluminum in arbuscular mycorrhizal fungal structures and glomalin using confocal laser scanning microscopy. *Soil Biology and Biochemistry*, *43*(12), 2427–2431. https://doi.org/10.1016/J.SOILBIO.2011.09.001

Almagro, M., López, J., Querejeta, J. I., & Martínez-Mena, M. (2009). Temperature dependence of soil CO_2 efflux is strongly modulated by seasonal patterns of moisture availability in a Mediterranean ecosystem. *Soil Biology and Biochemistry*, *41*(3), 594–605. https://doi.org/10.1016/J.SOILBIO.2008.12.021

Bååth, E., & Arnebrant, K. (1994). Growth rate and response of bacterial communities to pH in limed and ash treated forest soils. *Soil Biology and Biochemistry*, *26*(8), 995–1001. https://doi.org/10.1016/0038-0717(94)90114-7

Baraniya, D., Puglisi, E., Ceccherini, M. T., Pietramellara, G., Giagnoni, L., Arenella, M., Nannipieri, P., & Renella, G. (2016). Protease encoding microbial communities and protease activity of the rhizosphere and bulk soils of two maize lines with different N uptake efficiency. *Soil Biology and Biochemistry*, *96*, 176–179. https://doi.org/10.1016/J.SOILBIO.2016.02.001

Barton, L., Gleeson, D. B., Maccarone, L. D., Zúñiga, L. P., & Murphy, D. V. (2013a). Is liming soil a strategy for mitigating nitrous oxide emissions from semi-arid soils? *Soil Biology and Biochemistry*, *62*, 28–35. https://doi.org/10.1016/J.SOILBIO.2013.02.014

Barton, L., Murphy, D. V., & Butterbach-Bahl, K. (2013b). Influence of crop rotation and liming on greenhouse gas emissions from a semi-arid soil. *Agriculture, Ecosystems & Environment*, *167*, 23–32. https://doi.org/10.1016/J.AGEE.2013.01.003

Blanco-Canqui, H., & Ruis, S. J. (2018). No-tillage and soil physical environment. *Geoderma*, *326*, 164–200. https://doi.org/10.1016/J.GEODERMA.2018.03.011

Bolinder, M. A., Janzen, H. H., Gregorich, E. G., Angers, D. A., & VandenBygaart, A. J. (2007). An approach for estimating net primary productivity and annual carbon inputs to soil for common agricultural crops in Canada. *Agriculture, Ecosystems & Environment*, *118*(1–4), 29–42. https://doi.org/10.1016/J.AGEE.2006.05.013

Caires, E. F., Corrêa, J. C. L., Churka, S., Barth, G., & Garbuio, F. J. (2006). Surface application of lime ameliorates subsoil acidity and improves root growth and yield of wheat in an acid soil under no-till system. *Scientia Agricola*, *63*, 502–509.

Conant, R. T., Cerri, C. E. P., Osborne, B. B., & Paustian, K. (2017). Grassland management impacts on soil carbon stocks: A new synthesis. *Ecological Applications*, *27*(2), 662–668. https://doi.org/10.1002/eap.1473

Creamer, R. E., Schulte, R. P. O., Stone, D., Gal, A., Krogh, P. H., Lo Papa, G., Murray, P. J., Pérès, G., Foerster, B., Rutgers, M., Sousa, J. P., & Winding, A. (2014). Measuring basal soil respiration across Europe: Do incubation temperature and incubation period matter? *Ecological Indicators*, *36*, 409–418. https://doi.org/10.1016/J.ECOLIND.2013.08.015

Čuhel, J., Šimek, M., Laughlin, R. J., Bru, D., Chèneby, D., Watson, C. J., & Philippot, L. (2010). Insights into the effect of soil pH on N_2O and N_2 emissions and denitrifier community size and activity. *Applied and Environmental Microbiology*, *76*(6), 1870–1878. https://doi.org/10.1128/AEM.02484-09

Mariscal Sancho I de L. (2008). *Recuperación de la calidad de Ultisoles mediterráneos degradados, mediante la aplicación de enmiendas y formas alternativas de uso.* Doctoral thesis, E.T.S.I. Agrónomos (UPM). https://doi.org/10.20868/UPM.thesis.1597

Espejo, R. (1987). The soils and ages of the "raña" surfaces related to the Villuercas and Altamira mountain ranges (Western Spain). *Catena*, *14*(5), 399–418. https://doi.org/10.1016/0341-8162(87)90012-9

Fageria, N. K. (2014). *Nitrogen management in crop production.* CRC Press.

García, C., Gil, F., Hernández, T., & Trasar, C. (2003). Determinación de la actividad de la A-glucosidasa. In *Técnicas de Análisis de Parámetros Bioquímicos en Suelos: Medidas de Actividades Enzimáticas y Biomasa Microbiana* (pp. 173–174). Mundi-Prensa.

García-Marco, S., Abalos, D., Espejo, R., Vallejo, A., & Mariscal-Sancho, I. (2016). No tillage and liming reduce greenhouse gas emissions from poorly drained agricultural soils in Mediterranean regions. *Science of The Total Environment*, *566–567*, 512–520. https://doi.org/10.1016/J.SCITOTENV.2016.05.117

Giagnoni, L., Taiti, C., León, P., Costa, C., Meneasatti, P., Espejo, R., Gómez-Paccard, C., Hontoria, C., Vázquez, E., Benito, M., Mancuso, S., & Renella, G. (2020). Volatile organic compound emission and biochemical properties of degraded Ultisols ameliorated by no tillage and liming. *Pedosphere*, *30*(5), 597–606. https://doi.org/10.1016/S1002-0160(20)60024-8

Gómez-Paccard, C., Hontoria, C., Mariscal-Sancho, I., Pérez, J., León, P., González, P., & Espejo, R. (2015). Soil-water relationships in the upper soil layer in a Mediterranean Palexerult as affected by no-tillage under excess water conditions—Influence on crop yield. *Soil and Tillage Research*, *146*(Part B), 303–312. https://doi.org/10.1016/j.still.2014.09.012

Gómez-Paccard, C., Mariscal-Sancho, I., León, P., Benito, M., González, P., Ordóñez, R., Espejo, R., & Hontoria, C. (2013). Ca-amendment and tillage: Medium term synergies for improving key soil properties of acid soils. *Soil and Tillage Research*, *134*, 195–206. https://doi.org/10.1016/j.still.2013.08.009

González-Fernández, P., Ordóñez-Fernández, R., Mariscal-Sancho, I., & Espejo-Serrano, R. (2012). Economics of applications of sugar factory lime refuse in Mediterranean Ultisols. *Archives of Agronomy and Soil Science*, *58*(Suppl), S185–S189. https://doi.org/10.1080/03650340.2012.696779

Gulledge, J., Hrywna, Y., Cavanaugh, C., & Steudler, P. A. (2004). Effects of long-term nitrogen fertilization on the uptake kinetics of atmospheric methane in temperate forest soils. *FEMS Microbiology Ecology*, *49*(3), 389–400. https://doi.org/10.1016/j.femsec.2004.04.013

Hontoria, C., Gómez-Paccard, C., Mariscal-Sancho, I., Benito, M., Pérez, J., & Espejo, R. (2016). Aggregate size distribution and associated organic C and N under different tillage systems and Ca-amendment in a degraded Ultisol. *Soil and Tillage Research*, *160*, 42–52. https://doi.org/10.1016/j.still.2016.01.003

Hontoria, C., Gómez-Paccard, C., Vázquez, E., Mariscal-Sancho, I., Ordónez-Fernández, R., Carbonell-Bojollo, R., & Espejo, R. (2018). Mid-long term effects of no tillage and Ca-amendment on degraded acid soils under contrasting weather conditions. *Soil and Tillage Research*, *183*, 83–92. https://doi.org/10.1016/j.still.2018.06.002

Hontoria, C., Seseña R., Gómez-Paccard C., Benito M., González P., & Espejo R. (2012). *Land use and amendment effects on soil aggregation and organic carbon in acid soils of SW*. Proceedings of the 4th International Congress EUROSOIL, July 2–16, 2012, Bari, Italy (p. 1755). European Confederation of Soil Science Societies.

Jackson, R. B., Lajtha, K., Crow, S. E., Hugelius, G., Kramer, M. G., Piñeiro, G., & Piñeiro, P. (2017). The ecology of soil carbon: Pools, vulnerabilities, and biotic and abiotic controls. *Annual Review of Ecology Evolution*, *48*, 419–464. https://doi.org/10.1146/annurev-ecolsys-112414

Kabir, Z. (2005). Tillage or no-tillage: Impact on mycorrhizae. *Canadian Journal of Plant Sciences*, *85*, 23–29.

Kay, B. D., & VandenBygaart, A. J. (2002). Conservation tillage and depth stratification of porosity and soil organic matter. *Soil and Tillage Research*, *66*(2), 107–118. https://doi.org/10.1016/S0167-1987(02)00019-3

Knight, T. R., & Dick, R. P. (2004). Differentiating microbial and stabilized β-glucosidase activity relative to soil quality. *Soil Biology and Biochemistry*, *36*(12), 2089–2096. https://doi.org/10.1016/J.SOILBIO.2004.06.007

León, P., Espejo, R., Gómez-Paccard, C., Hontoria, C., Mariscal, I., Renella, G., & Benito, M. (2017). No tillage and sugar beet foam amendment enhanced microbial activity of degraded acidic soils in South West Spain. *Applied Soil Ecology*, *109*, 69–74. https://doi.org/10.1016/j.apsoil.2016.09.012

Lipiec, J., Kuś, J., Słowińska-Jurkiewicz, A., & Nosalewicz, A. (2006). Soil porosity and water infiltration as influenced by tillage methods. *Soil and Tillage Research*, *89*(2), 210–220. https://doi.org/10.1016/j.still.2005.07.012

Mariscal, I., Peregrina, F., Terefe, T., González, P., & Espejo, R. (2007). Evolution of some physical properties related to soil quality in the degraded ecosystems of "raña" formations from SW Spain. *Science of the Total Environment*, *378*(1–2), 130–132. https://doi.org/10.1016/j.scitotenv.2007.01.025

Mariscal-Sancho, I., Espejo, R., & Peregrina, F. (2009a). Potentially toxic effects of phosphogypsum on palexerults in Western Spain. *Soil Science Society of America Journal*, *73*(1), 146–153. https://doi.org/10.2136/sssaj2007.0394

Mariscal-Sancho, I., Peregrina, F., Espejo, R., & Hontoria, C. (2024). Rehabilitation of a degraded acid Mediterranean soil after 11 years of by-product calcium amendments and pasture improvement. *Soil Use and Management*, *40*(4), e13124. https://doi.org/10.1111/sum.13124

Mariscal-Sancho, I., Peregrina, F., Mendiola, M. A., Santano, J., & Espejo, R. (2009b). Exchange complex composition in Mediterranean Ultisols under various types of vegetation and soil uses. *Soil Science*, *174*(6), 339–345. https://doi.org/10.1097/SS.0b013e3181a974fe

Mariscal-Sancho, I., Santano, J., Mendiola, M. Á., Peregrina, F., & Espejo, R. (2010). Carbon dioxide emission rates and β-glucosidase activity in Mediterranean Ultisols under different soil management. *Soil Science*, *175*(9), 453–460. https://doi.org/10.1097/SS.0b013e3181f51704

Menzies, N., Edwards, D., & Bell, L. (1994). Effects of calcium and aluminum in the soil solution of acid, surface soils on root elongation of mungbean. *Australian Journal of Soil Research*, *32*, 721–737.

Munawar, A., Blevins, R. L., Frye, W. W., & Saul, M. R. (1990). Tillage and cover crop management for soil water conservation. *Agronomy Journal*, *82*(4), 773–777. https://doi.org/10.2134/agronj1990.00021962008200040024x

Muslera, P. E., & Ratera, G. C. (1991). Praderas y Forrajes Producción y Aprovechamiento 2ª. Edición Ed. Ediciones Mundi-Prensa.

Noble, A. D., Sumner, M. E., & Alva, A. K. (1988). The pH dependency of aluminum phytotoxicity alleviation by calcium sulfate. *Soil Science Society of America Journal*, *52*(5), 1398–1402. https://doi.org/10.2136/sssaj1988.03615995005200050036x

Paradelo, R., Virto, I., & Chenu, C. (2015). Net effect of liming on soil organic carbon stocks: A review. *Agriculture, Ecosystems & Environment*, *202*, 98–107. https://doi.org/10.1016/J.AGEE.2015.01.005

Peregrina Alonso, F., Santano Arias, J., Ordónez Fernández, R., González Fernández, P., & Espejo Serrano, R. (2006). Agronomic implications of the supply of lime and gypsum by-products to Palexerults from western Spain. *Soil Science*, *171*(1), 65–81. https://doi.org/10.1097/01.ss.0000200557.253069.50

Peregrina, F., Mariscal, I., & Espejo, R. (2008a). Agronomic implications of the application of lime plus gypsum by-products to a Hyperdistric Acrisol from western Spain. *Communications in Soil Science and Plant Analysis*, *39*(13–14), 2080–2101. https://doi.org/10.1080/00103620802135104

Peregrina, F., Mariscal, I., Ordóñez, R., González, P., Terefe, T., & Espejo, R. (2008b). Agronomic implications of converter basic slag as a magnesium source on acid soils. *Soil Science Society of America Journal*, *72*(2), 402–411. https://doi.org/10.2136/sssaj2006.0197

Peregrina, F., Mariscal, I., Terefe, T., Ordóñez, R., & Espejo, R. (2007a). Aluminum, fluorine, and sulfate species dynamics in the soil solution of a Palexerult amended with lime and gypsum by-products. *Soil Science*, *172*(8), 639–650. https://doi.org/10.1097/ss.0b013e31806db5d0

Peregrina, F., Mariscal, I., Terefe, T., Ordóñez, R., & Espejo, R. (2007b). Soil solution dynamics into Palexerult amended with gypsum and lime by-products: A laboratory test. *Soil Science*, *172*(5), 366–377. https://doi.org/10.1097/SS.0b013e31803bbba5

Plaza-Bonilla, D., Cantero-Martínez, C., Arrúe, J. L., Bareche, J., & Álvaro-Fuentes, J. (2017). Productividad y eficiencias en el uso del agua y del nitrógeno en un secano Mediterráneo. *Agricultura*, *254*, 37–42.

Plaza-Bonilla, D., Cantero-Martínez, C., Viñas, P., & Álvaro-Fuentes, J. (2013). Soil aggregation and organic carbon protection in a no-tillage chronosequence under Mediterranean conditions. *Geoderma*, *193*, 76–82. https://doi.org/10.1016/j.geoderma.2012.10.022

Prosser, J. I., & Nicol, G. W. (2012). Archaeal and bacterial ammonia-oxidisers in soil: The quest for niche specialisation and differentiation. *Trends in Microbiology*, *20*(11), 523–531. https://doi.org/10.1016/J.TIM.2012.08.001

Rillig, M. C. (2004). Arbuscular mycorrhizae, glomalin, and soil aggregation. *Canadian Journal of Soil Science*, *84*(4), 355–363. https://doi.org/10.4141/S04-003

Rowley, M. C., Grand, S., & Verrecchia, É. P. (2018). Calcium-mediated stabilisation of soil organic carbon. *Biogeochemistry*, *137*(1–2), 27–49. https://doi.org/10.1007/s10533-017-0410-1

Šimek, M., & Cooper, J. E. (2002). The influence of soil pH on denitrification: Progress towards the understanding of this interaction over the last 50 years. *European Journal of Soil Science*, *53*(3), 345–354. https://doi.org/10.1046/j.1365-2389.2002.00461.x

Six, J., Feller, C., Denef, K., Ogle, S., De Moraes, C., Sa, J., Albrecht, A., Carlos, J., & Ogle, S. M. (2002). Soil organic matter, biota and aggregation in temperate and tropical soils—Effects of no-tillage. *Agronomie*, *22*(7–8), 755–775. https://doi.org/10.1051/agro:2002043ï

Six, J., Ogle, S. M., Breidt, F. J., Conant, R. T., Mosiers, A. R., & Paustian, K. (2004). The potential to mitigate global warming with no-tillage management is only realized when practised in the long term. *Global Change Biology*, *10*(2), 155–160. https://doi.org/10.1111/j.1529-8817.2003.00730.x

Soane, B. D., Ball, B. C., Arvidsson, J., Basch, G., Moreno, F., & Roger-Estrade, J. (2012). No-till in northern, western and south-western Europe: A review of problems and opportunities for crop production and the environment. *Soil and Tillage Research*, *118*, 66–87. https://doi.org/10.1016/J.STILL.2011.10.015

Soil Survey Staff. (2014). *Keys to soil taxonomy* (12th ed.). USDA-Natural Resources Conservation Service.

Tamai, N., Takenaka, C., & Ishizuka, S. (2007). Water-soluble Al inhibits methane oxidation at atmospheric concentration levels in Japanese forest soil. *Soil Biology and Biochemistry*, *39*(7), 1730–1736. https://doi.org/10.1016/J.SOILBIO.2007.01.029

Tester, C. F. (1990). Organic amendment effects on physical and chemical properties of a sandy soil. *Soil Science Society of America Journal*, *54*(3), 827–831.

Teutscherova, N., Vazquez, E., Masaguer, A., Navas, M., Scow, K. M., Schmidt, R., & Benito, M. (2017). Comparison of lime- and biochar-mediated pH changes in nitrification and ammonia oxidizers in degraded acid soil. *Biology and Fertility of Soils*, *53*(7), 811–821. https://doi.org/10.1007/s00374-017-1222-0

Vázquez, E., Benito, M., Espejo, R., & Teutscherova, N. (2019a). Effects of no-tillage and liming amendment combination on soil carbon and nitrogen mineralization. *European Journal of Soil Biology*, *93*, Article 103090. https://doi.org/10.1016/j.ejsobi.2019.103090

Vázquez, E., Benito, M., Espejo, R., & Teutscherova, N. (2020a). No-tillage and liming increase the root mycorrhizal colonization, plant biomass and N content of a mixed oat and vetch crop. *Soil and Tillage Research*, *200*, Article 104623. https://doi.org/10.1016/j.still.2020.104623

Vázquez, E., Benito, M., Espejo, R., & Teutscherova, N. (2020b). Response of soil properties and microbial indicators to land use change in an acid soil under Mediterranean conditions. *Catena*, *189*, 104486. https://doi.org/10.1016/j.catena.2020.104486

Vázquez, E., Benito, M., Masaguer, A., Espejo, R., Díaz-Pinés, E., & Teutscherova, N. (2019b). Long-term effects of no tillage and Ca-amendment on the activity of soil proteases and β-glucosidase in a Mediterranean agricultural field. *European Journal of Soil Biology*, *95*, Article 103135. https://doi.org/10.1016/j.ejsobi.2019.103135

Vázquez, E., Benito, M., Navas, M., Espejo, R., Díaz-Pinés, E., & Teutscherova, N. (2019c). The interactive effect of no-tillage and liming on gross N transformation rates during the summer fallow in an acid Mediterranean soil. *Soil and Tillage Research*, *194*, Article 104297. https://doi.org/10.1016/j.still.2019.104297

Vázquez, E., Teutscherova, N., Almorox, J., Navas, M., Espejo, R., & Benito, M. (2017). Seasonal variation of microbial activity as affected by tillage practice and sugar beet foam amendment under Mediterranean climate. *Applied Soil Ecology*, *117–118*, 70–80. https://doi.org/10.1016/j.apsoil.2017.04.013

Vázquez, E., Teutscherova, N., Pastorelli, R., Lagomarsino, A., Giagnoni, L., & Renella, G. (2020c). Liming reduces N_2O emissions from Mediterranean soil after-rewetting and affects the size, structure and transcription of microbial communities. *Soil Biology and Biochemistry*, *147*, Article 107839. https://doi.org/10.1016/j.soilbio.2020.107839

Vázquez Garcia, E. (2019). *Effect of no-tillage and liming practices on soil microbial parameters in degraded acid soils of SW Spain*. Doctoral Thesis, E.T.S. de Ingeniería Agronómica, Alimentaria y de Biosistemas (UPM). https://doi.org/10.20868/UPM.thesis.57169

Vranova, V., Rejsek, K., & Formanek, P. (2013). Proteolytic activity in soil: A review. *Applied Soil Ecology*, *70*, 23–32. https://doi.org/10.1016/J.APSOIL.2013.04.003

Wright, S. F., Starr, J. L., & Paltineanu, I. C. (1999). Changes in aggregate stability and concentration of glomalin during tillage management transition. *Soil Science Society of America Journal*, *63*, 1825–1829. https://doi.org/10.2136/sssaj1999.6361825x

6

Can Irrigation Enhance Soil Health? Multi-Scale Analysis of Its Effects on Agricultural Soils in Navarre, Spain

Rodrigo Antón, Alberto Enrique, Isabel S. de Soto, Henar Urmeneta, Delphine Derrien, and Iñigo Virto

Chapter Overview

Irrigation has been vital for rural development and global food security throughout human history. Currently, irrigation is used for 23% of cultivated land and contributes to 45% of global food production worldwide. Irrigation has made agriculture the largest water-consuming sector in the world, and coupled with the intensification of cultivation, it has led to long-term consequences, including the progressive degradation of agricultural soils and competition for resources. In this context, soil health, defined as the capacity of soils to sustain productivity, diversity, and environmental services, is re-emerging as a new standard in assessing sustainable soil management. This chapter evaluates the impact of irrigation on soil functioning and soil health in the region of Navarre (northern Spain) by exploring the results of research projects conducted in the region. First, from an experimental field perspective, one study focused on understanding how the introduction of irrigation can affect the mechanisms that regulate soil organic carbon (SOC) stabilization. Results revealed that irrigation enhanced SOC storage primarily through increased carbon inputs and a more efficient incorporation of crop residues into SOC. Second, scaling up to a regional scale, other studies focused on evaluating the effect that irrigated agriculture can have on the agricultural soil health in this region. Results highlighted the potential of irrigation to provide key soil-based ecosystem services such as erosion control and climate regulation, enhancing both crop yields and net gains to farmers. However, the net effect depends on various factors, particularly those related to soil management and

Soil Health Series: Volume 5 Soil Health and Sustainability in Spain and Portugal, First Edition.
Edited by Iñigo Virto and Rodrigo Antón.

Published 2025 by Soil Science Society of America, Inc. and John Wiley & Sons, Inc.

cropping and irrigation intensity. Exploring an adequate combination of irrigation and crop intensification and understanding these dynamics in context-specific assessments is crucial for maximizing irrigation benefits on soil health and leading a transition to sustainable agriculture.

Introduction

Irrigation has been, and continues to be, one of the pillars of rural development and food security on a global scale, particularly in arid and semi-arid regions, where it is considered a fundamental component of agri-food systems. From the beginning of agriculture, early human societies recognized the crucial importance of water provision, implementing irrigation strategies that facilitated crop diversification and improved yields. As far back as 5,000 years BCE, farmers in Ancient Egypt devised methods to divert floods along the Nile to store water for use during drier periods in order to enhance agricultural productivity (Angelakis et al., 2020; Bazza, 2007). These advances in agriculture played a pivotal role in population settlement and the emergence of early civilizations by promoting growth and development.

The evolution and progress of agriculture have been key in the development of societies throughout history. This influence was especially decisive in the second half of the 20th century, when the advances associated with the Green Revolution, together with industrial development, led to an unprecedented demographic expansion worldwide. Technological developments during this period also significantly improved irrigation systems, contributing to an important expansion of irrigated land globally. Innovations in pump design and the introduction of pivot technologies and sprinkler distribution systems enhanced the efficiency and capacity of water distribution and supply (United Nations Food and Agriculture Organization [FAO], 2017). In 1900, the world area equipped for irrigation was estimated to be 53 million ha (Freydank & Siebert, 2008). By 1965, the FAO reported a global area equipped for irrigation of 169 million ha, which increased to 258 million ha in 1990 and reached 352 million ha by 2020 (FAO, 2024). At present, with an annual upward trend of 1.3% of global area under irrigated agriculture, irrigation is used on 23% of the cultivated land, and 45% of total food production is obtained through this system (Velasco-Muñoz et al., 2019).

The conversion of natural land to agriculture and the intensification of cultivation have long-term consequences, however, and have led to a progressive degradation of agricultural soils in many instances. At present, the European Commission estimates that 60–70% of soils in the European Union are degraded (European Commission, 2020), with the Mediterranean region being one of the areas most susceptible to this degradation in Europe (Ferreira et al., 2022). According to Prăvălie et al. (2021), agricultural soil degradation in Europe affects more than 56% of the arable land, involving a process of multiple pathways that includes aridification, soil erosion, vegetation decline, soil salinization, and soil organic carbon (SOC) decline. Most of these degradation problems are concentrated in Southern Europe, which is the most arid region (Peel et al., 2007), presenting the lowest values in SOC stocks (de Brogniez et al., 2014) and the highest estimated soil erosion rates (Panagos et al., 2020).

Irrigated agriculture is not exempt from degradation issues. For instance, although soil salinization/sodification are not exclusive to irrigated agriculture, they can be induced and be more severe and have worse consequences in irrigated compared with rainfed agriculture. Soil salinization/sodification can limit plant growth with consequences for crop yields, but it can also have other effects on the soil, such as structure degradation, sealing, and crusting, as well as serious consequences for the environment, such as salinization of ground and surface water.

The use of irrigation in agriculture generates additional concerns. The FAO considers agriculture both a major cause and a major casualty of water scarcity (FAO, 2019). Indeed, the global expansion of irrigation has resulted in agriculture becoming the world's largest water-consuming sector, accounting for 70% of water use on average, although the actual figures of freshwater consumption for agricultural irrigation are estimated to range from 60 to 90% of all water use, depending on the level of economic development and the climate of an area (Adeyemi et al., 2017). Furthermore, an increase in water demands for irrigation could occur with changes in temperatures (increased temperature and major occurrence of heat waves) and precipitation patterns (less rainfall and major occurrence of extreme events) associated with climate change predictions for the Mediterranean region by the Intergovernmental Panel on Climate Change (Ali et al., 2022). For example, Savé et al. (2012) estimated that irrigation requirements would increase between 40 and 250% by the end of the 21st century in the Fluvià watershed (Catalonia, Spain).

The FAO estimates that it will be necessary to produce an additional 70% more food by 2050 owing to population growth (FAO, 2011). In light of these challenges, the increasing environmental concern of modern societies, their growing demand for public involvement on the management of natural resources, and the number of stakeholders involved in water management, the governance of water use and the competition for water generate an inevitable discussion in many territories (Ricart & Clarimont, 2016).

Thus, the conversion of rainfed agricultural land to irrigated land is generally a sensitive issue from social, environmental, and economical perspectives. To address this situation, developing agricultural systems that ensure maximum efficiency in the use of water in agriculture is necessary in making the transition to sustainable agriculture. This transition should be supported by policies that promote good governance in relation to water, land, and environmental management. These policies should also recognize that the territory is an integrated matrix of common resources with different uses and demands.

Under the umbrella of sustainable land management practices, several initiatives and strategies have emerged for addressing these ambitious goals. These initiatives are based on the premise that an optimal use of natural resources at the scale of ecosystem and agricultural systems minimizes land degradation and rehabilitates degraded areas. Given the relevance and multifunctionality of soil organic matter (SOM), many efforts have been made in the past few decades in the design, implementation, and evaluation of different agricultural management strategies to reverse land degradation by recovering SOC reserves in agricultural soils (Paustian et al., 2019). It is imperative to develop sustainable land management strategies that are adapted to both local edaphoclimatic conditions and local management opportunities. These strategies will need to take into account site-specific trade-offs to enable rapid scale-up and implementation to contribute to improving soil organic matter (SOM) storage for ecosystem services in arable lands and thus controlling and reversing soil degradation (Amelung et al., 2020; Costantini et al., 2020). For instance, the World Overview of Conservation Approaches and Technologies (WOCAT) initiative is a global network established in 1992 with the purpose of documenting, sharing, and applying knowledge on sustainable land management, providing an innovative space for sharing and scaling up good practices to address land degradation, climate change, and biodiversity loss (Liniger et al., 2019).

In this framework, irrigated agriculture has received worldwide scientific interest in recent decades, not only because of its pivotal role in the global food and raw material supply chain, but also because of its potential to promote SOM storage. In terms of SOC storage, irrigation may have contrasting effects, especially in arid and semi-arid areas (Apesteguía et al., 2015; Lal, 2020; Pareja-Sánchez et al., 2020), with some areas showing positive effects and others showing negative or no effects (da Gama et al., 2019; Mudge et al., 2017; Nunes et al., 2007; Trost et al., 2013; Zhou et al., 2016). Differences in climate and soil characteristics seem to play a key role in this sense (Costantini et al., 2020). However, irrigation is known to increase greenhouse gas (GHG) emissions per unit land, compared with rainfed management, especially for GHGs associated with soil and soil management (Erb et al., 2017; Sanz-Cobena et al., 2017). This issue illustrates the need for studies on net carbon balances of irrigated agriculture when assessing the associated soil multifunctionality.

From an agrosystem perspective, the introduction of irrigation in arid or semi-arid regions implies a significant change in soil management and water regime. Apart from its impact on the economic and social context of the region, this change may also affect soil functioning and thus alter its dynamic properties. These alterations could have consequences on the biogeochemical and nutrient cycles of the agrosystem, as well as on the performance of other relevant functions such as climate regulation through atmospheric carbon sequestration.

Within this framework, the concept of soil health is re-emerging as a new standard in the assessment of soil management (Karlen et al., 2021). Its definition has been the subject of some discussion. In an attempt to compile the different approaches defined in recent decades, the Intergovernmental Technical Panel on Soils (ITPS) of the Global Soil Partnership led by FAO released a summary definition of soil health in 2020, describing it as the capacity of the soil to sustain the productivity, diversity, and environmental services of terrestrial ecosystems (ITPS, 2020). This definition aligns well with the one provided by Doran and Zeiss (2000), who described soil health as the capacity of soil to function as a vital living system within ecosystem and land-use boundaries to sustain plant and animal productivity, maintain or enhance water and air quality, and promote plant and animal health. The definitions imply the need for a holistic analysis based on soils multifunctionality, focusing on the functions and services that soil can provide to society.

To assess how the adoption of irrigation affects soil functioning and soil health, it is necessary to better understand the benefits and drawbacks of irrigation in terms of agricultural sustainability and soil degradation monitoring policies. The aim of this chapter is to explore the potential of irrigated agriculture to improve soil health in the agricultural systems of the region of Navarre (northern Spain). This region is an example of the expansion of irrigation in the semi-arid Mediterranean area. Since 2000, more than 22,000 ha have been converted from rainfed to irrigated land, resulting in a total area dedicated to irrigated agriculture in the region to over 100,000 ha; the total agricultural area is ~340,000 ha, or ~39% of the total surface of the region. Additionally, another 30,000 ha is expected to be converted to irrigation in the next three decades (Gobierno de Navarra, 2020b).

We thus provide an overview at different scales in this chapter, exploring the results of several research projects carried out in the region of Navarre, to shed light on the multifunctioning of irrigated agricultural soils.

Characterization of the Region of Navarre

The region of Navarre is located in northern Spain, at the western end of the Pyrenees. The region presents an important north-south gradient in terms of temperature, but the gradient is especially prominent in terms of rainfall, mostly owing to the Föhn effect exerted by the Pyrenees and Cantabric mountain ranges in the north, which traps rainfall from most fronts entering the territory from the Gulf of Biscay. As a consequence, annual rainfall ranges between >2,500 mm in the north and <350 mm in the southeast (Pejenaute Goñi, 2017), while mean annual temperatures vary from 14.5 °C in the southernmost point of the region (Buñuel, 41°58′47″N; 1°26′38″W) to 9.3 °C in the Pyrenean highlands in the north (Irabia, 42°59′07″N; 1°09′28″W; Gobierno de Navarra, 2020a). The territory is thus divided into two major biogeographical regions—the Eurosiberian in the north and the Mediterranean in the south (Peralta et al., 2013)—and presents several different climate zones. According to Köppen's classification, these zones range from an arid climate in the south, through to Mediterranean and Oceanic climates in the central and Atlantic (north) zone of the region, to a continental climate at the area close to the Pyrenees (Figure 6.1). Projected regional climate change scenarios point to a general trend toward an aridification of the region, with a clear increase in average temperatures and changes in the seasonal distribution of rainfall, as well as an increase in the number of heat waves and hot days and nights (NADAPTA, 2017).

In relation to the lithologic characteristics of the region, the most relevant lithologies are limestone massifs in the north and center of the region and carbonaceous materials of sedimentary origin in the south. Climate and lithology, together with topography, are thus the most relevant soil-forming factors in the region. The variability of these factors results in a wide diversity of soil types in the region. In the north, Inceptisols and Entisols with an *udic* water regime, combined with Alfisols and Ultisols according to the USDA Soil Taxonomy (Soil Survey Staff, 2022), dominate. In the south, Entisols and Inceptisols with calcaric traits and xeric and aridic water regimes, which would be classified as Calcisols in the FAO World Reference Base (IUSS Working Group WRB, 2015), and some Aridisols dominate. These areas are generally irrigated, together with the alluvial plains and terrace systems of major rivers (containing Fluvents and other soils with relatively high stone contents). The central region is dominated by Inceptisols, Entisols, and Mollisols with calcic (and in some cases, petrocalcic) horizons.

This climatic diversity is also manifested in the agricultural use of the region, which represents an important share of the region's economy. Currently, 39% of the total area is under agricultural use (341,835 ha), with 90.7% of this area being croplands and 9.3% grasslands (Figure 6.1). Irrigated agriculture, developed mostly in the central and south areas of the region, represents more than 30% of the total agricultural area, with more than 100,000 ha at present. Of this area, 22,000 ha has been converted from dryland to irrigated agriculture since 2000, and a further 30,000 ha is expected to be converted in the next three decades (Gobierno de Navarra, 2020b). This conversion is possible through the Canal de Navarra project (among other infrastructure developments), which allows

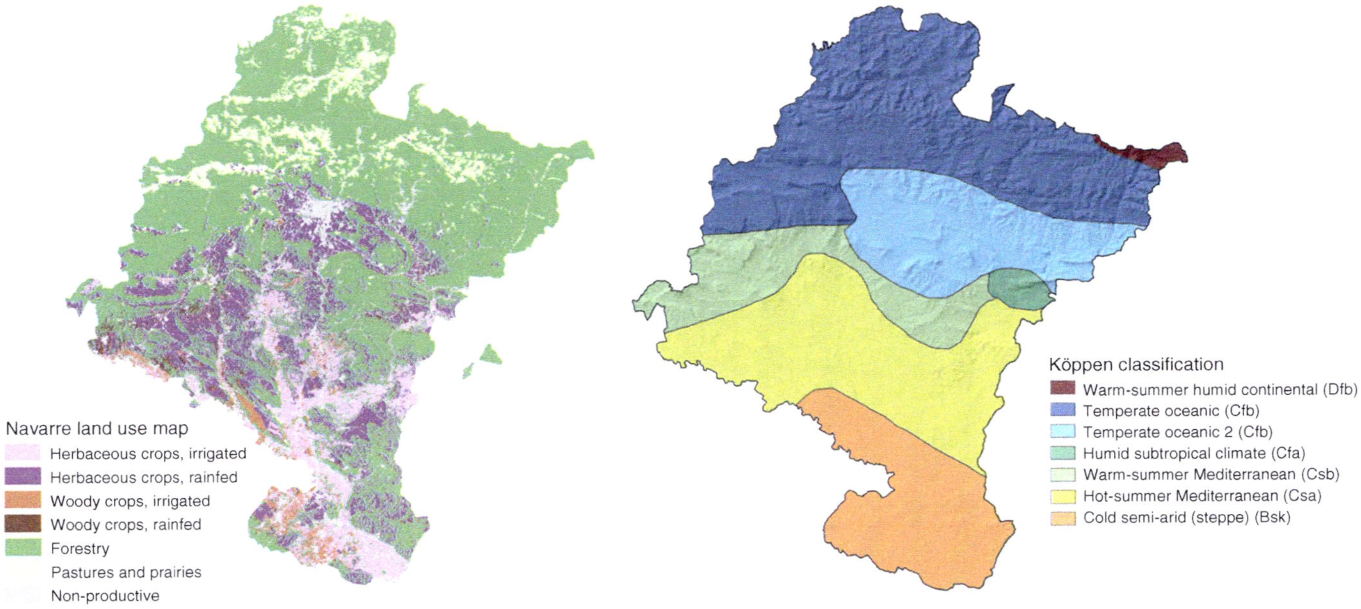

Figure 6.1 Agricultural land use in the region of Navarre (left) and Köppen climate classification (right). *Note*: Regional Statistical Service at https://nastat.navarra.es/es/subtema/-/tag/agricultura and Regional Meteorological Service at https://meteo.navarra.es/definiciones/koppen.cfm.

transporting water more than 190 km from the northern part of the region (Itoiz reservoir), with a more humid climate, to the southern part of Navarra, thus facilitating the expansion of irrigation in semi-arid and arid Mediterranean areas.

This chapter presents the results from several research projects conducted in the region of Navarre and explores the potential of irrigated agriculture to improve soil health at different scales. First, from an experimental field perspective, the chapter discusses results from a study focused on understanding the mechanisms that regulate the process of SOC stabilization in a characteristic soil of the region after the conversion to irrigation. Secondly, the chapter scales up to a regional perspective, based on two studies that evaluate the impact of irrigated agriculture on soil health across the region. Table 6.1 shows a list of the studies considered and the soil determinations and methods applied. Additional information on the analyses and methods applied can be found in the references included in the table.

Table 6.1 Soil Determinations and Methods.

Scale	Soil determinations	Method analyses	Reference
Experimental field	Soil organic carbon (SOC)	Wet oxidation	Antón et al. (2022)
	Particulate organic carbon (POC)	Wet oxidation, fraction >53 μm	
	Bulk density	Undisturbed sample rings	
	Natural abundance of ^{13}C	Mass spectrometry	
Regional	SOC	Wet oxidation	Antón et al. (2021a, 2021b)
	Soil available water-holding capacity	Pressure plate extractors, −33 and −1,500 kPa	
	Bulk density	Unaltered sample rings	
Regional	POC	Wet oxidation, fraction >53 μm	Antón et al. (2019)
	Greenhouse gas emission balances	C footprint calculator (EURENERS 3, 2017)	
	Microbial biomass carbon	Fumigation-extraction (BioLog® EcoPlates®)	
	Soil functional biodiversity index	Community-level physiological profiles (BioLog EcoPlates)	
	Erodibility	*K* factor (RUSLE)	

Assessment of the Effect of Irrigation at the Experimental Field Scale

Approach and Methodology

The first study described here explored the mechanisms and controls involved in SOC dynamics and plant residues incorporation in a calcareous cultivated Mediterranean soil (38% carbonate content at the 0 to 30 cm depth), and their interaction with the transition from rainfed to irrigated cultivation of extensive crops, as reported by Antón et al. (2022). The study is based on an ad hoc experimental design implemented in Enériz (Navarre, 42°40′18.74″N; 1°45′4.65″W) in 2009. This trial was designed to carry out a natural isotopic labeling study from a change from C3 to C4 plants. It is based on the discrimination of different carbon isotopes during photosynthesis in C3 and C4 plants and the consequent difference in the natural abundance of ^{13}C in the tissues of these two types of plants. Following the conversion from C3 to C4 plants grown in a soil, the rate of carbon loss derived from the original C3 vegetation and the rate of new carbon incorporation derived from the new C4 vegetation can be deduced from the resulting change in the natural abundance of ^{13}C in SOM and its fractions. Therefore, the singularity of this study lies in its design, which was developed not only to assess the soil response to a change in crop type from C3 to C4 plants, but also to evaluate the soil response to a change in the irrigation regime over time. It takes advantage of the conversion to irrigation in the area, thus considering cultivation and irrigation as factors in the trial (Antón et al., 2022).

The crops included were winter wheat (*Triticum aestivum* L., a C3 plant), which has been cultivated in the area under rainfed conditions for decades, and maize (*Zea mays* L., a C4 plant), which was introduced in the area with irrigation. Both crops represent two of the most widely and extensively grown crops in the region. The four possible combinations in this experimental field were the following: rainfed wheat, irrigated wheat, rainfed maize, and irrigated maize.

The monitoring included crop parameters, such as harvest index or crop yields, and soil parameters in the tilled layer (0–30 cm), such as bulk density, bulk SOC, and carbon content in particulate organic matter (POC), as well as the isotopic signature of ^{13}C in both bulk SOC and POC (Table 6.1). This approach allowed for the calculation of carbon stocks and the generation of a complete database with data covering a 7-year monitoring period. The sampling design and processing of samples are described in Antón et al. (2022). To analyze the evolution of bulk SOC and POC over the 7 years of the study, a repeated measures linear mixed effects model was considered, with crop, irrigation, year, and the interaction of the crop and year as categorical fixed effects, and a specific plot intercept as a random effect (Antón et al., 2022).

As described in Antón et al. (2022), this database was also used to develop and parameterize a mechanistic model based on the exponential kinetics of SOM decomposition. The model aimed to evaluate the differences in SOC incorporation dynamics into the soil when the irrigation factor was introduced under the study conditions. A General Likelihood Uncertainty Estimation (GLUE) calibration procedure was applied to optimize the model parameters. This procedure, based on the Monte Carlo approach, involves conducting a large number of runs of a model with different sets of parameter values that are randomly chosen from specified parameter distributions. From the comparison of predicted and observed responses, each set of parameter values is assigned a probability of accurately simulating the system (Beven & Binley, 1992). Each model parameter was assigned a random value within a range of possible values, and the model was run in a series of repeated cycles. The performance of the model output was determined using the Root Mean Square Error (RMSE; Antón et al., 2022). The GLUE optimization procedure allowed for the definition of intervals of possible model parameters for irrigated and rainfed systems, and differences were considered to be significant when these intervals did not overlap each other.

Results and Discussion

According to our results, the bulk SOC and POC stocks at year 0 (year of the onset of the experiment) showed no significant differences between the four crop-irrigation combination treatments, as depicted in Figures 6.2 and 6.3 (Antón et al., 2022).

A significant increasing trend was found in both bulk SOC and POC over the years in all four treatments (Antón et al., 2022). In contrast to other studies reporting losses of SOC upon the adoption of irrigation (da Gama et al., 2019; Mudge et al., 2017; Nunes et al., 2007), a significant mean effect associated with the adoption of irrigation was detected in the current study. This effect induced an overall greater increment in the irrigated systems in bulk SOC ($P = .08$), with a less significant increment in POC ($P = .18$), over the 7-year study period in the study soil (Antón et al., 2022). A greater and earlier evolution was observed over time in POC than in bulk SOC, as noted in other general and local studies (Apesteguía et al., 2017; Cotrufo et al., 2019; Poeplau et al., 2018).

From these data, SOC dynamics were simulated with a two-pool modeling approach (Figure 6.4), in which pool 1 (P1), corresponding to the POC fraction, consisted of rapidly decomposable organic matter with a fast decay rate (k_1) and, consequently, a short mean residence time ($\text{MRT}_1 = 1/k_1$). Pool 2 (P2) represented a pool with a slower decay rate (k_2) and a longer MRT ($\text{MRT}_2 = 1/k_2$), corresponding to the mineral-associated organic carbon fraction (Antón et al., 2022). A new parameter (incorporated fraction, *IF*) was added to the model to account for the

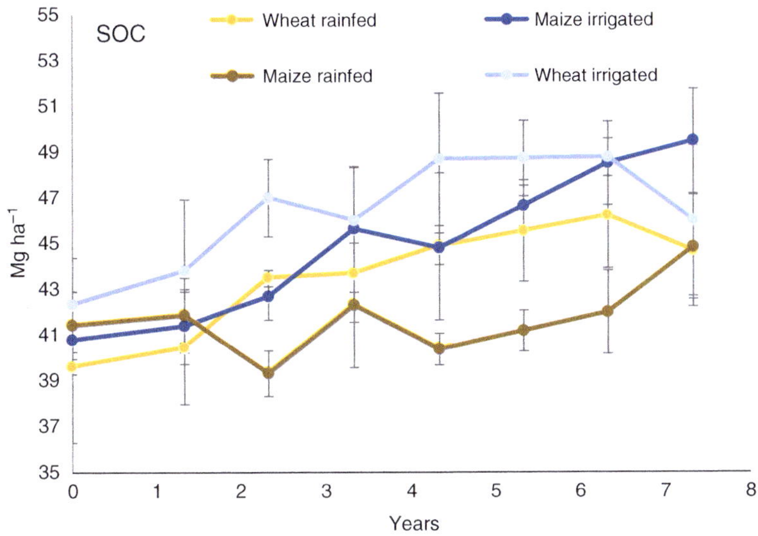

Figure 6.2 Bulk soil organic carbon (SOC) stocks (Mg ha^{-1}) per crop-irrigation combination (n = 6, mean and standard deviation): maize irrigated, maize rainfed, wheat irrigated, and wheat rainfed. *Note*: Adapted from Antón et al. (2022).

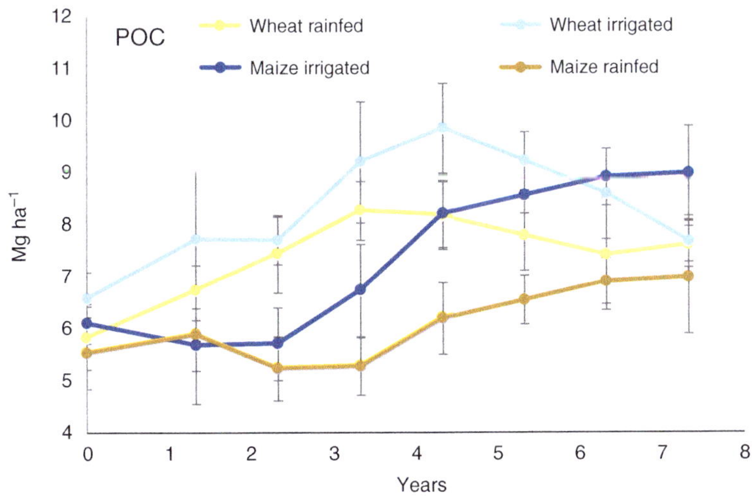

Figure 6.3 Particulate organic carbon (POC) stocks (Mg ha^{-1}) per crop-irrigation combination (n = 6, mean and standard deviation): maize irrigated, maize rainfed, wheat irrigated, and wheat rainfed. *Note*: Adapted from Antón et al. (2022).

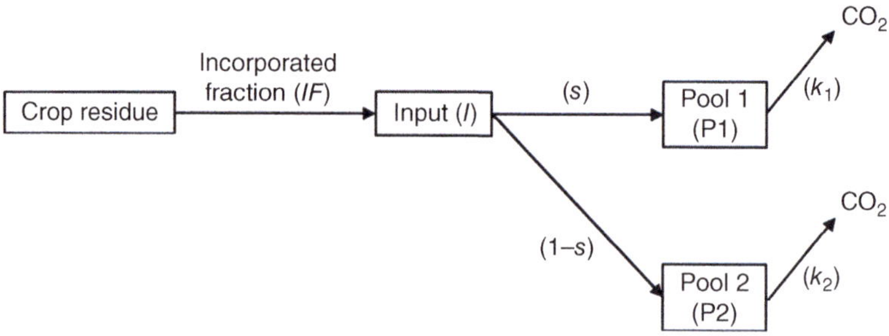

Figure 6.4 Flow chart of the two-parallel pools model. *IF* represents the proportion of crop residues actually incorporated as carbon inputs into the soil. The factor *s* represents the proportion of these incorporated crop residues allocated to pool P1, while (100 – *s*) indicates the proportion entering directly into pool P2. Parameters k_1 and k_2 represent the decay rate of pool P1 and P2, respectively. *Note*: Adapted from Antón et al. (2022).

possibility that only a proportion of crop residues would be effectively incorporated as carbon inputs into the soil.

As explained in Antón et al. (2022), after different model alternatives were tested, the results showed that the best approximation to the real SOC dynamics in the studied soil was a parallel model, in which the carbon inputs are directly transferred to the two pools, with no exchange of carbon between them. In this approach, a proportion (*s*) of the crop residues that are incorporated (*IF*) into the soil is allocated to pool P1, while the remainder (100 – *s*) directly enters pool P2 (Figure 6.4).

This result contrasts with the general understanding of the SOC stabilization process, which is based on the progressive decomposition of plant residues by the decomposer community from large residues to smaller molecules, which increases their reactivity with mineral surfaces and their incorporation into aggregates, and therefore, provides protection against further decomposition (Lehmann & Kleber, 2015; Six et al., 2000, 2004). This finding reveals the need for further investigations into SOC stabilization processes under different edaphoclimatic conditions. In this sense, as stated by Antón et al. (2022), the presence of carbonates in the soil of this study (38% in the upper studied layer) may have played an important role in soil functioning under our study conditions, possibly explaining this observation. It is known that carbonates can promote the rapid incorporation and stabilization of crop residues into stable aggregates of different sizes with slow cycling (Fernández-Ugalde et al., 2011, 2014; Rowley et al., 2018).

In relation to the differences between the irrigated and the rainfed systems, the modeling approach supported by the natural C3-C4 isotopic labeling study revealed that the greater carbon increment observed within the irrigation system was most likely due to an increase in carbon inputs from crop residues

(Antón et al., 2022). This outcome was not only because of the obvious higher yields and crop residues potentially incorporated into the soil in irrigated versus rainfed systems, but also because of a more efficient incorporation of these residues into SOC in the irrigated plots (Figure 6.5). As such, the proportion of carbon from crop residues incorporated into the soil (*IF*) was approximately 19% higher in the irrigated plots (from 36% in the rainfed system to 43% in the irrigated system) in the soil depth considered. This result also revealed that a large proportion of crop residues (>50%) were not actually incorporated into the soil in this study. This finding highlights the importance of the correct determination of carbon inputs from crop residues in this type of model, especially in cases in which steady-state conditions are not preserved, as a faulty estimate may result in errors in the estimation of mean turn-over rates (Derrien & Amelung, 2011).

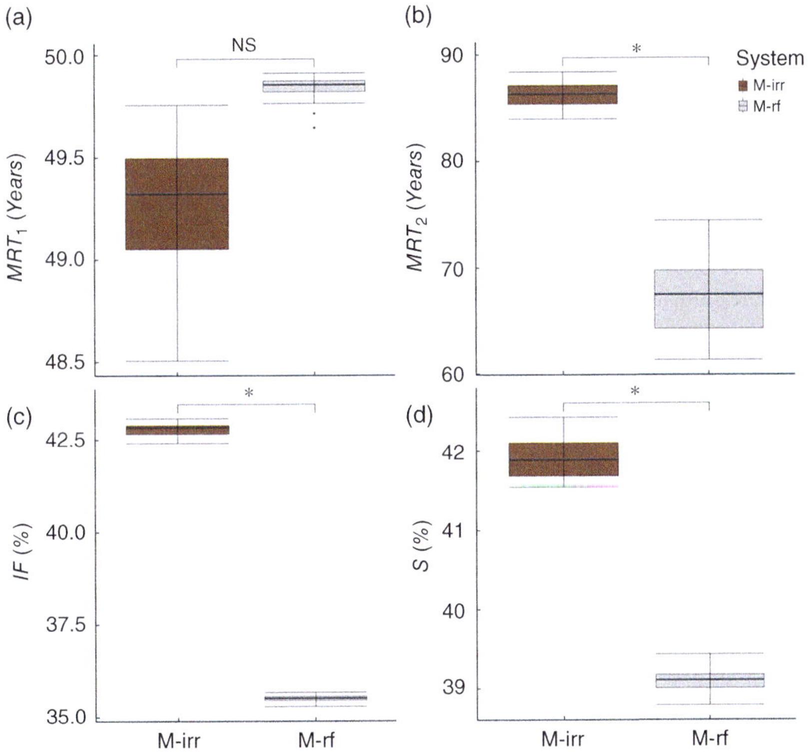

Figure 6.5 Values of MRT_1 (a), MRT_2 (b), the proportion from crop residues incorporated into the soil (IF) (c), and the split *s* parameter (d) for rainfed maize (M-rf) and irrigated maize (M-irr). The boxplots represent the median (n = 50), 25th and 75th percentiles. *Represents no overlapping of the intervals of possible model parameters; NS means some degree of overlapping, so no differences can be attributed.

An unexpected result was the difference in this parameter (*IF*) associated with irrigation, which was greater under this system (Figure 6.5). This outcome can be explained, in part, by the difference in root distribution in the studied soil layer of 0–30 cm, as irrigation tends to increase root density in the surface soil layer (Mahgoub et al., 2017). As remarked in Antón et al. (2022), the gap between conceptual understanding, measurement, and modeling represents a real challenge for understanding SOC dynamics and improving biogeochemical field projections. In addition, the importance of determining crop parameters such as the harvest index is noteworthy in this type of study, in which the determination of carbon residues is usually made from crop yields (Unkovich et al., 2010).

MRT_1 of fast-cycling pool was not significantly different between the irrigated system and the rainfed system (Figure 6.5), and contrary to expectation, MRT_2 of the slow-cycling pool was higher in the irrigated system than in the rainfed system (Figure 6.5). This outcome implies that the modeling results did not support the expectable acceleration of the organic matter turnover rate in the irrigated system owing to a possible enhancement of microbial activity, which opens the way for new research questions.

Finally, in relation to the distribution of plant inputs (*s*) from crop residues between P1 and P2, the results showed a slight but significant increase in the irrigated system (0.4%) compared with the rainfed system (0.39%, Figure 6.5). This result suggests that slightly less crop residues would be allocated to the slow-cycling pool (P2) in the irrigated system, but their protection would occur over a longer time, considering the higher MRT_2 associated with this pool.

This study suggests that further research is needed on the processes and mechanisms involved in SOC dynamics and incorporation in calcareous soils. The understanding of the interactions between soil minerals, organic matter, and microbial communities, and the impact that irrigation can have on them, may lead to improved soil management practices, optimization of nutrient cycling, and improved soil health in regions where these soil types are present.

Regional Assessment of the Effect of Irrigation on Soil Health in the Region of Navarre

Approach and Methodology

The exploration of the work developed in two regional-scale projects allows for evaluating the effect that irrigation can have on soil health of agricultural soils from a regional perspective. In short, the two projects in combination can provide a complete characterization of irrigated agriculture in the region.

In the first project, within the framework of the LIFE NADAPTA project (LIFE16 IPC/ES/000001), a territorial study was carried out in order to evaluate the adaptation of Navarra's agrosystems to climate change. As described by

Antón et al. (2021a), this assessment was based on the development of a regional zoning process, based on intrinsic and dynamic aspects of the territory such as climate, soil type, and land use (Figure 6.6). Subsequently, a network of demonstrative commercial plots under different management systems characteristic of the region was defined, in order to characterize the vulnerability of the agrosystems of the region to climate change.

Irrigation was included in the analysis to evaluate its performance as one of the most widespread agricultural practices in the region. In this sense, two contrasting tests related to irrigation were considered: (1) the transformation from rainfed to irrigated land per se, and (2) the different intensities of irrigation within irrigated land. This approach is relevant because irrigation expands the range of crops that can be grown in an area, as well as the intensity of land use, because it facilitates the cultivation of several crop cycles in the same season. Thus, the most common cropping systems in irrigated land in the region of Navarre were included in the assessment.

The network of commercial plots was defined based on groups of plots distributed throughout the irrigated areas in the region according to two criteria: contrasting use and soil type. Initially, the groups were defined by identifying plots with a known historical management, selecting a set of representative agricultural plots under contrasting management systems in each case.

Therefore, in the first proposed contrast test, rainfed plots were compared with irrigated plots. In this case, within each group, at least one plot under rainfed agricultural use and at least one plot under irrigated use were selected. In the second proposed contrasting test associated with irrigation, irrigated plots with different irrigation intensities were considered within each group (Table 6.2). Accordingly, annual irrigation plots (one crop cycle per year) were compared with intensively irrigated plots (two or more crop cycles per year).

At the same time, to ensure that the different contrasting uses considered in each case were evaluated on the same soil unit, all plots in each group were selected so that at least a portion of each plot was located on the same soil typology (Antón et al., 2019). This involved, when necessary, defining a sampling area within the plot by adjusting it to the area corresponding to the common soil typology within each group. An exhaustive cartographic analysis and fieldwork were carried out, including soil pit digging and analysis, when necessary, to guarantee the same soil unit.

Twelve groups of plots associated with irrigated agriculture were identified, six of them to assess the conversion to irrigation *per se* and six to evaluate different irrigation intensities (Table 6.2). Figure 6.6 illustrates the homogeneous zones and network of commercial plots associated with irrigated agriculture defined for this project, two examples of groups of plots associated to both proposed contrasting tests, and a sampling area definition example within a plot based on the common soil typology for this group.

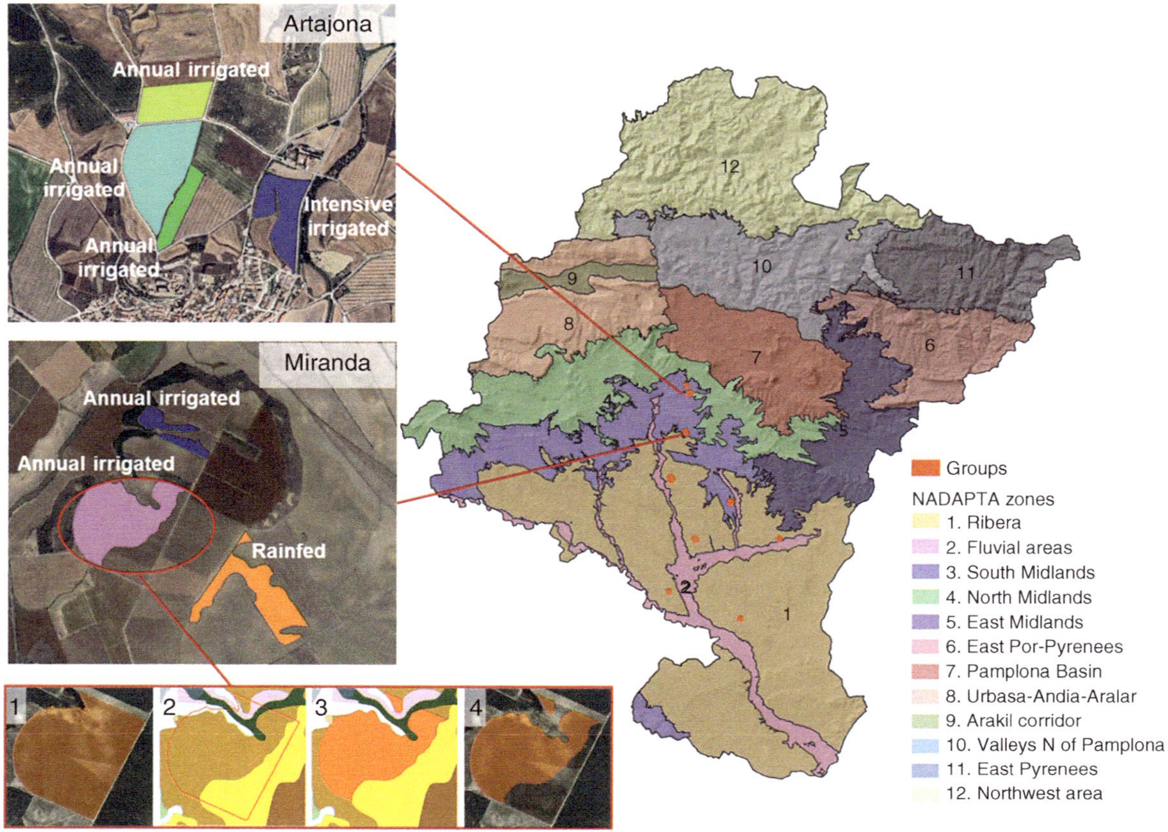

Figure 6.6 Homogeneous zones and network of groups of commercial agricultural plots defined for this study, example of groups of plots and sampling zones on homogeneous soil units, and sampling area definition example of a plot based on the common soil typology for this group.

Table 6.2 Characteristics of the Groups Included.

					Climate			
Group	Contrats use	Managements included in the study		Reference soil**	Mean annual precipitation (mm)	Mean annual temperature (°C)	Potential evapotranspiration (Thornthwaite) (mm)	Years since conversion to irrigation
Funes 1*	Irrigation transformation	Rainfed	Intensive irrigated	Xeric Haplocalcid	390	14.0	740	13
Valtierra*	Irrigation transformation	Rainfed/Organic rainfed	Intensive irrigated	Xeric Haplocalcid	379	13.9	740	20
Santacara	Irrigation transformation	Rainfed	Intensive irrigated	Typic Calcixerept	464	14.2	783	
Miranda*	Irrigation transformation	Rainfed	Annual irrigated	Typic Calcixerept	437	13.9	738	6
Peralta	Irrigation intensity	Annual irrigated	Intensive irrigated	Typic Calcixerept	450	14.4	785	—
Funes 2	Irrigation intensity	Annual irrigated	Intensive irrigated	Xeric haplocalcid	390	14.0	740	—
Funes 3	Irrigation intensity	Annual irrigated	Intensive irrigated	Xeric haplocalcid	390	14.0	740	—
Olite	Irrigation intensity	Annual irrigated	Intensive irrigated	Typic Calcixerept	473	13.7	759	—
Artajona	Irrigation intensity	Annual irrigated	Intensive irrigated	Typic Calcixerept	602	12,9	726	—
Muruzabal	Irrigation intensity	Annual irrigated	Intensive irrigated	Lithic Haplustolls	565	13.4	722	—
Eneriz 1	Irrigation transformation	Rainfed	Annual irrigated	Calcic Haploxerept	565	13.4	722	—
Eneriz 2	Irrigation transformation	Rainfed	Annual irrigated	Calcic Haploxerept	565	13.4	722	—

* Complementary analysis is provided for these groups. ** Soil Survey Staff (2022).

A set of soil indicators were studied on this network of plots to quantify the effect of the contrasted management types tested on soils. The indicators considered included SOC stocks (Antón et al., 2021a) and other soil physical indicators, namely bulk density (BD) and available water-holding capacity (AWHC; Table 6.1, Antón et al., 2021b).

As explained by Antón et al. (2021a), the effect of irrigation on the indicators considered was evaluated for each group of plots using the natural logarithm of the response ratio (RR), which was the ratio between the values of each indicator considered (SOC, AWHC, etc.) in plots with contrasting uses in each group. Then, following the approach commonly applied in meta-analyses comparing results on the same parameters but from different study areas, the overall effects were analyzed with an unweighted fixed effects model (Hedges et al., 1999). Results were represented in forest plots, transforming into RR for simplification, with the effect being considered significant when the 95% confidence interval of the RR did not overlap one ($\alpha = .05$; Hedges et al., 1999).

As the second project, the LIFE REGADIOX project (LIFE12 ENV/ES/000426) complemented the information for three of these groups: Funes 1, Valtierra, and Miranda (Table 6.2). This project provided a complementary dataset of soil physical, biological, and organic matter-related properties, as well as estimations of GHG emissions, allowing the assessment of a set of ecosystem services provided by irrigated agriculture in the region. The study focused on four key soil-based ecosystem services of agricultural systems, considering the characteristics of the region: biomass production, climate regulation, erosion control, and habitat for soil microorganisms.

As described by Antón et al. (2019), biomass production was evaluated based on the yield of each of the plots considered in each group, measured in kilograms of dry matter per hectare. Climate regulation was evaluated by estimating both carbon sequestration and GHG emissions rates for each soil type and management considered, which allowed for the calculation of an annual GHG balance. GHG emission rates were estimated using software developed in the framework of a regional project for carbon footprint calculations in agriculture (EURENERS 3, 2017). Erosion control was evaluated by calculating the *K* factor of the Revised Universal Soil Loss Equation (RUSLE), which represents the susceptibility of soil to erosion or erodibility. Finally, the ability of the soil to function as a habitat for soil microorganisms was determined by measuring both microbial biomass carbon (MBC), determined by the fumigation-extraction method of Vance et al. (1987), and soil microbial functional diversity index, expressed by the number of substrates used by the soil microbial community (NSU) in plates BioLog® EcoPlates® (Biolog, Hayward, CA, USA).

Results and Discussion

As described by Antón et al. (2021a), the observed effect on SOC associated with the transformation from rainfed to irrigation was generally positive, with a 25% increase on average (Figure 6.7). However, it presented considerable variability, with the effect being positive in four of the six groups tested (groups 1, 4, 5, and 6) but neutral in the other two groups (groups 2 and 3). This variability may be related to climatic conditions, with a slight trend in which aridity decreases SOC gains, as the average gain values ranged from 15 to 30% from the most arid zone to the least arid one. Crop intensity also seems to play an important role, since the three groups with annual irrigation showed positive results (groups 1, 5, and 6).

In contrast, when the crop intensification associated with irrigated agriculture was considered, the results were different (Figure 6.8). Although the overall effect on SOC was negative and none of the groups considered showed a positive effect, it was neutral in four of the six groups (groups 1, 2, 3, and 6). As explained by Antón et al. (2021a), when irrigation is introduced in semi-arid areas, more complex crop rotations can be adopted, and when this happens, irrigation is associated with an intensification of tillage and increased fertilization, which may in turn

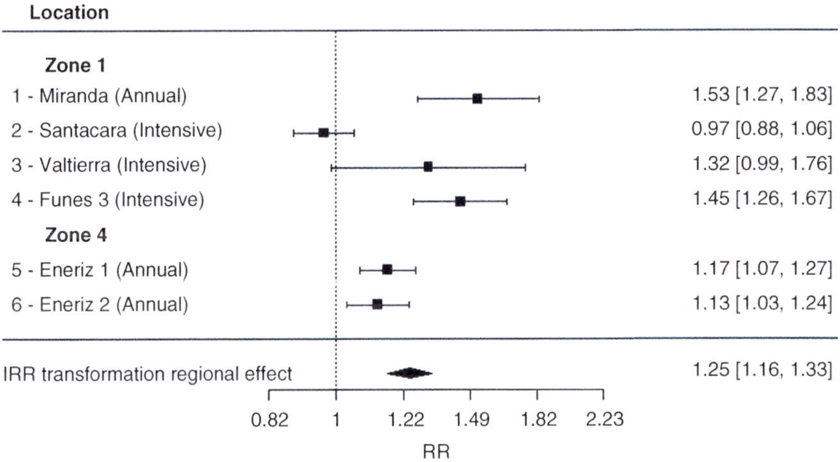

Figure 6.7 Response ratio (RR) of soil organic carbon (SOC) stocks (0–20 cm) for irrigation transformation evaluation. Zones correspond to those in Figure 6.6. The effect was considered significant when the 95% confidence interval (CI) of the RR did not overlap one (α = .05). "Annual" corresponds to groups contrasting rainfed versus annual irrigation, and "Intensive" corresponds to groups contrasting rainfed versus intensive irrigation. *Note*: Adapted from Antón et al. (2021a).

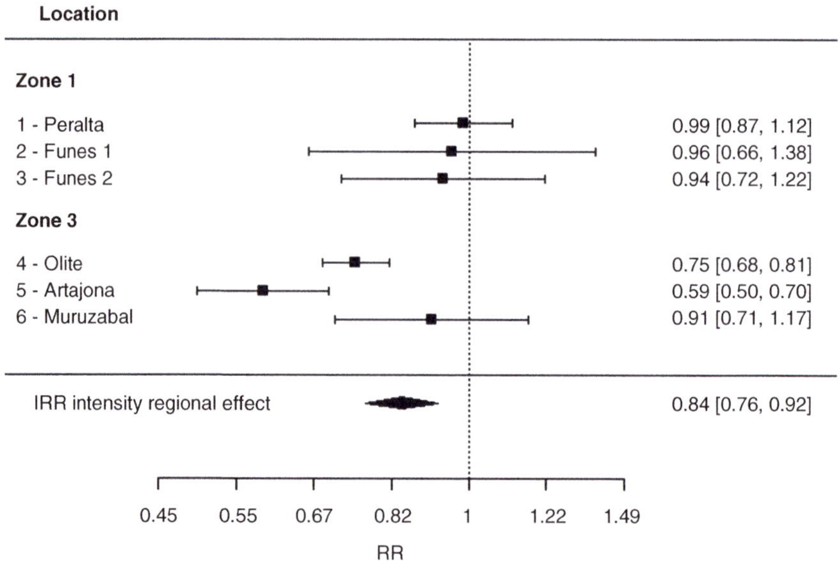

Figure 6.8 Response ratio (RR) of soil organic carbon (SOC) stocks (0–20 cm) for irrigation intensity evaluation. Zones correspond to those in Figure 6.6. The effect was considered significant when the 95% confidence interval (CI) of the RR did not overlap one ($\alpha = .05$). *Note*: Adapted from Antón et al. (2021a).

increase SOC losses (Meurer et al., 2018; Six et al., 1999). Even so, the neutral effect observed in four of six groups (groups 1, 4, 5, and 6) indicated that there was no SOC loss in these soils, while cropping intensification clearly led to increases in yield and profitability.

The analyses of bulk density associated with the transformation from rainfed cropland to irrigation showed a neutral regional effect (Figure 6.9; Antón et al., 2021b). Within the three groups under an annual irrigation regime (groups 1, 5, and 6), one of them (group 5) showed a positive effect (lower BD), while the other two groups (groups 1 and 6) showed a neutral effect. In the case of the three groups evaluating intensive irrigation (groups 2, 3, and 4), one (group 4) showed a positive effect (lower BD); one, a neutral effect (group 3); and one (group 2), a negative effect (higher BD). For the assessment of the effect of crop intensification associated with irrigated agriculture on bulk density (Figure 6.10), the overall effect was neutral, and only one of the groups (group 2) showed a positive effect (lower BD).

Finally, as reported by Antón et al. (2021b), in the analysis of the soil AWHC changes associated with the adoption of irrigation, the overall effect was neutral (Figure 6.11). In the groups with the annual irrigation regimen (groups 1, 5, and 6),

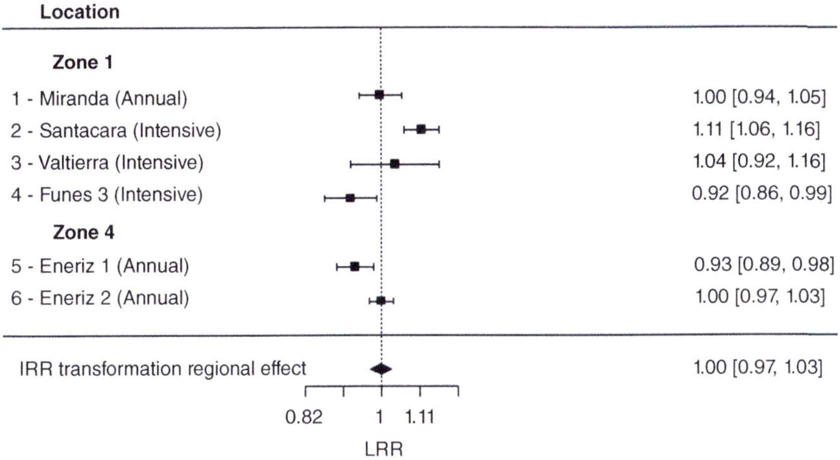

Figure 6.9 Response ratio (RR) of bulk density for irrigation transformation evaluation. Zones correspond to those in Figure 6.6. The effect was considered significant when the 95% confidence interval (CI) of the RR did not overlap one (α = .05). "Annual" corresponds to groups contrasting rainfed versus annual irrigation, and "Intensive" corresponds to groups contrasting rainfed versus intensive irrigation. *Note*: Adapted from Antón et al. (2021b).

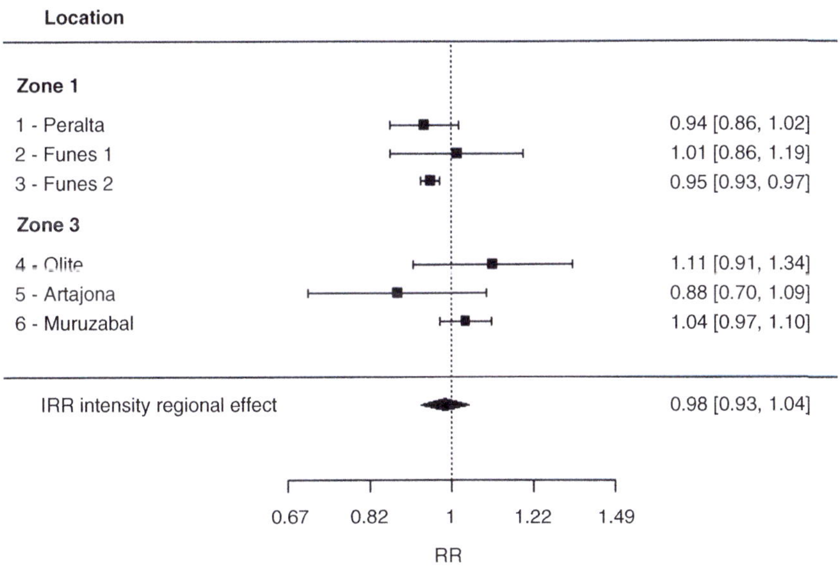

Figure 6.10 Response ratio (RR) of bulk density for irrigation intensity evaluation. Zones correspond to those in Figure 6.6. The effect was considered significant when the 95% confidence interval (CI) of the RR did not overlap one (α = .05). *Note*: Adapted from Antón et al. (2021b).

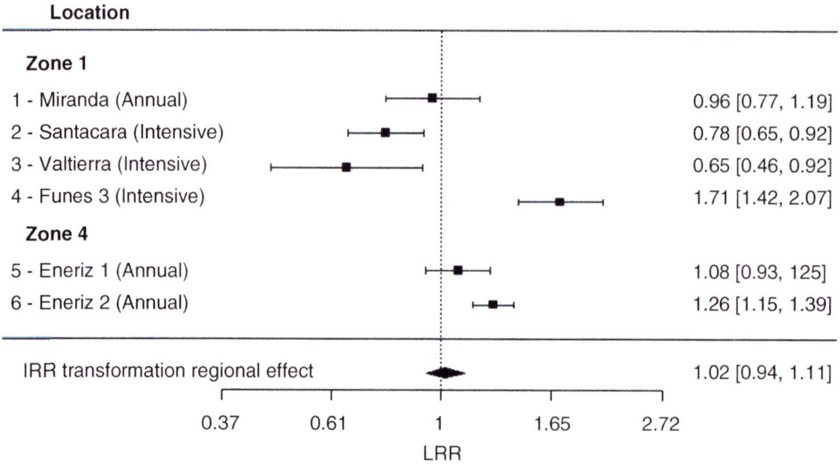

Figure 6.11 Response ratio (RR) of available water holding capacity for irrigation transformation evaluation. Zones correspond to those in Figure 6.6. The effect was considered significant when the 95% confidence interval (CI) of the RR did not overlap one (α = .05). "Annual" corresponds to groups contrasting rainfed versus annual irrigation, and "Intensive" corresponds to groups contrasting rainfed versus intensive irrigation. *Note*: Adapted from Antón et al. (2021b).

the results showed two groups with a neutral effect (groups 1 and 5) and one with a positive effect (group 6). In the three groups under an intensive irrigation regime (groups 2, 3, and 4), the results showed a negative effect in two of them (groups 2 and 3) and positive in the third (group 4). For AWHC, the results associated with crop intensification related to irrigated agriculture showed great variability within groups (Figure 6.12); one showed a positive effect (group 4), and the effect was neutral in the others.

In general, the analysis suggested that the conversion from rainfed to irrigated agriculture may be an effective strategy to promote carbon sequestration in the soils evaluated. Although this increase in SOC is usually correlated with an improvement in soil physical properties such as bulk density and AWHC, this analysis did not yield such as result. This difference can be explained by the fact that tillage, which is carried out regularly in the agricultural soils considered, and the consequent soil homogenization may counteract the positive effect on SOC on the soil physical status.

However, this positive effect on SOC was attenuated when we analyzed the crop intensification associated with irrigated agriculture. In this case, the overall effect on both SOC and the physical parameters considered was generally neutral, with high variability in some cases.

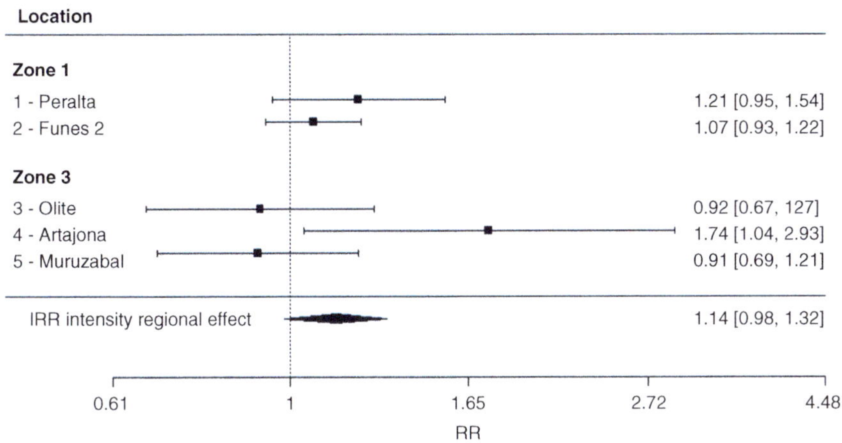

Figure 6.12 Response ratio (RR) of available water holding capacity for irrigation intensity evaluation. Zones correspond to those in Figure 6.6. The effect was considered significant when the 95% confidence interval (CI) of the RR did not overlap one ($\alpha = 0.05$) *Note*: Adapted from Antón et al. (2021b).

The variations in SOC stocks linked to the management practices defined as baseline within each group, may be decisive here. While the baseline for groups for assessing the transformation to irrigation comprised rainfed soils, the groups for examining management intensity were based on irrigated soils with the lowest intensity level observed. This implies that in the latter scenario, a new equilibrium in SOC levels may have been attained, indicating that the carbon stock cannot be further augmented, at least under the management and intensity parameters scrutinized in this study.

As stated previously, an adequate combination of irrigation and crop intensification should be explored in context-specific assessments, with particular consideration that the major climate change threats identified for this region are related to changes in temperatures (increased temperature and major occurrence of heat waves) and precipitation patterns (less rainfall and major occurrence of extreme events; NADAPTA, 2017). In this regard, a detailed investigation of the soil water dynamics and retention and possible measures to improve them is essential for developing an efficient irrigation system in the semi-arid part of this region, as well as in many other regions potentially facing limited water resources (Al-Rumikhani, 2002).

The detailed study conducted on three of the groups of plots described above, Funes 1, Miranda, and Valtierra, provided a deeper evaluation of the effect that irrigation can have on the provision of ecosystem services. As described at the beginning of this section, the study focused on four key soil ecosystem services in agricultural systems: biomass production, climate regulation, erosion control, and habitat for soil microorganisms.

Table 6.3 presents an overview of groups and data collected for each of these three groups, including biomass production, erodibility, annual carbon sequestration rate, GHG emissions rate, and annual GHG balance.

As expected, in relation to biomass production, the conversion to irrigation implied a remarkable increase in yields and dry biomass production (Table 6.3; Antón et al., 2019). Considering the erosion control service, the *K* factor, which represents the susceptibility of soil to erosion or erodibility, depends on both inherent and dynamic soil properties. It is usually reduced with SOC gains. However, since its calculation also depends on soil texture and structure, some interference with these factors implies that the observed gains in SOC translated into reduced erodibility only in Valtierra, for both organic rainfed and annual irrigation compared with the rainfed (conventional) reference system (Table 6.3; Antón et al., 2019).

Climate regulation can be assessed through a comparison of the annual SOC gain resulting from the irrigation transformation with the GHG emissions rate estimation generated by each agrosystem. The results indicated that irrigation increased SOC, but also GHG emissions overall (Table 6.3).

The estimated average annual carbon sequestration rates associated with annual irrigation management ranged from 0.84 ± 0.3 Mg C ha^{-1} y^{-1} in Funes to more than 2 Mg C ha^{-1} y^{-1} in Miranda and Valtierra. Irrigated fodder, with 3.20 ± 0.8 Mg C ha^{-1} y^{-1}, showed the highest annual sequestration rate. It would be expected that the sequestration rate would be reduced over time after the introduction of irrigation, as the SOC stock reaches a new steady state (Chenu et al., 2019). This outcome could not be confirmed in this study, as the estimated annual sequestration rates in the groups did not seem to show differences associated with the time since transformation to irrigation (Miranda, 6 years; Funes, 13 years; and Valtierra, 20 years).

With respect to GHG emissions, irrigation is known to be able to increase GHG emissions compared with rainfed management (Erb et al., 2017; Sanz-Cobena et al., 2017). In this study, GHG emissions were estimated from survey data associated with different inputs and the soil management system, and not from soil measurements. It was confirmed that soil and fertilization represented the most important sources of emissions in our study plots (Figure 6.13), which was similar to findings in other studies, based on measurements for annual irrigated maize (Álvaro-Fuentes et al., 2016; Franco-Luesma et al., 2019; Maris et al., 2018). In this sense, the most promising practices for GHG mitigation seem to be those associated with water management and nitrogen fertilization (Karimi et al., 2012; Maris et al., 2018; Patle et al., 2016; Zornoza et al., 2016).

Finally, considering the habitat for soil microorganism service, seasonal and site-dependent differences in the gains with irrigation were also observed in

Table 6.3 RUSLE *K* Factor (Soil Erosion Indicator), Biomass Production, Annual Soil Organic Carbon (SOC) Gains Rate, Annual Greenhouse Gas (GHG) Emissions Rate, and Annual GHG Balance

		Erodibility (*K*)	Biomass Production	SOC gains		GHG emissions	Annual GHG balance
Group	Management	(kg ha h ha^{-1} MJ^{-1} mm^{-1})	(kg dry matter ha^{-1})	(Mg C $ha^{-1}y^{-1}$)	(Mg $eqCO_2$ ha^{-1} y^{-1})	(Mg $eqCO_2$ ha^{-1} y^{-1})	(Mg $eqCO_2$ ha^{-1} y^{-1})
Miranda	Rainfed (reference)	44.9 ± 0.8	3,872	—		1.24	
	Annual irrigated	49.9 ± 0.5*	13,149.4	2.07 ± 0.4	7.59 ± 1.5	4.50	−3.09
	Irrigated fodder	57.9 ± 2.5*	11,440	3.20 ± 0.8	11.73 ± 2.9	0.83	−10.9
Funes 1	Rainfed (reference)	63.3 ± 0.1	3,918			0.67	
	Annual irrigated	61.8 ± 0. 5*	11,591	0.84 ± 0.3	3.08 ± 1.1	7.44	4.36
Valtierra	Rainfed (reference)	52.5 ± 0.5	4,840			0.67	
	Organic rainfed	49.5 ± 2.3*	290	0.86 ± 0.3	3.15 ± 1.01	0.77	−2.38
	Annual irrigated	44.3 ± 3.9*	11,610	2.12 ± 0.3	7.77 ± 1.01	5.36	−2.41

Note. Adapted from Antón et al. (2019). Annual GHG balance = GHG annual emissions – SOC annual gains. For erodibility (K), cases marked with * show significant differences ($P < .05$) in relation to the reference system within each group.

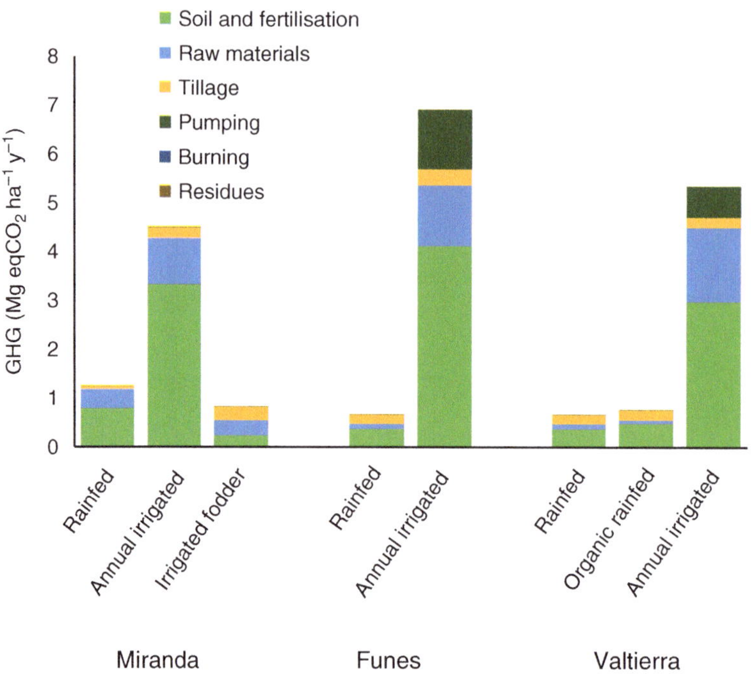

Figure 6.13 Annual GHG emissions associated to each system considered in the study. *Note*: Adapted from Antón et al. (2019).

both indicators considered, MBC and NSU. Only the organic rainfed management in Valtierra showed a positive evolution for MBC (Figure 6.14). Changes in the NSU were also uneven among sites (Figure 6.15), with the Miranda site showing no or negative effects, and Funes and Valtierra showing positive effects in annual crops.

Overall, these results highlight the potential of irrigation to provide key ecosystem services to society, such as erosion control and climate regulation, as well as to ensure an increase in crop yields and, therefore, net gains to farmers. The actual net effect will depend on multiple factors, especially those related to soil management strategies. The results also indicate that depending on the agricultural management, it is possible to compensate for the increase in GHG emissions associated with irrigation by SOC storage, at least in the first years after irrigation adoption. In this case, the system would be considered to perform as a carbon sink, acting as a climate change mitigation agent during this period. This possibility has been observed in other studies (Adviento-Borbe et al., 2007; Ghimire et al., 2017; Mosier et al., 2005, 2006; Sainju et al., 2014). However, it has to be stressed that this effect

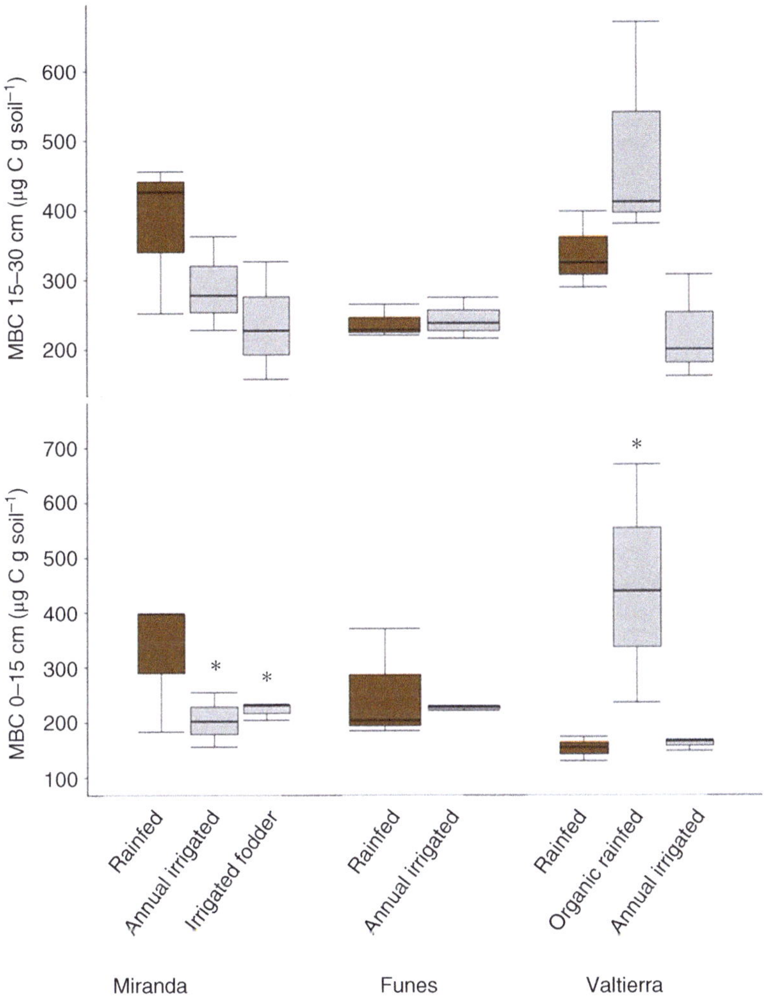

Figure 6.14 Microbial biomass carbon (MBC) measured in 0–15 cm and 15–30 cm in the network of agricultural plots (rainfed in brown). Cases marked with an asterisk (*) show significant ($P<.05$) differences in relation to the reference system within each group according to ANOVA. *Note*: Adapted from Antón et al. (2019).

will only occur until the SOC stock converges to a new steady state in the newly irrigated systems, thus approaching their maximum storage potential (Chenu et al., 2019; Follett, 2001), when the carbon accumulation rates may decrease over time or become nil. If emissions are not reduced in the meantime, the system would be revert to being a net GHG source.

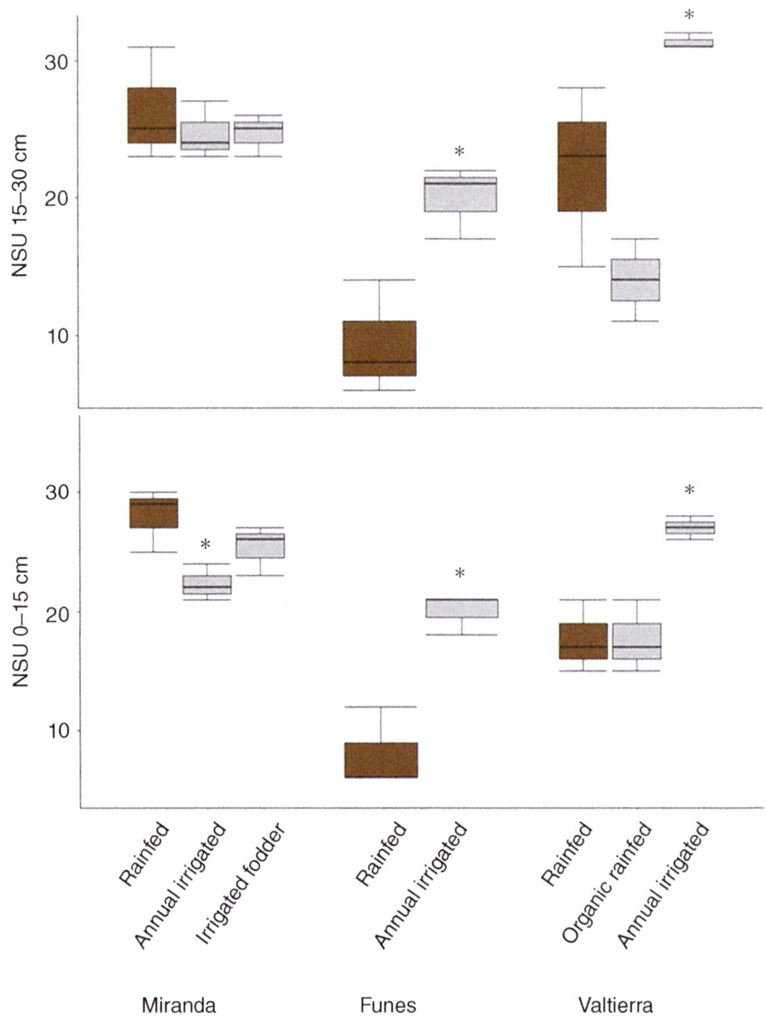

Figure 6.15 Number of substrates used by the soil microbial community (NSU) measured in 0–15 cm and 15–30 cm in the network of agricultural plots (rainfed in brown). Cases marked with an asterisk (*) show significant ($P < .05$) differences in relation to the reference system within each group according to ANOVA. *Note*: Adapted from Antón et al. (2019).

Conclusion

This work provides insights into the impact that irrigation can have on soil functioning and soil health in the Region of Navarre (northern Spain). The exploration of the resources developed in the region in the last few years has allowed for a multi-scale analysis approach of this issue.

The study of SOM dynamics at the experimental field scale demonstrated its effectiveness in assessing short- to mid-term changes resulting from the introduction of irrigation. It confirmed that the adoption of irrigation induced an increase both in bulk SOC and POC over the 7-year study period in the study soil.

This study also explored in detail the mechanisms and controls that regulate SOC dynamics in a calcareous cultivated Mediterranean soil and the interaction that the conversion from rainfed to irrigated land can have on this process. The modeling approach indicated that the greatest difference between irrigated and rainfed systems in SOC storage in the study field seemed to be related to incorporation of crop residues and not SOC mineralization rates. Carbon inputs would therefore be the major driver responsible for the observed and projected increase in SOC storage in the irrigated systems evaluated, as increasingly claimed (Chenu et al., 2019; Keel et al., 2019; Virto et al., 2012).

However, regional-scale studies highlight the potential of irrigation to provide essential soil-based ecosystem services to the local community. They also highlight the trade-offs and risks associated with it, as well as the importance of integrating this type of study into policy development and decision-making in relation to agricultural land management.

In general, the introduction of irrigation had an overall positive effect on SOC, resulting in an average 25% increase. However, improvements in other soil physical properties were less evident, likely owing to the intensive management practices associated with irrigated agriculture, including conventional tillage and in some cases intensive rotations. This positive effect on SOC is attenuated when crop intensification in irrigated agriculture is considered, resulting in a neutral overall effect on SOC and physical parameters.

Depending on agricultural practices, the increase in GHG emissions associated with irrigation can be offset by SOC storage, at least initially, after adopting irrigation. However, this effect is temporary, as SOC stocks eventually reach a new equilibrium, limiting further carbon accumulation rates. Without emission reduction measures, the system could revert to being a net GHG source over time.

Furthermore, when considering the potential climate change mitigating effect of irrigated agriculture (or other agricultural practices) based on the carbon sequestration potential of agricultural soils, it should be noted that, in addition to being limited in time, the storage process is reversible. Therefore, the carbon accumulated over the years in the soil can be re-emitted when the management (or climate) changes (Chenu et al., 2019; Follett, 2001).

Consequently, there is a (temporal) limit on the capacity of agricultural soils to offset GHG emissions. With the target of zero-emission agriculture in 2050 in mind, efforts should be clearly focused not only on enhancing the carbon sink capacity of agricultural systems but also, and especially, on reducing the emission factors associated with agricultural management.

According to Trost et al. (2013), although the irrigation type or intensity of irrigation may influence N_2O emissions, the intensive fertilization strategies normally associated with irrigation seem to be more relevant for N_2O release owing to the more reactive nitrogen compounds (and water) availability within the soil. According to Sanz-Cobena et al. (2017), optimizing nitrogen fertilization, not only in terms of dosage but also in terms of timing of application and type of fertilizer used, is crucial to reducing N_2O emissions. In Mediterranean agroecosystems, where irrigation is usually applied during the summer period, leading to optimal moisture and temperature conditions for N_2O production, water management must also be considered as an emission factor (Aguilera et al., 2013). There are numerous efforts focused on developing strategies for controlling and reducing GHG emissions, including crop residue management (Maris et al., 2018), the use of organic fertilizers (Aguilera et al., 2013), and approaches focused on efficient water management such as regulated deficit irrigation (Zornoza et al., 2016) and approaches related to irrigation timing and frequency (Franco-Luesma et al., 2019).

The multi-scale analysis revealed that irrigated soils in the region have similar soil health to rainfed systems, considering that, except for SOC and POC, no evident improvement (or worsening) has been observed in the physical and biological indicators considered.

As stated previously, assessing the potential of irrigation to enhance soil health requires exploration of an adequate combination of irrigation and crop intensification in context-specific assessments. Without ignoring the need to transition toward sustainable agricultural systems that ensure maximum efficiency in water use and promote good governance in relation to water and land management, the territory must be recognized as an integrated matrix of common resources with different uses and demands.

The interest in these types of studies not only depends on their fundamental role in climate change mitigation and adaptation strategies but also on their focus on improving soil health, supporting the ecosystem services provided by agricultural soils, increasing their resilience, and promoting the adaptation of the farming sector to global change.

Acknowledgments

The authors thank funding from INIA, Spanish Agency (Grant INIA RTA2009-00052-C02-02), and the European Commission Integrated Life Program (Life NADAPTA Project LIFE 16 IPC/ES/000001 and Life REGADIOX project LIFE12 ENV/ES/000426).

The authors would like to thank SILVATECH (Silvatech, INRAE, 2018. Structural and functional analysis of tree and wood Facility, doi: 10.15454/1.5572400113627854E12) from UMR 1434 SILVA, 1136 IAM, 1138 BEF and 4370

EA LERMAB from the research center INRAE Grand-Est Nancy for its contribution to isotopic analysis. SILVATECH facility is supported by the French National Research Agency through the Laboratory of Excellence ARBRE (ANR-11-LABX-0002-01).

The Instituto Navarro de Tecnologías e Innovación Agroalimentarias (INTIA) is thanked for the maintenance and crop management of the experimental field. We also thank the staff and students at the Soil Science group of the Department of Sciences (Universidad Pública de Navarra), for their assistance during soil and biomass sampling, and Prof. Esther González for valuable advice in C4-plants physiology.

References

Adeyemi, O., Grove, I., Peets, S., & Norton, T. (2017). Advanced monitoring and management systems for improving sustainability in precision irrigation. *Sustainability*, *9*(3), Article 353. https://doi.org/10.3390/su9030353

Adviento-Borbe, M. A. A., Haddix, M. L., Binder, D. L., Walters, D. T., & Dobermann, A. (2007). Soil greenhouse gas fluxes and global warming potential in four high-yielding maize systems. *Global Change Biology*, *13*(9), 1972–1988. https://doi.org/10.1111/j.1365-2486.2007.01421.x

Aguilera, E., Lassaletta, L., Sanz-Cobena, A., Garnier, J., & Vallejo, A. (2013). The potential of organic fertilizers and water management to reduce N_2O emissions in Mediterranean climate cropping systems. A review. *Agriculture, Ecosystems and Environment*, *164*, 32–52. https://doi.org/10.1016/j.agee.2012.09.006

Ali, E., Cramer, W., Carnicer, J., Georgopoulou, E., Hilmi, N. J. M., Le Cozannet, G., & Lionello, P. (2022). Cross-chapter paper 4: Mediterranean region. In H.-O. Pörtner, D. C. Roberts, M. Tignor, E. S. Poloczanska, K. Mintenbeck, A. Alegría, M. Craig, S. Langsdorf, S. Löschke, V. Möller, A. Okem, & B. Rama (Eds.), *Climate change 2022—Impacts, adaptation and vulnerability: Contribution of Working Group II to the Sixth Assessment Report of the Intergovernmental Panel on Climate Change* (pp. 2233–2272). Cambridge University Press. https://doi.org/10.1017/9781009325844.021

Al-Rumikhani, Y. A. (2002). Effect of crop sequence, soil sample location and depth on soil water holding capacity under center pivot irrigation. *Agricultural Water Management*, *55*(2), 93–104. https://doi.org/10.1016/S0378-3774(01)00190-1

Álvaro-Fuentes, J., Arrúe, J. L., Cantero-Martínez, C., Isla, R., Plaza-Bonilla, D., & Quílez, D. (2016). Fertilization scenarios in sprinkler-irrigated corn under Mediterranean conditions: Effects on greenhouse gas emissions. *Soil Science Society of America Journal*, *80*(3), 662–671. https://doi.org/10.2136/sssaj2015.04.0156

Amelung, W., Bossio, D., de Vries, W., Kögel-Knabner, I., Lehmann, J., Amundson, R., Bol, R., Collins, C., Lal, R., Leifeld, J., Minasny, B., Pan, G., Paustian, K., Rumpel, C., Sanderman, J., van Groenigen, J. W., Mooney, S., van Wesemael, B., Wander, M., & Chabbi, A. (2020). Towards a global-scale soil climate mitigation strategy. *Nature Communications*, *11*(1), Article 5427. https://doi.org/10.1038/s41467-020-18887-7

Angelakis, A. N., Zaccaria, D., Krasilnikoff, J., Salgot, M., Bazza, M., Roccaro, P., Jimenez, B., Kumar, A., Yinghua, W., Baba, A., Harrison, J. A., Garduno-Jimenez, A., & Fereres, E. (2020). Irrigation of world agricultural lands: Evolution through the millennia. *Water*, *12*(5), Article 1285. https://doi.org/10.3390/W12051285

Antón, R., Arricibita, F. J., Ruiz-Sagaseta, A., Enrique, A., de Soto, I., Orcaray, L., Zaragüeta, A., & Virto, I. (2021a). Soil organic carbon monitoring to assess agricultural climate change adaptation practices in Navarre, Spain. *Regional Environmental Change*, *21*(3), Article 63. https://doi.org/10.1007/s10113-021-01788-w

Antón, R., Derrien, D., Urmeneta, H., van der Heijden, G., Enrique, A., & Virto, I. (2022). Organic carbon storage and dynamics as affected by the adoption of irrigation in a cultivated calcareous Mediterranean soil. *Frontiers in Soil Science*, *2*(March), 1–18. https://doi.org/10.3389/fsoil.2022.831775

Antón, R., Ruiz-Sagaseta, A., Orcaray, L., Arricibita, F. J., Enrique, A., de Soto, I., & Virto, I. (2021b). Soil water retention and soil compaction assessment in a regional-scale strategy to improve climate change adaptation of agriculture in Navarre, Spain. *Agronomy*, *11*(3), Article 607. https://doi.org/10.3390/agronomy11030607

Antón, R., Virto, I., Gonzalez, J., Hernandez, I., Enrique, A., Bescansa, P., Arias, N., Orcaray, L., & Campillo, R. (2019). Extension of irrigation in semi-arid regions: What challenges for soil security? Perspectives from a regional-scale project in Navarre (Spain). In A. Richer de Forges, F. Carré, A. B. McBratney, & J. Bouma (Eds.), *Global soil security. Towards more science-society interfaces* (pp. 79–87). CRC Press.

Apesteguía, M., Virto, I., Orcaray, L., Bescans, P., Enrique, A., Imaz, M. J., & Karlen, D. L. (2017). Tillage effects on soil quality after three years of irrigation in northern Spain. *Sustainability*, *9*(8), Article 1476. https://doi.org/10.3390/su9081476

Apesteguía, M., Virto, I., Orcaray, L., Enrique, A., & Bescansa, P. (2015). Effect of the conversion to irrigation of semiarid Mediterranean dryland agrosecoystems on soil carbon dynamics and soil aggregation. *Arid Land Research and Management*, *29*(4), 399–414. https://doi.org/10.1080/15324982.2015.1016245

Bazza, M. (2007). Overview of the history of water resources and irrigation management in the Near East region. *Water Supply*, *7*(1), 201–209. https://doi.org/10.2166/ws.2007.023

Beven, K., & Binley, A. (1992). The future and distributed models: Model calibration and uncertainty prediction. *Hydrological Processes*, *6*, 279–298. https://doi.org/10.1002/hyp.3360060305

Chenu, C., Angers, D. A., Barré, P., Derrien, D., Arrouays, D., & Balesdent, J. (2019). Increasing organic stocks in agricultural soils: Knowledge gaps and potential innovations. *Soil and Tillage Research*, *188*, 41–52. https://doi.org/10.1016/j.still.2018.04.011

Costantini, E. A. C., Antichi, D., Almagro, M., Hedlund, K., Sarno, G., & Virto, I. (2020). Local adaptation strategies to increase or maintain soil organic carbon content under arable farming in Europe: Inspirational ideas for setting operational groups within the European innovation partnership. *Journal of Rural Studies*, *79*, 102–115. https://doi.org/10.1016/j.jrurstud.2020.08.005

Cotrufo, M. F., Ranalli, M. G., Haddix, M. L., Six, J., & Lugato, E. (2019). Soil carbon storage informed by particulate and mineral-associated organic matter. *Nature Geoscience*, *12*, 989–994. https://doi.org/10.1038/s41561-019-0484-6

da Gama, J. T., Nunes, J. R., Loures, L., Piñeiro, A. L., & Vivas, P. (2019). Assessing spatial and temporal variability for some edaphic characteristics of Mediterranean rainfed and irrigated soils. *Agronomy*, *9*(3), Article 132. https://doi.org/10.3390/agronomy9030132

de Brogniez, D., Ballabio, C., van Wesemael, B., Jones, R. J. A., Stevens, A., & Montanarella, L. (2014). Topsoil organic carbon map of Europe. In A. E. Hartemink & K. McSweeney (Eds.), *Soil carbon* (pp. 393–405). Springer International Publishing. https://doi.org/10.1007/978-3-319-04084-4_39

Derrien, D., & Amelung, W. (2011). Computing the mean residence time of soil carbon fractions using stable isotopes: Impacts of the model framework. *European Journal of Soil Science*, *62*(2), 237–252. https://doi.org/10.1111/j.1365-2389.2010.01333.x

Doran, J. W., & Zeiss, M. R. (2000). Soil health and sustainability: Managing the biotic component of soil quality. *Applied Soil Ecology*, *15*, 3–11.

Erb, K. H., Luyssaert, S., Meyfroidt, P., Pongratz, J., Don, A., Kloster, S., Kuemmerle, T., Fetzel, T., Fuchs, R., Herold, M., Haberl, H., Jones, C. D., Marín-Spiotta, E., McCallum, I., Robertson, E., Seufert, V., Fritz, S., Valade, A., Wiltshire, A., & Dolman, A. J. (2017). Land management: Data availability and process understanding for global change studies. *Global Change Biology*, *23*(2), 512–533. https://doi.org/10.1111/gcb.13443

EURENERS 3. (2017). *Huella de carbono de producto agroalimentario EURENERS 3.* https://teder.org/proyectos-finalizados/huella-de-carbono-de-producto-agroalimentario-eureners-3 [In Spanish].

European Commission. (2020). *Caring for soil is caring for life: ensure 75% of soils are healthy by 2030 for food, people, nature and climate: report of the Mission board for Soil health and food.* Publications Office of the EU.

Fernández-Ugalde, O., Virto, I., Barré, P., Apesteguía, M., Enrique, A., Imaz, M. J., & Bescansa, P. (2014). Mechanisms of macroaggregate stabilisation by carbonates:

Implications for organic matter protection in semi-arid calcareous soils. *Soil Research*, *52*(2), 180–192. https://doi.org/10.1071/SR13234

Fernández-Ugalde, O., Virto, I., Barré, P., Gartzia-Bengoetxea, N., Enrique, A., Imaz, M. J., & Bescansa, P. (2011). Effect of carbonates on the hierarchical model of aggregation in calcareous semi-arid Mediterranean soils. *Geoderma*, *164*(3–4), 203–214. https://doi.org/10.1016/j.geoderma.2011.06.008

Ferreira, C. S. S., Seifollahi-Aghmiuni, S., Destouni, G., Ghajarnia, N., & Kalantari, Z. (2022). Soil degradation in the European Mediterranean region: Processes, status and consequences. In *Science of the Total Environment* (Vol. 805). Elsevier B.V. https://doi.org/10.1016/j.scitotenv.2021.150106

Follett, R. F. (2001). Soil management concepts and carbon sequestration in cropland soils. *Soil and Tillage Research*, *61*(1–2), 77–92. https://doi.org/10.1016/S0167-1987(01)00180-5

Food and Agriculture Organization. (2011). *How to feed the world in 2050. Archiv Fur Kriminologie*, *228*(5–6), 151–159.

Food and Agriculture Organization. (2017). *Does improved irrigation technology save water? A review of the evidence*. https://www.fao.org/3/I7090EN/i7090en.pdf

Food and Agriculture Organization. (2019). *Water Scarcity — One of the greatest challenges of our time*. https://www.fao.org/newsroom/story/Water-Scarcity-One-of-the-greatest-challenges-of-our-time

Food and Agriculture Organization. (2024). Aquastat database. Retrieved November 2023 from https://www.fao.org/faostat

Franco-Luesma, S., Álvaro-Fuentes, J., Plaza-Bonilla, D., Arrúe, J. L., Cantero-Martínez, C., & Cavero, J. (2019). Influence of irrigation time and frequency on greenhouse gas emissions in a solid-set sprinkler-irrigated maize under Mediterranean conditions. *Agricultural Water Management*, *221*, 303–311. https://doi.org/10.1016/j.agwat.2019.03.04

Freydank, K., & Siebert, S. (2008). *Towards mapping the extent of irrigation in the last century: time series of irrigated area per country (Frankfurt Hydrology Paper 08)*. University of Frankfurt (Main). https://doi.org/10.13140/2.1.1743.1687

Ghimire, R., Norton, U., Bista, P., Obour, A. K., & Norton, J. B. (2017). Soil organic matter, greenhouse gases and net global warming potential of irrigated conventional, reduced-tillage and organic cropping systems. *Nutrient Cycling in Agroecosystems*, *107*(1), 49–62. https://doi.org/10.1007/s10705-016-9811-0

Gobierno de Navarra. (2020a). *Meteorología y climatolofía de Navarra (Metherology and Climatology of Navarre)*. Retrieved November 25, 2023, from http://meteo.navarra.es/estaciones/mapadeestaciones.cfm

Gobierno de Navarra. (2020b). *Canal de Navarra. Datos proyecto (Navarra Canal. Project information)*. Retrieved November 15, 2023, from https://obraspublicas.navarra.es/es/aerofer/departamento/canalnavarra

Hedges, L. V., Gurevitch, J., & Curtis, P. S. (1999). The meta-analysis of response-ratios in experimental ecology. *Ecology*, *80*(4), 1150–1156.

Intergovernmental Technical Panel on Soils. (2020). *Towards a definition of soil health*. FAO.

IUSS Working Group WRB. (2015). *World reference base for soil resources 2014: International soil classification system for naming soils and creating legends for soil maps. Update 2015* (World Soil Resources Reports No. 106). FAO.

Karimi, P., Qureshi, A. S., Bahramloo, R., & Molden, D. (2012). Reducing carbon emissions through improved irrigation and groundwater management: A case study from Iran. *Agricultural Water Management*, *108*, 52–60. https://doi.org/10.1016/j.agwat.2011.09.001

Karlen, D. L., De, M., McDaniel, M. D., & Stott, D. E. (2021). Evolution of the soil health movement. In D. L. Karlen, D. E. Stott, & M. M. Mikha (Eds.), *Approaches to soil health analysis* (pp. 21–48). Wiley. https://doi.org/10.1002/9780891189817.ch2

Keel, S. G., Anken, T., Büchi, L., Chervet, A., Fliessbach, A., Flisch, R., Huguenin-Elie, O., Mäder, P., Mayer, J., Sinaj, S., Sturny, W., Wüst-Galley, C., Zihlmann, U., & Leifeld, J. (2019). Loss of soil organic carbon in Swiss long-term agricultural experiments over a wide range of management practices. *Agriculture, Ecosystems and Environment*, *286*, Article 106654. https://doi.org/10.1016/j.agee.2019.106654

Lal, R. (2020). Food security impacts of the “4 per Thousand” initiative. *Geoderma*, *374*, Article 114427. https://doi.org/10.1016/j.geoderma.2020.114427

Lehmann, J., & Kleber, M. (2015). The contentious nature of soil organic matter. *Nature*, *528*(7580), 60–68. https://doi.org/10.1038/nature16069

Liniger, H., Harari, N., van Lynden, G., Fleiner, R., de Leeuw, J., Bai, Z., & Critchley, W. (2019). Achieving land degradation neutrality: The role of SLM knowledge in evidence-based decision-making. *Environmental Science and Policy*, *94*, 123–134. https://doi.org/10.1016/j.envsci.2019.01.001

Mahgoub, N. A., Ibrahim, A. M., & Ali, O. M. (2017). Effect of different irrigation systems on root growth of maize and cowpea plants in sandy soil. *Eurasian Journal of Soil Science*, *6*(4), 374–379. https://doi.org/10.18393/ejss.319952

Maris, S. C., Lloveras, J., Vallejo, A., & Teira-Esmatges, M. R. (2018). Effect of stover management and nitrogen fertilization on N_2O and CO_2 emissions from irrigated maize in a high nitrate Mediterranean soil. *Water, Air, and Soil Pollution*, *229*(1), Article 11. https://doi.org/10.1007/s11270-017-3660-6

Meurer, K. H. E., Haddaway, N. R., Bolinder, M. A., & Kätterer, T. (2018). Tillage intensity affects total SOC stocks in boreo-temperate regions only in the topsoil—A systematic review using an ESM approach. *Earth-Science Reviews*, *177*, 613–622. https://doi.org/10.1016/j.earscirev.2017.12.015

Mosier, A. R., Halvorson, A. D., Peterson, G. A., Robertson, G. P., & Sherrod, L. (2005). Measurement of net global warming potential in three agroecosystems.

Nutrient Cycling in Agroecosystems, *72*(1), 67–76. https://doi.org/10.1007/s10705-004-7356-0

Mosier, A. R., Halvorson, A. D., Reule, C. A., & Liu, X. J. (2006). Net global warming potential and greenhouse gas intensity in irrigated cropping systems in northeastern Colorado. *Journal of Environmental Quality*, *35*(4), 1584–1598. https://doi.org/10.2134/jeq2005.0232

Mudge, P. L., Kelliher, F. M., Knight, T. L., O'Connell, D., Fraser, S., & Schipper, L. A. (2017). Irrigating grazed pasture decreases soil carbon and nitrogen stocks. *Global Change Biology*, *23*(2), 945–954. https://doi.org/10.1111/gcb.13448

NADAPTA. (2017). *Evolución de indicadores climáticos en Navarra (Evolution of climate indicators in Navarra)*. Delegación Territorial de AEMET en Navarra. Retrieved on November 11, 2023, from https://monitoring.lifenadapta.eu/pages/clima

Nunes, J. M., López-Piñeiro, A., Albarrán, A., Muñoz, A., & Coelho, J. (2007). Changes in selected soil properties caused by 30 years of continuous irrigation under Mediterranean conditions. *Geoderma*, *139*(3–4), 321–328. https://doi.org/10.1016/j.geoderma.2007.02.010

Panagos, P., Ballabio, C., Poesen, J., Lugato, E., Scarpa, S., Montanarella, L., & Borrelli, P. (2020). A soil erosion indicator for supporting agricultural, environmental and climate policies in the European union. *Remote Sensing*, *12*(9), 1365. https://doi.org/10.3390/RS12091365

Pareja-Sánchez, E., Cantero-Martínez, C., Álvaro-Fuentes, J., & Plaza-Bonilla, D. (2020). Soil organic carbon sequestration when converting a rainfed cropping system to irrigated corn under different tillage systems and N fertilizer rates. *Soil Science Society of America Journal*, *84*(4), 1219–1232. https://doi.org/10.1002/saj2.20116

Patle, G. T., Singh, D. K., Sarangi, A., & Khanna, M. (2016). Managing CO_2 emission from groundwater pumping for irrigating major crops in trans Indo-Gangetic Plains of India. *Climatic Change*, *136*(2), 265–279. https://doi.org/10.1007/s10584-016-1624-2

Paustian, K., Collier, S., Baldock, J., Burgess, R., Creque, J., DeLonge, M., Dungait, J., Ellert, B., Frank, S., Goddard, T., Govaerts, B., Grundy, M., Henning, M., Izaurralde, R. C., Madaras, M., McConkey, B., Porzig, E., Rice, C., Searle, R., . . . Jahn, M. (2019). Quantifying carbon for agricultural soil management: From the current status toward a global soil information system. *Carbon Management*, *10*(6), 567–587. https://doi.org/10.1080/17583004.2019.1633231

Peel, M. C., Finlayson, B. L., & McMahon, T. A. (2007). Updated world map of the Köppen-Geiger climate classification. *Hydrology and Earth System Sciences*, *11*, 1633–1644. https://doi.org/10.1002/ppp.421

Pejenaute Goñi, J. M. (2017). *Navarra. Geografía*. Nafarroako Gobernua/Gobierno de Navarra.

Peralta, J., Biurrum, I., García-Mijangos, I., Remón, J. L., Olano, J. M., López, M. L., Loidi, J., & Campos, J. A. (2013). *Manual de Hábitats de Navarra*. Gobierno de Navarra.

Poeplau, C., Don, A., Six, J., Kaiser, M., Benbi, D., Chenu, C., Cotrufo, M. F., Derrien, D., Gioacchini, P., Grand, S., Yevdokimov, I., & Nieder, R. (2018). Isolating organic carbon fractions with varying turnover rates in temperate agricultural soils—A comprehensive method comparison. *Soil Biology and Biochemistry*, *125*, 10–26. https://doi.org/10.1016/j.soilbio.2018.06.025

Prăvălie, R., Patriche, C., Borrelli, P., Panagos, P., Roșca, B., Dumitrașcu, M., Nita, I. A., Săvulescu, I., Birsan, M. V., & Bandoc, G. (2021). Arable lands under the pressure of multiple land degradation processes. A global perspective. *Environmental Research*, *194*, Article 110697. https://doi.org/10.1016/j.envres.2020.110697

Ricart, S., & Clarimont, S. (2016). Modelling the links between irrigation, ecosystem services and rural development in pursuit of social legitimacy: Results from a territorial analysis of the Neste System (Hautes-Pyrénées, France). *Journal of Rural Studies*, *43*, 1–12. https://doi.org/10.1016/j.jrurstud.2015.09.012

Rowley, M. C., Grand, S., & Verrecchia, É. P. (2018). Calcium-mediated stabilisation of soil organic carbon. *Biogeochemistry*, *137*(1–2), 27–49. https://doi.org/10.1007/s10533-017-0410-1

Sainju, U. M., Stevens, W. B., Caesar-TonThat, T., Liebig, M. A., & Wang, J. (2014). Net global warming potential and greenhouse gas intensity influenced by irrigation, tillage, crop rotation, and nitrogen fertilization. *Journal of Environmental Quality*, *43*(3), 777–788. https://doi.org/10.2134/jeq2013.10.0405

Sanz-Cobena, A., Lassaletta, L., Aguilera, E., del Prado, A., Garnier, J., Billen, G., Iglesias, A., Sánchez, B., Guardia, G., Abalos, D., Plaza-Bonilla, D., Puigdueta-Bartolomé, I., Moral, R., Galán, E., Arriaga, H., Merino, P., Infante-Amate, J., Meijide, A., Pardo, G., . . . Smith, P. (2017). Strategies for greenhouse gas emissions mitigation in Mediterranean agriculture: A review. *Agriculture, Ecosystems and Environment*, *238*, 5–24. https://doi.org/10.1016/j.agee.2016.09.038

Savé, R., de Herralde, F., Aranda, X., Pla, E., Pascual, D., Funes, I., & Biel, C. (2012). Potential changes in irrigation requirements and phenology of maize, apple trees and alfalfa under global change conditions in Fluvià watershed during XXIst century: Results from a modeling approximation to watershed-level water balance. *Agricultural Water Management*, *114*, 78–87. https://doi.org/10.1016/j.agwat.2012.07.006

Six, J., Bossuyt, H., Degryze, S., & Denef, K. (2004). A history of research on the link between (micro)aggregates, soil biota, and soil organic matter dynamics. *Soil and Tillage Research*, *79*(1), 7–31. https://doi.org/10.1016/j.still.2004.03.008

Six, J., Elliott, E. T., & Paustian, K. (1999). Aggregate and soil organic matter dynamics under conventional and no-tillage systems. *Soil Science Society of America Journal*, *63*, 1350–1358.

Six, J., Merckx, R., Kimpe, K., Paustian, K., & Elliott, E. T. (2000). A re-evaluation of the enriched labile soil organic matter fraction. *European Journal of Soil Science, 51*(2), 283–293. https://doi.org/10.1046/j.1365-2389.2000.00304.x

Soil Survey Staff. (2022). *Keys to soil taxonomy* (13th ed.). USDA Natural Resources Conservation Services.

Trost, B., Prochnow, A., Drastig, K., Meyer-Aurich, A., Ellmer, F., & Baumecker, M. (2013). Irrigation, soil organic carbon and N_2O emissions. *A review. Agronomy for Sustainable Development, 33*(4), 733–749. https://doi.org/10.1007/s13593-013-0134-0

Unkovich, M., Baldock, J., & Forbes, M. (2010). Variability in harvest index of grain crops and potential significance for carbon accounting: Examples from Australian agriculture. In D. L. Sparks (Ed.), *Advances in agronomy* (pp. 173–219). Elsevier Inc. https://doi.org/10.1016/S0065-21131005005-4

Vance, E. D., Brookes, P. C., & Jenkinson, D. S. (1987). An extraction method for measuring soil microbial biomass C. *Soil Biology and Biochemistry, 19*, 703–707. https://doi.org/https://doi.org/10.1016/0038-0717(87)90052-6

Velasco-Muñoz, J. F., Aznar-Sánchez, J. A., Batlles-delaFuente, A., & Fidelibus, M. D. (2019). Sustainable irrigation in agriculture: An analysis of global research. *Water, 11*(9), Article 1758. https://doi.org/10.3390/w11091758

Virto, I., Barré, P., Burlot, A., & Chenu, C. (2012). Carbon input differences as the main factor explaining the variability in soil organic C storage in no-tilled compared to inversion tilled agrosystems. *Biogeochemistry, 108*(1–3), 17–26. https://doi.org/10.1007/s10533-011-9600-4

Zhou, X., Zhou, L., Nie, Y., Fu, Y., Du, Z., Shao, J., Zheng, Z., & Wang, X. (2016). Similar responses of soil carbon storage to drought and irrigation in terrestrial ecosystems but with contrasting mechanisms: A meta-analysis. *Agriculture, Ecosystems and Environment, 228*, 70–81. https://doi.org/10.1016/j.agee.2016.04.030

Zornoza, R., Rosales, R. M., Acosta, J. A., de la Rosa, J. M., Arcenegui, V., Faz, Á., & Pérez-Pastor, A. (2016). Efficient irrigation management can contribute to reduce soil CO2 emissions in agriculture. *Geoderma, 263*, 70–77. https://doi.org/10.1016/j.geoderma.2015.09.003

7

Applicability of Visual Assessment for Evaluating Soil Health in Managed Planted Forests in Northern Spain

Nahia Gartzia-Bengoetxea, Inazio Martínez de Arano, and Ander Arias-González

Chapter Overview

Forest operations can cause substantial soil disturbance, including compaction, ruts, and displacement, with varied impacts on soil health. In northern Spain (Galicia, Asturias, Cantabria, Basque Country, and north of Navarre) managed planted forests cover approximately 1.5 million ha. These forests, predominantly pine and eucalyptus plantations, are managed in a clear-cut regime using heavy machinery for harvesting and site preparation. Accurately describing the impact of this management on soil health at a stand level based on soil strength, void space, bulk density, or organic matter content can be challenging and costly. Our study explored the effectiveness of visual assessments for describing soil health changes caused by forest management activities.

Six manually prepared and six mechanically prepared commercially managed *Pinus radiata* and *Fagus sylvatica* planted stands in the Basque Country were studied. Visual assessments were conducted to evaluate the forest soil health after site preparation. At each monitoring point, the presence or absence of predefined visual disturbance category (undisturbed soil, compacted area, gouge, rut, soil displacement, scalp, and trail) was noted, along with the presence or absence of the forest floor. Soil samples were taken from the most prevalent visual categories (undisturbed soil as a reference point, compacted areas, and gouges) to determine traditional physico-chemical attributes of the soils and evaluate the usefulness of the visual assessment. Bulk density, saturated hydraulic conductivity, and water holding capacity at field capacity and permanent wilting point were measured using

Soil Health Series: Volume 5 Soil Health and Sustainability in Spain and Portugal, First Edition.
Edited by Iñigo Virto and Rodrigo Antón.

Published 2025 by Soil Science Society of America, Inc. and John Wiley & Sons, Inc.

undisturbed soil cores at a depth of 5 cm. Additionally, resistance to penetration, soil moisture, soil texture, and soil organic carbon and nitrogen were determined in the top 10 cm of soil. Visual soil disturbance categories were found to be an efficient means for differentiating disturbances based on both physical and fertility properties. Thus, visual assessment of soil health might be a simple, user-friendly, and affordable tool for evaluating the degree of alteration from natural or pre-activity states to those resulting from management techniques in Northern Spain.

Author contribution: Nahia Gartzia-Bengoetxea, Inazio Martínez de Arano, and Ander Arias-González were responsible of conception and design of the work. Nahia Gartzia-Bengoetxea and Ander Arias-González were responsible also for the acquisition and analysis of data; Nahia Gartzia-Bengoetxea was the responsible of drafting, and Ander Arias-González and Inazio Martínez de Arano contributed to the interpretation of data and the critical revision of the manuscript.

Funding: EUSKOBASOA2050- Research, Development, and Innovation Plan for Enhancing Forestry in the Autonomous Community of the Basque Country.

Competing interests: The authors declare that no competing interests exist.

Introduction

Forests are one of the best green infrastructures to prevent soil degradation and protect watercourses (Horswell & Quinn, 2003). Their dense canopy and continuous supply of organic matter through litterfall and root exudates, paired with their long lifespans, protect soils and prevent degradation. In that sense, forests have been managed for the purpose of soil protection as far back as the 14th century when the felling of mountain forests was prohibited in Switzerland in order to prevent avalanches and landslides (Farrell, 2015). In Spain, forest cover was as low as 12.5% in the mid-19th century (de Armenteras, 1903, cited in Vadell et al., 2022) owing to overgrazing, forest clearing for agriculture, and charcoal production for the iron industry and fuel for fires. The creation of several forest management administrations and policies since 1850 aimed at halting this degradation process (Montero, 2018) and resulted in a sharp increase in forest cover in the past 150 years. Forest area currently covers almost 40% of the country (Eurostat, 2021). However, this large-scale reforestation and afforestation were not uniform throughout the country.

With its temperate Atlantic climate, northern Spain (Galicia, Asturias, Cantabria, Basque Country, and north of Navarre) is optimal for even-aged stands of fast-growing, frequently exotic tree species, which are managed through clearcutting for wood or biomass production. Currently, managed planted forests cover approximately 1.5 million ha in northern Spain, accounting for approximately 40% of the forested area (Ministry for Ecological Transition and Demographic

Challenge, 2008). These forests, predominantly pine and eucalyptus plantations, have been managed with the use of heavy machinery for logging operations over the past four to five decades. To be economically efficient, modern forestry relies on machinery, but the inappropriate use of such machinery, especially on steep slopes, may degrade the forest soil health (Labelle et al., 2022).

Within forests, appropriate assessment of soil health can be very difficult because it requires knowledge of soil chemical, physical, and biological properties to gauge a site's capacity to sustainably provide ecosystem services, such as clean water and air, food, fiber, and bioenergy, and numerous valuable forest services, such as the maintenance and promotion of biodiversity. Even though there is currently no universally accepted protocol for assessing soil health, several key soil chemical, physical, and biological properties are widely used to assess forest soil health (Page-Dumroese et al., 2021).

Soil compaction has been widely used for assessing soil health (Shestak & Busse, 2005). It occurs when mechanical pressures applied to soil result in the reorganization of its particles and thus reduce the void space and increase the bulk density (BD; Labelle and Jaeger, 2012; Picchio et al., 2018). Compaction represents the primary consequence of forest harvesting operations for soil (Hansson et al., 2018a, 2018b; Picchio et al., 2020). Mechanical site preparation aims to reduce soil and vegetation constraints that limit seedling establishment and early growth (Collet et al., 2021). However, the use of heavy machinery during site preparation can cause forest floor removal (i.e., gouges, or the removal of topsoil due to excavation), soil nutrient depletion, a negative impact on mulching, and soil compaction (Gartzia-Bengoetxea et al., 2009). The compaction produced by ground-based traffic of forest machines can adversely affect key forest soil properties and ecosystem services provided by forest soils. For example, increased soil strength and soil BD can impede the growth of roots and trees and thereby negatively affect site productivity (Canga et al., 2018; Gartzia-Bengoetxea et al., 2021; Solgi et al., 2019). Further, the decrease in macroporosity (Naghdi et al., 2017) and the consequent decrease in oxygen availability for roots and the decrease of infiltration (Jourgholami et al., 2019) and saturated hydraulic conductivity (Hansson et al., 2019) increase runoff and erosion, at least during the first years after the forest operations (Gartzia-Bengoetxea et al., 2021). In addition, these impacts can hinder colonization of the soil by fauna, thus reducing soil biodiversity (Frey et al., 2009; Jourgholami et al., 2019).

The extent and severity of soil disturbance during mechanized logging operations and site preparation are influenced by several factors. These include the characteristics of the mechanized system employed, the frequency of machine passes, the slope of the terrain, soil type, soil moisture content during silvicultural treatments, and how carefully treatments are conducted. Therefore, describing or defining soil disturbance in terms of variables such as soil strength, pore space, or BD at the stand level presents challenges, and evaluating soil changes resulting from management activities can carry high costs (Labelle et al., 2022). Soil variability and variation in equipment operations further complicate the assessment process. Moreover, the assessments become more complex with the occurrence of multiple forms of disturbance in a single location, such as forest floor removal, mineral soil displacement, and compaction. As an example, in most forest soils, BD increases with depth. In soils where the topsoil has been excavated (gouges), the natural increase in BD could be confused with compaction.

One approach to simplifying and standardizing evaluations of soil disturbance is to use visual classifications to describe the degree of alteration from natural or pre-activity states to those resulting from management techniques (British Columbia Ministry of Forests, 2001). The scope of using visual methods is broader than ocular alone and may also include observations on biological, hydraulic, or chemical properties, making them multi-attribute systems (Collet et al., 2021). This method is also affordable. Moreover, soil disturbance is a continuous process that occurs throughout the terrain. The visual categorization technique permits an observer to divide this continuum into significant and describable segments that can be replicated (Page-Dumroese et al., 2009). To ensure the efficacy of these visual evaluations, they must be comparable to a reference point ("undisturbed soil"), applicable over a vast area, and both economical and user friendly. These evaluations may also serve as a valuable means of communicating the influence of land management on the resilience and productivity of the soil. Moreover, the visual assessment of soil health could potentially be beneficial in assessing compliance with forest certification programs and determining the necessity for corrective measures. The true correlations between visually determined categories of soil disturbance and various physical/chemical properties of the soil and site characteristics and the significance of such correlations have yet to be established.

The objective of this study was to investigate the usefulness of a visual assessment of soil health in association with traditional methods for assessing soil disturbance in forest operations at a stand level in commercially managed planted forests.

Materials and Methods

Study Area and Experimental Plots

The study was conducted on 10 commercially managed *Pinus radiata* and two *Fagus sylvatica* planted forests in the Basque Country, as representative of northern Spain. The climate in the area is temperate and humid, with annual mean temperature and annual mean precipitation of around 9 °C and 1,200 mm, respectively, and 190 rainy days per year. Precipitation is distributed rather evenly throughout the year, with maximum levels in autumn and spring. The sites are thus representative of Atlantic temperate conditions, and the soil moisture regime can be defined as udic or perudic (Soil Survey Staff, 2022).

Pinus radiata is the tree par excellence in the Basque Country, accounting for 50% of the planted forests with around 100,000 ha (Gobierno Vasco-Eusko Jaurlaritza, 2022). *Fagus sylvatica* is not of great commercial importance in the region, but it is managed similarly and occupies approximately 55,000 ha. The rotation length for radiata pine is around 35–40 years and around 100 years for beech. Timber has been traditionally harvested with chainsaw cutting and limbing, after which the logs are moved, with the aid of skidders, to a landing area or forest road where they can be collected by trucks for transportation to mills. However, in the past several years, harvesting with forest-processors is being widely used. For establishing a new plantation, site preparation is performed with heavy machinery to clear forest residues and facilitate planting. The purposes of site preparation are to reduce soil and vegetation constraints that limit seedling establishment and early growth (Collet et al., 2021) and to remove slash and logging debris to facilitate further planting. Subsoiling or ripping is sometimes also carried out to loosen compacted soil during forest harvesting operations. Bulldozers have often been used for site preparation in the region, although foresters are now replacing bulldozers with excavators. When excavators are used during site clearance, they are also usually used to dig plantation holes. Direct planting or manual site preparation without the use of heavy machinery for clearing and site preparation is very uncommon because of high labor costs. The manual soil preparation technique involves digging holes with a shovel for planting each seedling, after clearing understory vegetation and slash and logging debris with a manual brush cutter.

Twelve of these managed planted forests, which were clear-cut and harvested using skidders, were selected. Site characteristics are shown in Table 7.1. Forest stands differed in the soil preparation technique applied. Six were prepared by direct planting, and the other six were prepared using heavy machinery (i.e., mechanical site preparation). Mechanical site preparation was performed with a Komatsu D65EX-12 bulldozer or similar with a K-G blade attached to the front and a ripper used for subsoiling in furrows (0.5-m depth) in favor of the slope (machine characteristics in Gartzia-Bengoetxea et al., 2021). After the mechanical operations, the planting holes were dug manually with a shovel.

Visual Assessment of Soil Health

In each stand, a visual assessment of soil health was performed by adapting guidelines from the British Columbia Ministry of Forests (2001) and Page-Dumroese et al. (2009) for planted forests in Atlantic temperate conditions.

Table 7.1 Site Characteristics of 10 Radiata Pine Stands and 2 Beech Stands Studied in the Basque Country.

Name	UTM_X	UTM_Y	UTM_Z (masl)	Forest species	Texture class (USDA)	Slope (%)	Site preparation
Arbaliza I	483,654	4,781,268	270	*Pinus radiata*	Loam	18	Direct planting
Kolitxa	479,932	4,783,301	682	*P. radiata*	Loam	29	Direct planting
La Herbosa II	483,763	4,785,707	453	*P. radiata*	Loam	25	Direct planting
Aretxabalagane	520,125	4,792,200	314	*P. radiata*	Silty clay loam	15	Direct planting
Sarasolalde	514,000	4,776,450	253	*P. radiata*	Silty clay loam	10	Direct planting
Entzia I	559,366	4,741,959	1,040	*Fagus sylvatica*	Clay	25	Direct planting
La Guinea	481,433	4,784,039	581	*P. radiata*	Clay loam	40	Mechanical site preparation
La Sancha I	481,387	4,786,796	293	*P. radiata*	Clay loam	13	Mechanical site preparation
La Sancha II	481,458	4,786,635	297	*P. radiata*	Clay loam	11	Mechanical site preparation
Santa Lucia	523,300	4,795,550	154	*P. radiata*	Silty clay loam	46	Mechanical site preparation
Baluga	488,175	4,784,250	147	*P. radiata*	Silty clay loam	25	Mechanical site preparation
Entzia II	559,224	4,741,929	150	*F. sylvatica*	Clay	15	Mechanical site preparation

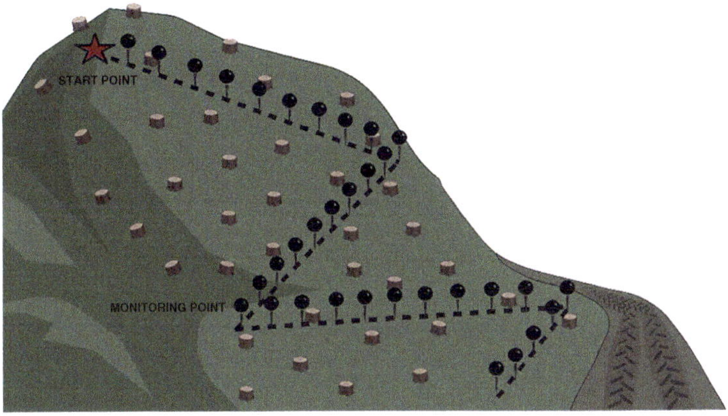

Figure 7.1 Illustration of a random transect with a randomly located a start point that covers the entire unit so that the monitoring points are spaced to provide an adequate assessment of the site.

The assessment required randomly locating a start point and traversing a transect that covered the entire forest stand so that monitoring points were spaced to provide an adequate assessment of the site (Figure 7.1; Page-Dumroese et al., 2009). A monitoring point was defined as a 15-cm-diameter circular area around the tip of the assessor's foot. The presence or absence of each disturbance category at the point was noted.

The visual disturbance category of each sample point was determined using the most limiting visual indicator at the monitoring point (Table 7.2). A monitoring point located in a stump, rock, or log was noted as "no evaluation," and the category of "others" was used to note a monitoring point in harvesting residues. In addition, the presence or absence of the forest floor was noted at each monitoring point.

The minimum number of monitoring points for an assessment was 50, but the optimum number of monitoring points was calculated from the stand surface (Table 7.3). The visual assessment provided estimates of soil disturbance and confidence intervals around the monitoring results.

Relationship of Visual Categories with Physicochemical Characteristics of the Soil

In every manually prepared stand, three subplots were arranged within the soil category classified as undisturbed (18 subplots). Similarly, in every mechanically prepared stand, subplots were arranged in both undisturbed (18 subplots) and gouges (18 subplots) categories, while subplots for compacted areas were arranged

Table 7.2 Soil Disturbance Categories Used in the Visual Assessment of Soil Health in Planted Forest Soils. Photos by Ander Arias-González/NEIKER.

Soil disturbance	Description	Implications for soil quality	Example picture
Undisturbed	No evidence of disturbance, either compaction ruts, or skid trails. No management-generated soil erosion Forest floor layers present and intact.	None	
Compacted areas	Areas with evidence of compaction at the monitoring point caused by repeated machine traffic. Forest floor may be present or absent.	Compaction generates a rearrangement of soil aggregates that can strongly affect the pore geometry of soils. These soil structural changes directly affect soil water movement by altering the hydraulic properties such like water holding capacity and infiltration rate (e.g., Alaoui et al., 2011). Usually, water runoff increases and soils are more vulnerable to soil erosion (e.g., Fullen, 1985; Prats et al., 2019).	

Table 7.2 (Continued)

Soil disturbance	Description	Implications for soil quality	Example picture
Ruts	Wheel or track ruts are impressions or ruts in the soil created during the operation of ground-based equipment on wet soils of limited load-bearing strength. Ruts vary in depth but are defined as machine-generated soil displacement or compression. Often soil puddling is also present within the rut.	Rutting is caused by soil compression and displacement, which become preferential flow paths (Peng & Horn, 2008). Rut formation suggests that the load imposed on the soil has exceeded its bearing capacity. Soil compaction is the primary result of rutting.	
Gouges	Evidence of topsoil loss owing to excavation.	Gouges are mainly caused by the blade or the digging bucket of heavy machinery cutting into the soil, which creates deep grooves and displaces significant amounts of soil. They are characterized by similar properties to those of compaction. It is also characterized by a decrease in soil organic carbon (SOC) due to the exposure of soil from lower depths.	

(*Continued*)

Table 7.2 (Continued)

Soil disturbance	Description	Implications for soil quality	Example picture
Soil displacement	Soil that is not originally from the monitoring point and is management generated.	Soil that has been excavated will end up in a different place from where it previously was. Normally this category may appear where the forest residues are collected to ease the plantation. The soil in these piles may be very compacted (with the characteristics shown in compacted areas) or be very loose and thus easily erodible.	
Scalps	Scalps are areas where the forest floor has been removed at the monitoring point, but no alteration has occurred in the mineral soil.	Removal of forest floor has a negative impact on mulching, with two major consequences: rain drops may compact the area, and the exposed soil may be easily eroded. The loss of SOC may have consequences on nutrient depletion.	

Table 7.2 (Continued)

Soil disturbance	Description	Implications for soil quality	Example picture
Trails	Included in this category are skid trails or backspar trails and bladed trails. Skid trails are trails that have been ballasted with logs, and excavated or bladed trails are constructed trails.	In a study performed by (Zemke et al., 2019), soil compaction was increased by 21% on average in skid trail wheel ruts. As a consequence, observed runoff was 8.5 times higher on skid trails, while saturated hydraulic conductivity was diminished by 36%. Preferential flow paths were created enhancing the sediment export to water courses	

in three stands (nine subplots). These categories—undisturbed soils, gouges, and compacted areas—were chosen for study owing to their prevalence in each respective soil preparation method.

In each subplot, three undisturbed and non-stony soil cores (diameter = 53 mm, height = 50 mm) were sampled from the first 5 cm of the soil profile (after removal of the forest floor) to determine BD, saturated hydraulic conductivity (K_s), and water holding capacity at −33 kPa (θ_{fc}) and −1,500 kPa (θ_{pwp}). Undisturbed field moist soil cores were prepared for saturation as described by Flint and Flint (2002), and placed in a bucket with a porous bottom, so that all pores were filled by capillary action without affecting the soil structure. BD was measured by the Archimedes' displacement method (Flint & Flint, 2002). The still saturated soil cores were then transferred to a laboratory constant head permeameter (Eijkelkamp Agrisearch equipment, Netherlands) to determine K_s by measuring the difference in water levels and by collecting the drained water in a burette during a fixed period of time when the water was flowing constantly. Four determinations were performed to obtain a reliable value of hydraulic conductivity, which was expressed in units of centimeters per hour. Once the K_s was measured, the

Table 7.3 Transect Length, Distance Between Monitoring Points and the Optimum Monitoring Points for Visual Assessment of Soil Quality in Forest Soils.

Area (ha)	Transect length (m)	Distance between monitoring points (m)	Optimum monitoring points
0.25	200	4	50
0.5	400	4	100
0.75	400	4	100
1	800	4	100
1.25	1,000	5	100
1.5	1,500	5	150
1.75	1,500	5	150
2	2,000	5	200
2.5	2,000	5	200
3	2,000	5	200
3.5	2,000	5	200
4	2,500	5	250
4.5	2,500	5	250
5	2,500	5	250
5.5	2,500	5	250
6	2,500	5	250
6.5	2,500	5	250
7	2,500	5	250
7.5	2,500	5	250
8	2,500	5	250
8.5	2,500	5	250
9	2,500	5	250
9.5	2,500	5	250
10	2,500	5	250
11	2,500	5	250
12	2,500	5	250
13	3,000	6	250
14	3,000	6	250
15	3,000	6	250
16	3,500	7	250
17	3,500	7	250
18	3,500	7	250
19	3,500	7	250
20	3,500	7	250

samples were transferred to a pressure plate apparatus (Soilmoisture Equipment Corp., USA) to determine water holding capacity at −33 kPa. The undisturbed soil cores were kept in the pressure plate extractor for 11 days to reach equilibrium. The water holding capacity at −1,500 kPa, or permanent wilting point, was determined with air-dried sieved (<2 mm) soil samples. The samples were placed in rubber cores of height 1.5 cm and saturated by capillary action, and pressure was applied at −1,500 kPa. Pressure was kept until no more water was extracted from the chambers (4 days). Afterward, all samples were weighed and oven-dried for 48 h at 105 °C. Available water content (AWC) was calculated as the difference between θ_{fc} and θ_{pwp}.

Around each soil core, four resistance to penetration (RP) measurements in the first 10 cm of the soil profile were performed using a hand penetrometer for top layers (Eijkelkamp Agrisearch equipment, Netherlands), and four soil samples were taken for moisture measurements. The mean of four RP measurements was calculated.

In addition, a composite sample of the first 10 cm was also collected. The soil samples were air-dried and sieved to <2 mm. Soil organic carbon (SOC) and nitrogen analysis was performed in a LECO TruSPEC analyzer. After removal of the organic matter by oxidation with H_2O_2 and dispersion of the sample with sodium hexametaphosphate, the soil texture was determined by laser diffractometry (Mastersizer 2000 diffractometer, Malvern Instruments, Worcestershire, UK) correcting the clay value according to Konert and Vandenberghe (1997).

Data Processing and Statistical Analysis

Mean values and standard errors are depicted in figures, tables, and within the text. The site preparation effects on the ground determined by visual assessment were compared for each soil disturbance category between direct planting and mechanical site preparation with a two-sample *t*-test using R (R Core Team, 2022). First, normality was assessed using Shapiro's test and equality of variances was examined using Bartlett's test. The Welch *t*-test was used when homogeneity of variances was not achieved and Mann–Whitney U test when non-normality was observed.

Afterwards, soil properties in the undisturbed soils category were compared between direct planting and mechanical site preparation using a *t*-test or Mann–Whitney *U* test. This comparison was done to determine whether the reference point was independent of the soil preparation method. The extent of treatment response was evaluated by means of a 95% confidence interval, presented as a percentage for improved interpretability. Confidence intervals were computed using the bootstrap percentile method with 1,000 resamples. Three of the K_s values were excluded as they were considered very high (>36 cm/h; Soil Science Division Staff, 2017) related to preferential flows (Allaire et al., 2009).

Subsequently, an additional comparison was conducted, specifically focusing on the soil properties of the undisturbed soils, compacted, and gouges categories within plots subjected to mechanical site preparation. In this case, a paired *t*-test or Wilcoxon signed-rank test was employed, depending on whether there was a violation of the equality of variances and normality.

Results

The visual assessment of soil disturbance revealed significant differences between direct planting and mechanical site preparation (Figure 7.2). The average percentage of undisturbed area in manually prepared plots was 50.7% ± 9.06, while in mechanized plots the undisturbed area was 11.0% ± 4.46 ($t = 3.93$, $P < .01$). The main disturbance categories in manually prepared plots were compacted areas with an average occupation of 12.9% ± 7.17 and scalps with average occupation of 12.5% ± 5.02 of the stands. In contrast, in mechanically prepared plots, the main soil disturbance present was gouges, with an average occupation of 40.9% ± 9.14 of the stands, while soil

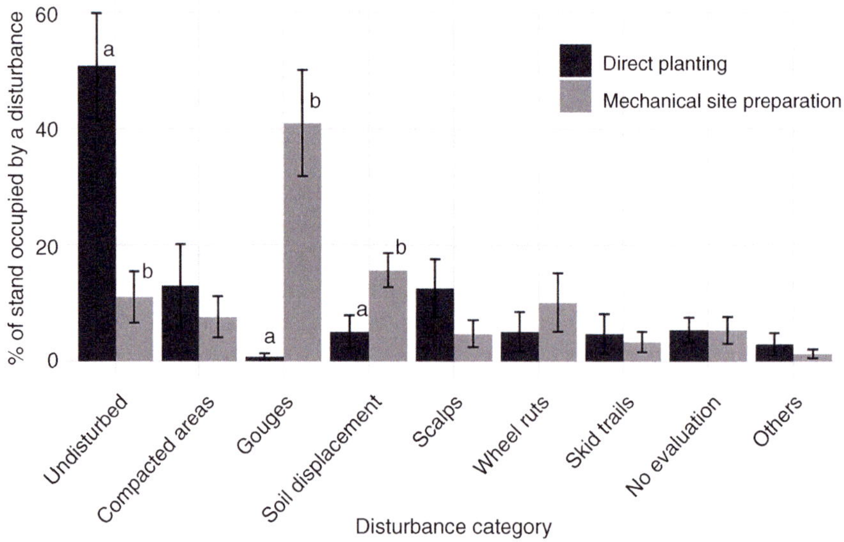

Figure 7.2 Percentage of stand occupied by each soil disturbance in manually prepared plots (Direct planting) and mechanically prepared plots (Mechanical site preparation). Means and standard errors are shown in the figure. Bar plots followed by different letters mean significant difference between direct planting and mechanical site preparation for each disturbance category performed by a two-sample *t*-test with the confidence interval of 95% for the difference of means.

displacement was also significant with an average occupation of 15.6% ± 2.94 of the stands. Significant differences were observed between direct planting and mechanical site preparation for the following soil disturbance categories: gouges ($U = 0$, $P < .01$) and soil displacement ($t = -2.60$, $P < .05$; Figure 7.2). In addition, the presence of the forest floor was only noted in 31.7% ± 13.56 of the monitoring points in mechanically prepared plots, while forest floor was present in 61.6% ± 8.70 of the monitoring points in manually prepared plots.

To determine whether the reference point was independent of the soil preparation method, soil properties of soil samples from the undisturbed category in direct planting and mechanical preparation sites were compared. No significant differences were found (Table 7.4). BD of undisturbed soils in direct planting stands was 0.95 ± 0.04 g/cm^3, while that of undisturbed soils from mechanically prepared plots was slightly higher 1.10 ± 0.05 g/cm^3 ($t = -2.23$, $P = .05$). No significant differences were observed either in RP, soil moisture at sampling time, saturated hydraulic conductivity, AWC, or SOC or nitrogen concentration in soil categories visually determined as undisturbed in sites that differed by soil preparation technique.

Later, soil samples gathered from gouges were juxtaposed with the undisturbed soils in mechanically prepared plots, using a paired comparison. All soil physical properties were affected in soil samples from gouges in comparison with soil

Table 7.4 Resistance to Penetration (RP), Soil Moisture (SM), Saturated Hydraulic Conductivity (K_S), Bulk Density (BD), Available Water Content (AWC), and Soil Organic Carbon (SOC) and Nitrogen (N) of Soils Collected from "Undisturbed" Category in Direct Planting (DP) and Mechanized (MP) Stands

Soil property	DP	MP	95% CI for the difference of medians	*t*-value/ *U* value	*P* value
PR (MPa)	1.27 (0.17)	1.68 (0.29)	(−98.1%; 98.9%)	−1.11	>.05
SM (%)	31.4 (4.25)	35.1 (3.88)	(−1,247%; 1,228%)	−0.64	>.05
K_S (cm/h)	8.58 (4.05)	6.06 (1.56)	(−656%; 627%)	17	>.05
BD (g/cm^3)	0.95 (0.04)	1.10 (0.05)	(−15.7%; 16.78%)	−2.24	.05
AWC (%)	25.7 (2.64)	19.1 (2.29)	(−747%; 718%)	1.87	>.05
SOC (g C/kg soil)	45.3 (7.8)	39.9 (6.5)	(−268%; 270%)	0.48	>.05
N (g N/kg soil)	2.2 (0.3)	2.1 (0.3)	(−11%; 11%)	0.13	>.05

Soils from 0- to 5-cm depth for K_S, BD, and AWC and from 0- to 10-cm depth for PR, SM, SOC, and N. Means with standard errors in parenthesis are shown ($n = 6$). Two-sample t-test or Mann–Whitney U test results are also shown with the 95% confidence interval (CI) for the difference of medians in percentages.

Table 7.5 Resistance to Penetration (RP), Soil Moisture (SM), Saturated Hydraulic Conductivity (K_S), Bulk Density (BD), Available Water Content (AWC), and Soil Organic Carbon (SOC) and Nitrogen (N) of Soil Samples Collected from "Undisturbed" and "Gouges" Categories in Mechanized (MP) Stands.

Soil property	Undisturbed	Gouges	95% CI for the difference of medians	*t*-value/ W-value	*P* value
PR (MPa)	1.68 (0.29)	3.07 (0.60)	(24%; 253%)	3.35	<.05
SM (%)	36.0 (4.22)	23.0 (3.06)	(−2,436%; −172%)	−3.20	<.05
K_S (cm/h)	6.06 (1.56)	0.46 (0.11)	(−798%; −134%)	15	<.05
BD (g/cm^3)	1.10 (0.05)	1.27 (0.05)	(3%; 34%)	3.19	<.05
AWC (%)	19.15 (2.29)	13.97 (1.04)	(−1,205%; 169%)	−1.94	>.05
SOC (g C/kg soil)	39.9 (6.5)	18.0 (2.6)	(−392%; −46%)	−3.51	<.05
N (g N/kg soil)	2.1 (0.3)	0.9 (0.1)	(−21%; −3%)	−3.60	<.05

Soils from 0- to 5-cm depth for K_S, BD, and AWC and from 0- to 10-cm depth for RP, SM, SOC, and N. Means with standard errors in parenthesis are shown ($n = 6$). Two-sample *t*-test or Wilcoxon sum rank test results are also shown with the 95% confidence interval (CI) for the difference of medians in percentages.

samples from undisturbed sites (Table 7.5). BD in soil samples from subplots representing the undisturbed category was 1.10 ± 0.05 g/cm^3 and 1.27 ± 0.04 g/cm^3 in gouges subplots. The 95% confidence interval for the median difference between gouges and undisturbed for BD was between 3% and 34%. Soil moisture at sampling time was also significantly lower in samples taken from gouges category than from undisturbed soil samples collected at the same day; a reduction between 1,234% and 106% might be observed in the median differences between both categories. In contrast, mechanized stands showed a mean penetration resistance above 3 MPa, the threshold at which the growth of most roots are restricted (Sinnett et al., 2008) and at which radiata pine root penetration is also severely damaged (Sands et al., 1979). The 95% confidence interval for the median difference between the groups (undisturbed and gouges) in the variable RP was 24–253%. There was evidence of a median difference between the groups.

The saturated hydraulic conductivity in samples taken from gouges was 0.46 ± 0.11 cm/h, significantly lower than in undisturbed soil samples taken from the same stand. While the AWC did not differ significantly between gouges and undisturbed soil samples, the water holding capacity at −33 kPa was significantly reduced in gouges compared with undisturbed samples from the same stand (θ_{fc} = 37.62% ± 1.73 in undisturbed and 30.92% ± 1.37 in gouges, $t = -3.03$, $P < .05$). On average, θ_{fc} was between 1,162% and 178% lower in gouges than in undisturbed soil.

Table 7.6 Resistance to Penetration (RP), Soil Moisture (SM), Saturated Hydraulic Conductivity (K_S), Bulk Density (BD), Available Water Content (AWC), and Soil Organic Carbon (SOC) and Nitrogen (N) of Soil Samples Collected from "Undisturbed" and "Compacted areas" Categories in Mechanized (MP) Stands.

Soil property	Undisturbed	Compacted areas	95% CI for the difference of medians	*t*-value/ W-value	*P* value
PR (MPa)	1.94 (0.88)	3.47 (1.25)	(46%; 259%)	6.15	<.05
SM (%)	32.30 (11.65)	24.91 (8.44)	(−2,439%; −960%)	−1.87	<.05
K_S (cm/h)	6.87 (2.79)	0.58 (0.19)	(−1,213%; −248%)	6	<.05
BD (g/cm^3)	1.02 (0.05)	1.24 (0.06)	(3.9%; 4.1%)	3.53	<.1
AWC (%)	19.33 (1.93)	18.88 (4.36)	(−1,258%; 1,169%)	−0.16	>.05
SOC (g C/kg soil)	48.3 (13.9)	39.1 (13.9)	(−325%; 139%)	−1.72	>.05
N (g N/kg soil)	2.6 (0.6)	2.2 (0.8)	(−16%; 7%)	−1.59	>.05

Soils from 0- to 5-cm depth for K_S, BD, and AWC and from 0- to 10-cm depth for RP, SM, SOC, and N. Means with standard errors in parenthesis are shown (n = 3). Two-sample t-test or Wilcoxon sum rank test results are also shown with the 95% confidence interval (CI) for the difference of medians in percentages.

Organic carbon and nitrogen levels in soil were diminished in areas referred to as gouges compared with undisturbed soils. Specifically, the average SOC decreased from 39.9 g C/kg ± 6.5 in undisturbed to 18.0 g C/kg ± 2.6 in gouges. The confidence interval for the median difference indicates a reduction in SOC ranging from 46% to 392% and a reduction in nitrogen ranging from 3% and 21%.

Finally, compacted areas were compared with undisturbed areas in mechanically prepared stands using a paired comparison (Table 7.6). Compacted areas revealed significantly higher RP and BD in comparison with their paired undisturbed soil samples and lower soil moisture content at sampling time and saturated hydraulic conductivity. Again, soil physical properties were diminished in those categories identified as compacted areas visually; however, SOC and nitrogen were not affected in this case.

Discussion

Protection of forest soil is of increasing concern in European forestry (Thees & Olschewski, 2017). The current focus is on the impact of forest harvesting machinery on soil and the resulting risks to the forest ecosystem in general and to soil fertility in particular. Plantation forestry in southwestern Europe must also address this issue. Soil protection is a central issue in the European soil strategy for 2030 (European Commission, 2021a), and it is a relevant input for the European forest strategy for 2030 (European Commission, 2021b). In addition, it may represent a climate change adaptation strategy (European Commission, 2021c) that maintains forest productivity and associated ecosystem services, while also being very relevant to the protection of adjacent ecosystems such as watercourses.

The comparison between soil properties assessed through traditional methods in the undisturbed soils categories arising from both direct planting and mechanical site preparation revealed no significant differences, either in physical soil properties (RP, soil moisture, saturated hydraulic conductivity, BD, and AWC) or in soil fertility properties (SOC and nitrogen). Sinnett et al. (2008) concluded that tree root development is significantly impeded when soil penetration resistance is above 2 MPa. The soils classified as undisturbed in this study showed average RP values below 2 MPa. Soil BD is the single most useful parameter of soil physical structure, but it is seldom measured. It is a direct measure of soil compaction (or loosening) and is essential to assess total available void space within a soil (i.e., total porosity; Hernanz et al., 2000). Those undisturbed soils showed a soil BD of around 1 g/cm^3 revealing a good level of volumetric pore space, which is fundamental for sustainable soil use, both in terms of productivity and environmental health. These values for BD might be related to the good SOC levels observed in these categories (around 40 g C/kg soil; Table 7.4). Indeed, for all soil groups in England and Wales, Loveland and Webb (2003) found that increased SOC levels improved soil structure by decreasing BD, improving aggregate stability, increasing pore size, and increasing the proportion of air-filled pore space.

However, BD is an insensitive measurement of pore size distribution and pore continuity parameters, which are critical factors affecting soil hydrology and biochemistry (Hernanz et al., 2000). Saturated hydraulic conductivity (K_S) was also assessed in this study. Undisturbed soil samples exhibited values exceeding 3.6 cm/h, categorizing them as high hydraulic conductivity in terms of K_S classification (Soil Science Division Staff, 2017). High saturated hydraulic conductivity allows water to move rapidly through soil, which can be advantageous for drainage and water movement in the soil profile, reducing runoff and erosion risk (Prats et al., 2019).

In summary, soil categories identified visually as undisturbed in a managed planted forest in the Basque Country were characterized by a BD of approximately 1 g/cm^3, a SOC concentration of around 40 g C/kg, high saturated hydraulic conductivity, and RP below 2 MPa. These values may be considered reference points for healthy forest soils with similar texture in plantation forestry in northern Spain.

The effect of topsoil compaction was observed in soil categories that were visually classified as gouges and compacted areas in this study. They also revealed a change in the hydrological response. RP was above the threshold of 3 MPa in both disturbance categories (Tables 7.5 and 7.6). Consequently, the growth and health of forest vegetation could be disturbed as roots struggle to reach deeper soil layers, and thus, suffer impaired nutritional, aeration, and hydric status (Bonnaud et al., 2019; Gartzia-Bengoetxea et al., 2021). The USDA Forest Service Pacific Southwest Region determined that following a management activity, a change of 0.5 MPa or more in soil resistance should be considered detrimental to soil structure and processes (Landsberg et al., 2003). In gouges and compacted areas, average increases of 1.39 MPa $\pm$ 0.37 and 1.52 MPa $\pm$ 0.2 in RP were observed, respectively. These increases represent an almost threefold rise compared with the threshold set by the USDA Forest Service.

BD was also significantly increased in both categories, reflecting the evidence of compaction caused by mechanized site preparation. Several studies have shown that tractor passes can form an anisotropic soil pore system owing to simultaneous movement of aggregates or particles forward and downward and to wheel slippage (Pagliai et al., 2003; Peng & Horn, 2008). The changes can form a platy structure in the upper few centimeters with elongated pores that are oriented parallel to the soil surface. These pores are not vertically continuous and induce mainly horizontal fluxes of water and gas (Horn et al., 2003; Pagliai et al., 2003), reducing soil infiltration and increasing runoff and soil erosion risk. Indeed, very low saturated hydraulic conductivities were observed in gouges (0.46 cm/h $\pm$ 0.11) and compacted areas (0.58 cm/h $\pm$ 0.19) in comparison with undisturbed soil, which showed K_S higher than 3.6 cm/h.

While both gouges and compacted areas showed negative impacts on the soil physical condition, their primary distinction was associated with soil fertility. In contrast to compacted areas, which maintained SOC and nitrogen concentrations similar to those of undisturbed soils, gouges exhibited a significant reduction in carbon and nitrogen levels. The SOC and nitrogen concentrations in undisturbed soils were approximately 40 g C/kg and 2 g N/kg, respectively, whereas the average SOC concentration in gouges was 18 g C/kg $\pm$ 2.6, with a nitrogen concentration of 0.9 g N/kg $\pm$ 0.1. According to Loveland and Webb (2003), a major threshold for critical concentrations of SOC in agricultural temperate soils is widely believed to be 20 g C/kg, below which a potentially serious decline in soil health will occur. Soils labeled as gouges showed SOC concentrations below this threshold. The SOC stock in the top 10 cm of soils from managed forest areas categorized as undisturbed soil was estimated to be approximately 43 Mg C/ha based on this study, whereas the SOC stock in categories classified as gouges was estimated to be around 23 Mg C/ha. SOC is a global environmental issue, especially in the framework of the three multilateral environmental agreements—the United Nations Framework Convention on Climate Change, the Convention on Biological Diversity, and the United Nations Convention to Combat Desertification—and these data show that forest management has the potential to affect all of them based on the site preparation technique selected.

The results from the current study seem to prove that visually classified soil disturbance categories are linked to soil properties determined by traditional methods. Thus, visual assessment might be a simple and standardized useful tool to describe the degree of alteration from natural or pre-activity states to those resulting from management techniques. Mechanized site preparation resulted in a higher disturbance in comparison with direct planting or manual site preparation, as the undisturbed area represented only 11.0% ± 4.46 of the stands (Figure 7.2). In mechanized stands, disturbances such as gouges, soil displacement, and compacted areas occupied nearly two-thirds of the forest area, with gouges being the most prevalent disturbance, averaging 40.9% ± 9.14. This finding indicates that mechanical site preparation negatively affects forest soil health, which can be visually assessed, in the conditions of this study. Areas identified visually as gouges in managed planted forest in the Basque Country might be characterized by a BD above 1.2 g/cm, a SOC concentration of below 20 g C/kg, low saturated hydraulic conductivity, and RP above 3 MPa and may occupy more than one-third of the forest area.

This type of forest operation could significantly threaten soil health and could potentially affect the full spectrum of ecosystem services that planted forests provide in northern Spain. Opting for manual site preparation or other low-impact forestry methods may be preferable when preserving soil integrity is a primary objective. Regulations and subsidies could help sustain soil health in planted forests in northern Spain, while maintaining forest activity in these mountainous regions where, owing to steep terrain, poor accessibility, and high labor costs, the economic benefit of forestry is very difficult to obtain without these subsidies.

Conclusion

Visual assessment of soil health might be a simple, user-friendly, and affordable tool to evaluate the degree of alteration from natural or pre-activity states to those resulting from mechanical management techniques in northern Spain. Visual soil disturbance categories effectively differentiate the soil condition based on variations in physical properties (RP, soil moisture, saturated hydraulic conductivity, BD, and AWC) as well as in fertility properties (SOC and nitrogen). BD increased between 3% and 34% in gouges and in compacted areas in comparison with undisturbed soil or reference points, while saturated hydraulic conductivity decreased by up to 1,000%, resulting in areas prone to soil erosion. Although SOC was not affected by compaction, it significantly decreased in gouges, showing a reduction of 46–392% compared with undisturbed soils. SOC undoubtedly plays a key role in the health of temperate forest soils, as it positively influences their physical, chemical, and biological properties.

This approach revealed that mechanized site preparation led to greater disturbance than direct planting or manual site preparation in planted forests from northern Spain. Gouges were the primary disturbance observed in mechanized stands occupying 40.9% ± 9.14 of the forest area. SOC removal and compaction are the main characteristics of this disturbance causing declines in temperate forest soil health. In conclusion, these surveys may also serve as a valuable and cost-effective means of communicating the influence of forest management on soil health at the stand level. Moreover, the visual assessment of soil health could potentially be beneficial in assessing compliance with forest certification programs and determining the necessity for corrective measures.

Acknowledgments

We thank Ibai Segurola, Pablo Gamiz, and Sergio Zamora for helping on the sampling and lab work. We are grateful to Iñaki Garmendia, forest ranger from the Provincial Council of Bizkaia, for his valuable contribution to the project identifying forest stands. This study was part of EUSKOBASOA2050- Research, Development, and Innovation Plan for Enhancing Forestry in the Autonomous Community of the Basque Country.

References

Alaoui, A., Lipiec, J., & Gerke, H. H. (2011). A review of the changes in the soil pore system due to soil deformation: A hydrodynamic perspective. *Soil and Tillage Research*, *115–116*, 1–15. https://doi.org/doi.org/10.1016/j.still.2011.06.002

Allaire, S. E., Roulier, S., & Cessna, A. J. (2009). Quantifying preferential flow in soils: A review of different techniques. *Journal of Hydrology*, *378*, 179–204. https://doi.org/10.1016/j.jhydrol.2009.08.013

Bonnaud, P., Santenoise, P., Tisserand, D., Nourrisson, G., & Ranger, J. (2019). Impact of compaction on two sensitive forest soils in Lorraine (France) assessed by the changes occurring in the perched water table. *Forest Ecology and Management*, *437*, 380–395. https://doi.org/10.1016/j.foreco.2019.01.029

British Columbia Ministry of Forests. (2001). *Soil conservation surveys guidebook* (2nd ed.). Forest Practices Code of British Columbia.

Canga, E., Arias-González, A., Majada, J., & Gartzia-Bengoetxea, N. (2018). Effects of forwarding on soil hydrological properties in thinning operations in northern Spain. In *Book of Abstracts* (pp. 5767–5779). Presented at the VIII Congreso Ibérico De Las Ciencias Del Suelo/VIII Congresso Ibérico De Ciências Do Solo, Donostia/San Sebastián,.

Collet, C., Vast, F., Richter, C., & Koller, R. (2021). Cultivation profile: A visual evaluation method of soil structure adapted to the analysis of the impacts of mechanical site preparation in forest plantations. *European Journal of Forest Research*, *140*, 65–76. https://doi.org/10.1007/s10342-020-01315-2

de Armenteras, A. A. (1903). *Árboles y montes: Curiosidades artística e históricas de los montes*. Imprenta Ricardo Rojas.

European Commission. (2021a). *EU soil strategy for 2030, Reaping the benefits of healthy soils for people, food, nature and climate*. COM (2021) 699 final.

European Commission. (2021b). *New EU Forest Strategy for 2030*. COM (2021) 572 final.

European Commission. (2021c). *Forging a climate-resilient Europe—The new EU strategy on adaptation to climate change*. SWD (2021) 25 final.

Eurostat. (2021). *Statistics explained—Forests, forestry and logging*. Retrieved July 30, 2020, from https://ec.europa.eu/eurostat/statistics-explained/index.php?title=Forests,_forestry_and_logging

Farrell, E. P. (2015). Forward. In A. A. González & N. G. Bengoetxea (Eds.), *Soil degradation risks in planted forests* (pp. 13–14). Eusko Jaurlaritzaren Argitalpen Zerbitzua, Servicio Central de Publicaciones del Gobierno Vasco. Legal Record: VI 410-2015

Flint, L. E., & Flint, A. L. (2002). Porosity. In J. H. Dane & G. C. Topp (Eds.), *Methods of soil analysis* (pp. 241–254). John Wiley & Sons, Ltd. https://doi.org/10.2136/sssabookser5.4.c11

Frey, B., Kremer, J., Rüdt, A., Sciacca, S., Matthies, D., & Lüscher, P. (2009). Compaction of forest soils with heavy logging machinery affects soil bacterial community structure. *European Journal of Soil Biology*, *45*, 312–320. https://doi.org/10.1016/j.ejsobi.2009.05.006

Fullen, M. A. (1985). Compaction, hydrological processes and soil erosion on loamy sands in east Shropshire, England. *Soil and Tillage Research*, *6*, 17–29. https://doi.org/10.1016/0167-1987(85)90003-0

Gartzia-Bengoetxea, N., González-Arias, A., Merino, A., & Martínez de Arano, I. (2009). Soil organic matter in soil physical fractions in adjacent semi-natural and cultivated stands in temperate Atlantic forests. *Soil Biology and Biochemistry*, *41*, 1674–1683. https://doi.org/10.1016/j.soilbio.2009.05.010

Gartzia-Bengoetxea, N., Martínez de Arano, I., & Arias-González, A. (2021). Forest productivity and associated soil ecosystem services remain altered 15years after mechanized site preparation for reforestation with *Pinus radiata*. *Soil and Tillage Research*, *213*, Article 105150. https://doi.org/10.1016/j.still.2021.105150

Gobierno Vasco-Eusko Jaurlaritza. (2022). *Mapa forestal 2022-2022ko Baso mapa*.

Hansson, L., Šimůnek, J., Ring, E., Bishop, K., & Gärdenäs, A. I. (2019). Soil compaction effects on root-zone hydrology and vegetation in boreal forest clearcuts. *Soil Science Society of America Journal*, *83*, S105–S115. https://doi.org/10.2136/sssaj2018.08.0302

Hansson, L. J., Koestel, J., Ring, E., & Gärdenäs, A. I. (2018a). Impacts of off-road traffic on soil physical properties of forest clear-cuts: X-ray and laboratory analysis. *Scandinavian Journal of Forest Research*, *33*, 166–177. https://doi.org/10.1080/02827581.2017.1339121

Hansson, L. J., Ring, E., Franko, M. A., & Gärdenäs, A. I. (2018b). Soil temperature and water content dynamics after disc trenching a sub-xeric Scots pine clearcut in central Sweden. *Geoderma*, *327*, 85–96. https://doi.org/10.1016/j.geoderma.2018.04.023

Hernanz, J. L., Peixoto, H., Cerisola, C., & Sánchez-Girón, V. (2000). An empirical model to predict soil bulk density profiles in field conditions using penetration resistance, moisture content and soil depth. *Journal of Terramechanics*, *37*, 167–184. https://doi.org/10.1016/S0022-4898(99)00020-8

Horn, R., Way, T., & Rostek, J. (2003). Effect of repeated tractor wheeling on stress/strain properties and consequences on physical properties in structured arable soils. *Soil and Tillage Research*, *73*, 101–106. https://doi.org/10.1016/S0167-1987(03)00103-X

Horswell, M., & Quinn, N. (2003). *Minimising sediment delivery to rivers: A spatial modelling approach developed for commercial forestry*. In Bruen, M. (ed.) Diffuse Pollution and Basin Management. Proceedings of the 7th International Specialised IWA Conference, Dublin, Ireland. ISBN 1902277767

Jourgholami, M., Ghassemi, T., & Labelle, E. R. (2019). Soil physio-chemical and biological indicators to evaluate the restoration of compacted soil following reforestation. *Ecological Indicators*, *101*, 102–110. https://doi.org/10.1016/j.ecolind.2019.01.009

Konert, M., & Vandenberghe, J. (1997). Comparison of laser grain size analysis with pipette and sieve analysis: A solution for the underestimation of the clay fraction. *Sedimentology*, *44*, 523–535. https://doi.org/10.1046/j.1365-3091.1997.d01-38.x

Labelle, E. R., Hansson, L., Högbom, L., Jourgholami, M., & Laschi, A. (2022). Strategies to mitigate the effects of soil physical disturbances caused by forest machinery: A comprehensive review. *Current Forestry Reports*, *8*, 20–37. https://doi.org/10.1007/s40725-021-00155-6

Labelle, E. R., & Jaeger, D. (2012). Quantifying the use of brush mats in reducing forwarder peak loads and surface contact pressures. *Croatian Journal of Forest. Engineering*, *33*, Article 249

Landsberg, J. D., Miller, R. E., Anderson, H. W., & Tepp, J. S. (2003). *Bulk density and soil resistance to penetration as affected by commercial thinning in northeastern Washington* (No. PNW-RP-551). U.S. Department of Agriculture, Forest Service, Pacific Northwest Research Station. https://doi.org/10.2737/PNW-RP-551

Loveland, P., & Webb, J. (2003). Is there a critical level of organic matter in the agricultural soils of temperate regions: A review. *Soil and Tillage Research*, *70*, 1–18. https://doi.org/10.1016/S0167-1987(02)00139-3

Ministry for Ecological Transition and Demographic Challenge. (2008). *IFN4, 4th Spanish Forest Inventory*.

Montero, G. (2018). *Breve resumen del proceso repoblador en España*. Distrito Forestal. https://distritoforestal.es/historia-forestal/breve-resumen-del-proceso-repoblador-en-espana

Naghdi, R., Solgi, A., Zenner, E. K., Najafi, A., Salehi, A., & Nikooy, M. (2017). Compaction of forest soils with heavy logging machinery. *Silva Balcanica*, *18*, 25–39.

Page-Dumroese, D. S., Abbott, A. M., & Rice, T. M. (2009). *Forest soil disturbance monitoring protocol: Volume I: Rapid assessment* (General Technical Report WO-GTR-82a). USDA Forest Service.

Page-Dumroese, D. S., Sanchez, F. G., Udawatta, R. P., Perry, C. H., & González, G. (2021). Soil health assessment of forest soils. In D. L. Karlen, D. E. Stott, & M. M. Mikha (Eds.), *Soil health series: Approaches to soil health analysis* (pp. 100–138). Soil Science Society of America, Inc.

Pagliai, M., Marsili, A., Servadio, P., Vignozzi, N., & Pellegrini, S. (2003). Changes in some physical properties of a clay soil in Central Italy following the passage of rubber tracked and wheeled tractors of medium power. *Soil and Tillage Research*, *73*, 119–129. https://doi.org/10.1016/S0167-1987(03)00105-3

Peng, X., & Horn, R. (2008). Time-dependent, anisotropic pore structure and soil strength in a 10-year period after intensive tractor wheeling under conservation and conventional tillage. *Journal of Plant Nutrition and Soil Science*, *171*, 936–944. https://doi.org/10.1002/jpln.200700084

Picchio, R., Mederski, P. S., & Tavankar, F. (2020). How and how much, do harvesting activities affect forest soil, regeneration and stands? *Current Forestry Reports*, *6*, 115–128. https://doi.org/10.1007/s40725-020-00113-8

Picchio, R., Mercurio, R., Venanzi, R., Gratani, L., Giallonardo, T., Lo Monaco, A., & Frattaroli, A. (2018). Strip clear-cutting application and logging typologies for renaturalization of pine afforestation—A case study. *Forests*, *9*, Article 366. https://doi.org/10.3390/f9060366

Prats, S. A., Malvar, M. C., Coelho, C. O. A., & Wagenbrenner, J. W. (2019). Hydrologic and erosion responses to compaction and added surface cover in post-fire logged areas: Isolating splash, interrill and rill erosion. *Journal of Hydrology*, *575*, 408–419. https://doi.org/10.1016/j.jhydrol.2019.05.038

R Core Team. (2022). *R: A language and environment for statistical computing*. R Foundation for Statistical Computing. https://www.R-project.org/

Sands, R., Greacen, E. L., & Gerard, C. J. (1979). Compaction of sandy soils in radiata pine forests. I A penetrometer study. *Aust. J. Soil Res.*, *17*, 101–113.

Shestak, C. J., & Busse, M. D. (2005). Compaction alters physical but not biological indices of soil health. *Soil Science Society of America Journal*, *69*, 236–246. https://doi.org/10.2136/sssaj2005.0236

Sinnett, D., Morgan, G., Williams, M., & Hutchings, T. R. (2008). Soil penetration resistance and tree root development. *Soil Use and Management*, *24*, 273–280. https://doi.org/10.1111/j.1475-2743.2008.00164.x

Soil Science Division Staff. (2017). *Soil survey manual* (Agriculture Handbook No. 18). USDA-Natural Resources Conservation Service.

Soil Survey Staff. (2022). *Keys to soil taxonomy* (13th ed.). USDA-Natural Resources Conservation Service.

Solgi, A., Naghdi, R., Marchi, E., Laschi, A., Keivan Behjou, F., Hemmati, V., & Masumian, A. (2019). Impact assessment of skidding extraction: Effects on physical and chemical properties of forest soils and on maple seedling growing along the skid trail. *Forests*, *10*, Article 134. https://doi.org/10.3390/f10020134

Thees, O., & Olschewski, R. (2017). Physical soil protection in forests - insights from production-, industrial- and institutional economics. *For. Policy Econ.*, *80*, 99–106. https://doi.org/10.1016/j.forpol.2017.01.024

Vadell, E., Pemán, J., Verkerk, P. J., Erdozain, M., & de-Miguel, S. (2022). Forest management practices in Spain: Understanding past trends to better face future challenges. *Forest Ecology and Management*, *524*, Article 120526. https://doi.org/10.1016/j.foreco.2022.120526

Zemke, J. J., Enderling, M., Klein, A., & Skubski, M. (2019). The influence of soil compaction on runoff formation. A case study focusing on skid trails at forested Andosol sites. *Geosciences*, *9*, Article 204. https://doi.org/10.3390/geosciences9050204

8

Soil Health After >25 Years of Sewage Sludge Application in Arable Crops in Northern Spain

Ana Simoes-Mota, Armelle Zaragüeta, Rodrigo Antón, Luis Orcaray, Alberto Enrique, Isabel S. de Soto, and Iñigo Virto

Chapter Overview

The application of urban sewage sludge (SS) to agricultural soils is considered an economical and sustainable disposal method, as it can help prevent environmental pollution while recycling nutrients and decreasing the use of mineral fertilizers. Usually, SS is rich in organic carbon and plant nutrients and can promote soil microbial activity. Therefore, using SS as a fertilizer can improve some soil properties related to soil organic carbon (SOC) cycling, such as aggregation or water retention. However, SS can also include some trace metals, soluble salts, and emerging contaminants and may display nutrient imbalances, which implies both environmental and crop production risks.

To study such consequences and to assess soil health in relation to the use of different doses and frequencies of SS application to agricultural sub-humid rainfed agrosystems, an experimental field was established near Pamplona (Spain) in 1992. The plot supports a 4-year rainfed rotation on a calcareous soil. This chapter summarizes the results of the monitoring of several soil physical, chemical and biological indicators and their relation to crop yields over more than 25 years.

The results showed contrasting trends for the various indicators. Surface compaction was reduced with the use of SS, while the content of major nutrients (nitrogen and phosphorus) and organic carbon increased with increasing doses of SS compared with an unfertilized control and conventional mineral fertilization. At the same time, aggregate stability, porosity >9 µm, and the abundance and size of earthworms were reduced, likely in relation to the observed increase in

Soil Health Series: Volume 5 Soil Health and Sustainability in Spain and Portugal, First Edition.
Edited by Iñigo Virto and Rodrigo Antón.

Published 2025 by Soil Science Society of America, Inc. and John Wiley & Sons, Inc.

electrical conductivity and some trace metals. Yields were higher with mineral fertilization and moderate-to-low doses of SS than with higher doses of SS or no fertilization at all. Other indicators considered sensitive, such as microbial functional diversity and bulk density, were not affected by the type or dose of fertilization. A dose-dependent effect of SS application was also observed, with the highest doses having the most acute positive (i.e., organic C storage) and negative (i.e., loss of aggregate stability) effects. The effects of intermediate and low doses were closer to those obtained by using agronomic recommendations, yielding adequate results in soil indicators and good yields.

In conclusion, the effect of SS on soil health is dependent on the dose and frequency of its application, and it can differ depending on the soil functions being assessed in the agrosystem under study. While climate regulation through organic carbon stabilization might be better achieved with the highest doses and frequencies, similar success would not be seen for crop productivity (as measured from yields), erosion and pollution control (as expressed in soil structural stability or trace metals accumulation), or habitat provision (as observed from earthworms' biomass and density). Intermediate doses and/or frequencies (equivalent to 10–20 Mg ha^{-1} $yr.^{-1}$) closer to the ones used agronomically in the region (20 Mg/ha) performed better in those functions, while also storing carbon.

Introduction

Soils represent a fundamental part of life on Earth. When used for agriculture, they provide important services that have value beyond food production, such as the control of landscape degradation and the vulnerability to (and regulation of) climate (Adhikari & Hartemink, 2016; Dominati et al., 2010, 2014). However, many agricultural soils, and many of those in the Mediterranean region in particular, face threats that increase soil susceptibility to degradation processes, hampering the adequate provision of soil ecosystem services. Some of these threats are related to agricultural management (Virto et al., 2015).

In this framework, the slogan used by the European Union to spread awareness about soils within communities—"Caring for soils is caring for life" (Veerman et al., 2020)—is a well-chosen analogy with a sentimental impact. It shortens the distance between soil degradation in agriculture and the effect on human communities. Unfortunately, the complexity of the processes and interactions that occur in soil limits determining whether a soil is healthy to a specific time and conditions. Extrinsic factors (especially climate, parent material, topography, and hydrology) influence the potential value of soil properties and their response to short-term changes such that there is still no accepted universal method for measuring soil health (Bünemann et al., 2018).

In general, one crucial step in assessing soil health is the identification of adequate indicators that are able to represent key soil functions that can be affected by changes in management or other local conditions. Global studies on these indicators (e.g., Bai et al., 2018; Karlen et al., 2013) highlight that the indicators most frequently proposed for soil health assessment studies are related to soil porosity, nutrients, water retention, and SOC storage and dynamics (Bongiorno et al., 2019). The relationship between SOC and soil health explains the important role of C-related indicators in these assessments (Coll et al., 2011; Drobnik et al., 2018; Lorenz et al., 2019). Still, any indicator lacks relevance if it is not contextualized in terms of management practices and intrinsic properties. For example, the soil health assessment in agrosystems under conservation agriculture in Navarre (Spain; Apesteguía et al., 2017; Imaz et al., 2010) showed that the most sensitive indicators can vary rapidly with changes in the agrosystem management (in this case, from rainfed to irrigated cultivation).

In this context, the management of soil organic matter seems highly relevant for maintaining or improving soil functions in general and in Mediterranean agricultural areas in particular (Romanyà & Rovira, 2011). In agricultural soils, increasing the organic carbon content is known to improve their fertility and, in some cases, their water retention capacity (Yagüe et al., 2016; Yost & Hartemink, 2019), indirectly enhancing agricultural productivity and food security and eventually promoting a virtuous carbon cycle, in which effective policies for SOC storage are promoted beyond socioeconomic barriers (Bruni et al., 2021, 2022). The application of different types of organic residues from urban and agricultural subproducts (Torri et al., 2014) and the use of conservation agriculture (Lal, 2016), organic agriculture (Altieri et al., 2015), crop rotations (Land et al., 2017), and intercropping (Martínez-Mena et al., 2021; Sánchez-Navarro et al., 2022) are some of the strategies that have been considered with the aim of achieving a positive SOC balance in the medium-long term.

The application of SS to agricultural soils is therefore seen as an economical and sustainable sludge management option. When managed adequately, it can enable a controlled deposition of residues, while preventing environmental pollution, recycling nutrients, and reducing the use of mineral fertilizers (Simoes-Mota et al., 2021). As a source of carbon and other nutrients, SS can promote soil biological activity, including that of soil fauna and microorganisms. As a result, other properties related to SOC cycling, such as aggregation, can also be improved. These improvements can result in favorable changes in the soil resistance to erosion or water retention, among other benefits.

However, the composition of SS can also include some trace metals; different emerging pollutants, including microplastics and pharmaceuticals; and soluble salts. In addition, SS may display nutrient imbalances. These factors represent risks not only for the environment as a whole, but also to soil functioning and, ultimately, to crops (Zaragüeta et al., 2021). As such, some long-term studies on the effect of carbon-rich residues on soil health (Obriot et al., 2016) have reported an overall positive effect but have also noted negative effects related to soil contamination or grain quality that vary with the type of residue used as an amendment. Awareness of these risks has resulted in a large number of studies focusing on the potential consequences of SS for soil chemical properties and the accumulation of pollutants (e.g., Hechmi et al., 2020; Urra, Alkorta, Mijangos, et al., 2019b). These studies report heterogeneous results depending on the amount and quality of SS used, the type of soil and soil management, and the crops considered.

Aware of these risks, the European Commission implemented a specific EU Directive (Directive 86/278/EEC) that includes regulation on the use of SS on agricultural land and limits the allowable concentration of some toxic elements (European Commission, 1986). Other national and regional regulations exist worldwide. However, less attention has been paid to the impacts on other soil health indicators or their relation to soil functions (Obriot et al., 2016). It can be expected that, as established for other agricultural practices altering soil health, the consequences of SS application can vary depending on the soil, climate, and agronomic characteristics of specific agrosystems, as well as the composition and dose of SS (Doran & Parkin, 1996; Guo et al., 2019). Therefore, all the dimensions of soil properties (chemical, physical, and biological) must be considered in any study of this kind (Adhikari & Hartemink, 2016).

In this sense, long-term experiments (LTEs) can provide valuable data for soil health assessment, as they allow for an adequate monitoring of different indicators that might be affected by changes in management but respond at different paces and/or have different trends in time (Alhameid et al., 2019; Carof et al., 2022; Guo et al., 2019; Karlen et al., 2013; Rodrigues et al., 2022; Sousa et al., 2022; Valdez et al., 2020). In addition, LTEs allow for an adequate calibration of models aimed at understanding and simulating the response of different soil components such as SOC, nutrients, or chemicals (Bruni et al., 2022; Clivot et al., 2019).

The main objective of the current work was to make a holistic assessment of the state of an agricultural soil in a sub-humid Mediterranean area after >25 years of continuous fertilization with SS at different doses and frequencies, which resulted in a gradient in the total amount of SS added to the soil. In particular, we aimed to (a) test the effect of long-term SS application on a number of physical, chemical, biological, and organic matter–related soil indicators related to different soil functions, and (b) analyze how some of these indicators were related to crop yields. We hypothesized that the amount of SS used was a major factor in soil health after >25 years, and that the use of one or another combination of dose and frequency could be recommended, depending on the soil functions considered of interest at the study site.

Materials and Methods

In 1992, 2 years after the Arazuri Wastewater Treatment Plant started operating in Pamplona (northern Spain), the Pamplona Commonwealth established a number of experimental fields next to the facility itself. The goal was to use these trials as LTEs to enable collecting data to evaluate the effects of using SS on cropping fields in the long term. The LTEs using SS were carried out in an area of 2 ha (Figure 8.1).

From the beginning of the trials, INTIA (Regional Institute for Transfer and Innovation in the Food Chain in Navarre), together with universities and research centers, provided technical advice for practitioners and extensionists that has also served as a source of information for farmers interested in using SS in the region.

The experimental design used in the LTE is a randomized factorial block design with eight treatments and six replicates (n = 6, Figure 8.1), consisting of plots of

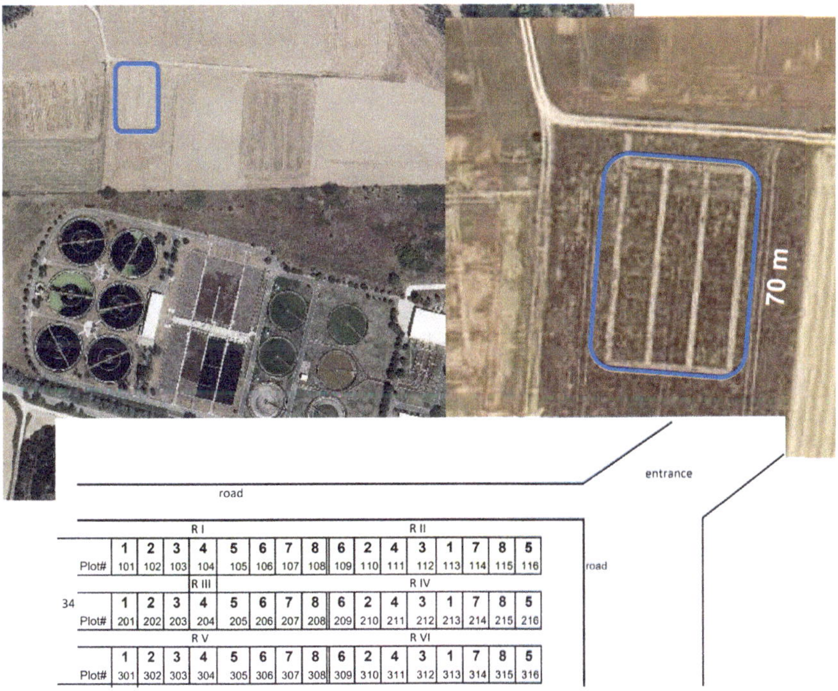

Figure 8.1 Arazuri Wastewater Treatment Plant and Arazuri long-term experiment field used in this study (top), and schematic representation of the Arazuri experimental field, with the three replicates used in this study highlighted in green (bottom).
Source: SITNA (2023).

$35\,m^2$ ($10\,m \times 3.5\,m$). For most of the studies conducted on soil properties, the first three replicates have been used.

The six treatments tested are the result of combining two different SS doses (40 and 80 Mg/ha) and three frequencies of application (every year, every 2 years, and every 4 years), plus two control treatments: one that uses the conventional mineral fertilization (MF, with 46% urea and ammonium sulfate) and a baseline control without SS application or mineral fertilization (C0), as shown in Table 8.1. These doses and frequencies represent a range that includes the common practices in the region (plots with low doses or frequencies) and the highest possible rates (plots with high doses and frequencies). As such, the treatments representing an average annual dose of 20 Mg/ha were considered representative of the usual agronomic dose, whereas the treatment adding 80 Mg/ha annually represented the most excessive dose manageable.

In a study aiming to determine the fate of the added organic carbon 28 years after the onset of the experiment, the equivalent total accumulated input of carbon for treatments corresponding to average annual doses of 10, 20, 40, and 80 Mg/ha, plus MF and C0 (see Table 8.1) was calculated. This calculation included carbon from SS and from crop residues for each average annual dose (10SS, 20SS, 40SS, 80SS, C0, and MF, respectively; Simões-Mota et al., 2024).

Table 8.1 Experimental Design for Fertilization With and Without SS at the Arazuri LTE.

Treatment and plots	Type of fertilization	Frequency (seasons)	Average SS annual dose (tag)	Cumulative SS dose	Accumulated C inputs (Simoes-Mota et al., 2024)
			(Mg/ha)	(Mg C/ha, average over 28 yr)	
40-1 (1)	40 Mg SS/ha	All seasons	40 (40SS)	1,000	110
40-2 (2)	40 Mg SS/ha	Every 2	20 (20SS)	540	79.1
40-4 (3)	40 Mg SS/ha	Every 4	10 (10SS)	250	57.9
80-1 (4)	80 Mg SS/ha	All seasons	80 (80SS)	2,000	181
80-2 (5)	80 Mg SS/ha	Every 2	40 (40SS)	1,000	110
80-4 (6)	80 Mg SS/ha	Every 4	20 (20SS)	540	79.1
MF (7)	Mineral fertilization	All seasons	0 (0SS)	0	38.8
C0 (8)	No fertilization	Never	0 (0SS)	0	26.8

Note. The treatments and plots are shown in Figure 8.1. The accumulated carbon inputs include accumulated carbon from SS and from crop residues.

The crops included in the most common rainfed 3-year rotation in the study area (cereal-cereal-cereal-noncereal; Table 8.2) were used in the LTE. These crops were wheat (*Triticum aestivum* L.) and barley (*Hordeum vulgare* L.) as cereals and sunflower (*Helianthus annuus* L.) as the noncereal. Crop management (phytosanitary treatments, tillage, harvesting) was done conventionally and adapted to crop needs. Wheat sampling was carried out in June in the year of study at harvest.

To determine carbon inputs from crops, crop yield sampling was carried out at complete physiological maturity (5 days before harvest) in each year of study. A combine adapted to the plot size was used to record yield, avoiding the edges of the plot, and grain yields were recorded. To obtain yield data on a dry-mass basis, grain weights were taken directly from the combine and grain samples were collected to analyze their water content. The carbon inputs from crop residues (Table 8.1) were then calculated following Bruni et al. (2021, 2022) with the approach first described by Bolinder et al. (2007) and later adapted by Clivot et al. (2019) for crops in France. This methodology allocates carbon in different pools, splitting carbon inputs from crop residues after harvest into aboveground and belowground, using measured dry matter yields and estimations of the shoot-to-root ratio and harvest indexes of the crops. For correlation with indicators and to

Table 8.2 Crop Rotation over the Years in the Arazuri LTE.

Year	Crop	Year	Crop
1992–1993	Barley	2006–2007	Wheat
1993–1994	Barley	2007–2008	Barley
1994–1995	Barley	2008–2009	Sunflower
1995–1996	Pea	2009–2010	Wheat
1996–1997	Wheat	2010–2011	Barley
1997–1998	Wheat	2011–2012	Oats
1998–1999	Barley	2012–2013	Wheat
1999–2000	Barley	2013–2014	Wheat
2000–2001	Sunflower	2014–2015	Barley
2001–2002	Wheat	2015–2016	Oats
2002–2003	Wheat	2016–2017	Wheat
2003–2004	Barley	2017–2018	Wheat
2004–2005	Sunflower	2018–2019	Wheat
2005–2006	Wheat	2019–2020	Sunflower

take the cumulative effect of SS on soil properties into consideration, yields for wheat (the crop most repeated in the rotation) in year 25 were used.

The properties of SS are described in Simoes-Mota et al. (2021, 2022) and Zaragüeta et al. (2021) and are summarized in Table 8.3. The doses were calculated in 1992 based on the restrictive regulation of 250 nitrogen fertilizer units (NFUs) per hectare and year, without considering the nitrogen yield for the year of application. As such, the treatment using 40 Mg/ha per year (40-1) was set as the one corresponding to this dose, and those receiving different SS amounts were used to follow the evolution of soil and crops above and below the recommended

Table 8.3 Physicochemical Properties of the SS Used in This Study in 2018.

pH	8.16 ± 0.03
Electrical conductivity (μS/cm³)	1,795 ± 28.0
Dry material (%)	18.1 ± 0.40
C/N	5.35 ± 0.08
Nutrients (% of dry mass)	
Total N	5.85 ± 0.13
Ammonium-N	0.75 ± 0.02
Phosphorus (P_2O_5)	5.59 ± 0.22
Potassium (K_2O)	0.62 ± 0.05
Iron (Fe)	1.68 ± 0.04
Calcium (CaO)	7.98 ± 0.29
Magnesium (MgO)	1.10 ± 0.05
Trace metals (mg/kg dry mass)	
Cadmium (Cd)	0.88 ± 0.09
Copper (Cu)	187 ± 11.0
Nickel (Ni)	32.1 ± 0.77
Lead (Pb)	39.0 ± 1.20
Zinc (Zn)	874 ± 38.0
Mercury (Hg)	0.03 ± 0.03
Chromium (Cr)	58.3 ± 3.20

Note. Adapted from Simoes-Mota et al. (2021, 2022), and Zaragüeta et al. (2021). Values are the mean annual amounts ± standard deviation (n = 3). NA, not analyzed.
The nitrogen equivalence of the added doses varied owing to the SS composition changes over time (Table 8.4).

Table 8.4 Nitrogen Fertilizer Units (NFUs) Provided by the Study Doses According to the Nitrogen Concentration in SS as Used in 1992 (Study Start) and in 2018.

SS dose	NFU 1992	NFU 2018
40 Mg/ha	250	480
80 Mg/ha	500	960

dose at the time. The frequencies of 2 and 4 years were established according to the common practices of farmers in the area and the recommendations of the entity that supplied the SS.

These characteristics accorded with the Spanish national regulations on the maximum concentration of trace metals in SS used as organic amendments in agriculture. The SS used in this study came from the wastewater treatment plant corresponding to the city of Pamplona and its metropolitan area, which have approximately 335,000 inhabitants. Wastewater underwent primary treatment, followed by biological treatment and nitrification and denitrification treatment. The SS was applied each campaign in September, 3–4 weeks before sowing, using a 3.5-m-wide spreader trailer, followed by moldboard plowing down to 30 cm. MF was applied before sowing.

Climate in the area is temperate Mediterranean (Papadakis, 1961). The mean annual rainfall is 750 mm/yr and the mean annual Thornthwaite's evapotranspiration is 687 mm/yr, according to the Navarra government data (Gobierno de Navarra, 2021). The mean annual temperature is 13.3 °C.

The soil in this field is a well-drained Calcaric Cambisol (IUSS Working Group WRB, 2022). It contains ~20% calcium carbonate, and the topsoil has a clay-loam texture (31% clay, 30% silt, 39% sand). The most relevant soil characteristics in the upper 0- to 30-cm layer are described in Table 8.5.

For the results reported in this chapter, soil was sampled at each replicate ($n = 3$) at 0–15 cm and at 15–30 cm between 25 and 28 years after the onset of the experiment. Most samplings were carried out in September, at 2 months after harvesting, and at the furthest moment from the last soil alteration. Undisturbed samples were collected in June (i.e., before harvesting) to minimize the effect of harvesting machinery.

As described by Simoes-Mota et al. (2021, 2022), disturbed and undisturbed samples were collected for analyses of soil properties encompassing physical, chemical, and biological indicators (Figure 8.2). A shovel or Edelman type auger (Ø = 5 cm) was used for collecting disturbed soil samples. To ensure representativeness at each replicate plot, three subsamples were collected and

Table 8.5 Soil Physical and Chemical Properties (0–30 cm).

Properties	Mean annual values ± standard deviation (n = 3)
pH (water 1:2.5)	8.67 ± 0.03
Electrical conductivity (μS/cm at 25 °C) (soil: water extract 1:2.5)	169 ± 10.0
Bulk density (g/cm^3)	1.59 ± 0.08
Carbonates (%)	16.0 ± 2.10
Clay (%)	27.7 ± 1.03
Organic C (%)	1.35 ± 0.02

Note. Adapted from Simoes-Mota et al. (2021).

Figure 8.2 Sampling in Arazuri experimental field for disturbed samples (top left) and earthworms (top center and right) and the view of the field and plots in June (bottom left), when undisturbed samples were collected using Kubiena boxes (bottom center and bottom right) for preparation of thin sections.

then combined into a composite sample. Part of the composite sample was stored at 4 °C for biological analyses. Another portion of the sample was gently pushed through a 6-mm sieve to obtain natural aggregates that were then air-dried. The rest of the sample was air-dried and ground to pass a 2-mm sieve for chemical analyses. Bevel-edged steel rings (Ø = 5 cm, total volume = 100 cm^3) were used to collect undisturbed core samples for soil bulk density (ρ_b), permeability, and water retention analysis.

Undisturbed samples were also collected with Kubiena boxes for thin section preparation. Earthworms were collected from two 20 cm × 20 cm × 30 cm soil blocks extracted from each plot with a shovel (Figure 8.2).

The soil physical indicators analyzed included those related to aggregation, water storage, and transit, as well as compaction and porosity. The methods used are described in detail in Simoes-Mota et al. (2021). In short, dry aggregate stability was determined for the 0- to 30-cm depth (by mixing in equal parts samples from the 0- to 15-cm and 15- to 30-cm depths), from 100 g of air-dried aggregates <6 mm using shaking in a rotary Retsch VS 100 device (Retsch GmbH & Co., Haan, Germany) and recovering stable aggregates in sieves with 4-, 2-, 1-, 0.5-, and 0.25-mm openings. After addition of a shower-like flux (6 L/min) of distilled water to samples, the same device was used to study wet aggregate stability. Samples were previously homogenized in a mechanical sample divisor (Retsch GmbH & Co., Haan, Germany). Two indexes of aggregate dry and wet stability were calculated: the mean weight diameter (MWD) after dry and wet sieving (as in Bosch-Serra et al., 2017) and the mass proportion of water-stable aggregates (WSAs) > 0.25 mm (Franzluebbers, 2002).

Bulk density (ρ_b) was determined with the core method. Soil water retention (SWR) was determined using pressure plate extractors (Soil Moisture Equipment Corp., Santa Barbara, CA) at −33, −50, and −90 kPa from undisturbed samples and at −1,500 kPa from air-dried sieved (<2 mm) samples. From these data, the soil available water-holding capacity was calculated using water retention at −33 and −1,500 kPa. From these data, the model proposed by Rose (1966) was used to estimate the equivalent pore diameter corresponding to each of the water potentials, as described by Bescansa et al. (2006) and Fernández-Ugalde et al. (2009). According to this model, the equivalent pore diameter is 9 μm for −33 kPa and 0.2 μm for −1,500 kPa.

Finally, penetration resistance (PR) was measured using a field penetrometer (Rimik CP20, Agridy Rimik Pty Ltd., Toowoomba, QLD, Australia) at 9 points per field replicate to a depth of 60 cm. Measurements were recorded every 15 mm, and even soil moisture conditions were verified. Values of PR for fixed depths were calculated as weighted depth averages.

Thin sections of soil were prepared as described in Benyarku and Stoops (2005; Figure 8.2). The protocol for the image analysis described by Simoes-Mota et al. (2022) was used on a digitalized binary image per section to determine macroporosity, using ImageJ (Rasband, 2008) and selecting five random images (10 mm × 10 mm) thin sections as in the study by Virto et al. (2013; Figure 8.3). From these data, the pore-size distribution (in sizes corresponding to apparent

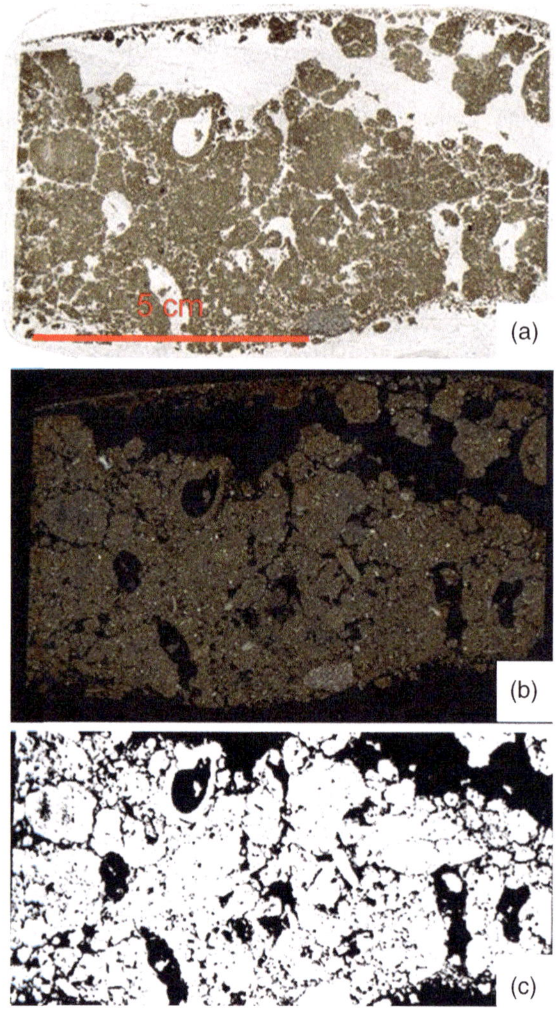

Figure 8.3 Thin section (SS-fertilized treatment): (a) under plain polarized light, (b) under crossed polarizer light, and (c) binarized with ImageJ (A. Simoes-Mota, A. Zaragüeta, R. Antón, L. Orcaray, A. Enrique, I. de Soto, I. Virto).

diameters of 100–200 µm, 200–400 µm, and >400 µm) was determined using the Quantim4 library (Simoes-Mota et al., 2022; Vogel, 2008).

Chemical analyses were done on air-dried and sieved (<2 mm) samples. Total nitrogen was analyzed using the Kjeldahl digestion method. Available phosphorus was determined as described by Olsen and Sommers (1982). Exchangeable potassium and sodium were quantified using atomic absorbance after extraction with NH_4OAc 1 N (Knudsen et al., 1982). The soil electrical conductivity (EC) and soil pH were measured in distilled water (1 : 2.5). Soil pH was determined with a Crison GLP22 pH meter (Crison Instruments, S.A., Barcelona, Spain). Conductivity was read with a Crison GLP32 conductivity meter (Crison Instruments, S.A.).

Carbonate concentration was measured in a modified Bernard's calcimeter as described by (Sherrod et al., 2002). DTPA($C_{14}H_{23}N_3O_{10}$)-extractable concentrations of copper, manganese, nickel, zinc, cadmium, and lead were analyzed at 0–15 cm, following the standard procedure of ISO 14870 : 2001 Zaragüeta et al. (2021).

The concentration in total organic carbon (TOC) was determined by wet oxidation (Nelson & Sommers, 1982). Particulate organic matter (POM; Cambardella & Elliot, 1992) was isolated based on its size (>53 μm) by dispersion and sieving of 10 g of air-dried soil. Organic carbon in the form of POM (POM-C) was determined by wet oxidation. The 20 cm × 20 cm × 30 cm soil blocks were crumbled by hand to collect earthworms (Figure 8.2), which were weighed and counted for each field replicate (Baker & Lee, 1993). Total biomass of earthworms (g/m^2), average size (grams per individual), and abundance (number of individuals per square meter) were determined from these data.

The study of soil microbial populations included analyzing their mass and functional diversity, as described by Simoes-Mota et al. (2021), and microbial biomass carbon (MBC) was determined following Vance et al. (1987). Community-level physiological profiles in fresh samples were determined from carbon-source utilization patterns with a Biolog Ecoplates™ microplating system (Biolog, Hayward, CA, USA) incubated at 30 °C. Color development was read at 595 nm using a microplate reader (Thermo Scientific Multiskan® EX). The mean reading of each well's absorbance value served to determine the average well color development index. The number of wells with corrected absorbance values >0.25 at the exponential phase of microbial growth (59 h from the start) represented the number of substrates used by the community, equivalent to species richness according to Zak et al. (1994).

Statistical Analysis

The effect of SS application on soil health was assessed at different levels. First, for the treatments described as 40-1, 40-2, 40-4, 80-1, 80-2, 80-4, MF, and C0 in Table 8.1, a univariate analysis of variance (ANOVA) was done to examine the treatment effect on the general physical, chemical, and biological indicators for the two soil depths studied. Duncan's test ($P < 0.05$) was used to detect homogeneous groups.

Second, the most sensitive soil health indicators under the study conditions were selected, as described by Imaz et al. (2010) and Simoes-Mota et al. (2021). This process started with principal component analysis (PCA) of the variables with a significant effect ($P < 0.05$) from SS application or fertilization (Shukla et al., 2006). The determinant of the correlation matrix of the standardized values of each variable then served to detect correlations among variables. Factor scores for each sample point were calculated, and ANOVA was performed on the scores' variables. Soil attributes were then assigned to factors based on the highest loadings for each (Shukla et al., 2006). For each retained factor, highly weighted attributes were selected as possible soil health indicators. We considered attributes to be highly weighted when they were within 10% of the highest factor loading, as in Andrews et al. (2004). The correlation of these indicators with yield was tested using a simple correlation test (Pearson's coefficient).

Results

The use of SS significantly affected nearly all the soil properties studied. For the physical quality indicators, differences were found related to SWR_{-50kPa} and PR in 0–15 cm; WSAs and MWD for 0–30 cm (Tables 8.6 and 8.7); and SWR_{-50kPa}, PØ < 0.2, and PØ > 0.9 µm for 15–30 cm (Table 8.7). No direct relationship was found between the dose and/or frequency of SS addition, and an effect was not found for all parameters. For instance, at 0–15 cm, while PR was clearly reduced by the use of SS compared with C0 or MF, WSAs showed the lowest values with 40-1, and the highest ones corresponded to C0 and MF or to the treatments with higher doses, similar to SWR_{-50kPa}. MWD was smaller in 80-1 and 80-2 than in any other treatment, and C0 had the highest values together with 40-2 and 40-4. At 15–30 cm, MF and C0 had the lowest values for SWR_{-50kPa}, and the highest proportion of pores PØ > 0.9. The smallest quantified pores (PØ < 0.2) were most abundant in 80-2, followed by C0 and the other treatments with SS, with the exception of 80-1, which had the smallest proportion together with MF. Pores of >400 µm were the most abundant ones in all treatments compared (C0, MF, 20SS, and 80SS; Figure 8.4), but no differences were observed between these contrasting treatments for any pore category analyzed using thin sections (Figure 8.4).

Under the studied conditions, fertilization with SS significantly affected almost all the selected chemical soil health indicators (Tables 8.8 and 8.9). In particular, major nutrients (nitrogen, phosphorus, and potassium) were found to be affected by the type of fertilization, although at different rates and depths. For total nitrogen, the lowest values were observed for C0 and the highest for 80-1 (80 Mg/ha SS every year) at 0–15 and 15–30 cm. The other treatments showed intermediate values, with more variability at 15–30 than 0–15 cm. For available phosphorus,

Table 8.6 Results for Bulk Density (BD), Soil Water Retention (SWR) at Different Potentials, and Available Water-Holding Capacity (AWHC).

Treatment	BD	SWR_{-33kPa}	SWR_{-50kPa}	SWR_{-90kPa}	$SWR_{-1,500kPa}$	AWHC
	(g/cm^3)	(g/g)	(g/g)	(g/g)	(g/g)	(L/m^2)
0–15 cm	$P = 0.7063$	$P = 0.1210$	$P = 0.0084$	$P = 0.7405$	$P = 0.1017$	$P = 0.2040$
40-1	1.45 ± 0.01	0.19 ± 0.00	0.11 ± 0.02 (B)	0.17 ± 0.00	0.12 ± 0.00	15.4 ± 1.30
40-2	1.51 ± 0.03	0.18 ± 0.01	0.13 ± 0.00 (B)	0.18 ± 0.00	0.12 ± 0.00	13.4 ± 1.79
40-4	1.52 ± 0.09	0.27 ± 0.01	0.20 ± 0.00 (A)	0.18 ± 0.00	0.13 ± 0.00	19.5 ± 1.90
80-1	1.46 ± 0.07	1.43 ± 0.01	0.21 ± 0.01 (A)	0.18 ± 0.01	0.13 ± 0.01	22.0 ± 3.57
80-2	1.60 ± 0.03	1.59 ± 0.03	0.21 ± 0.00 (A)	0.18 ± 0.01	0.13 ± 0.01	17.2 ± 0.85
80-4	1.43 ± 0.04	1.43 ± 0.04	0.13 ± 0.00 (B)	0.17 ± 0.01	0.12 ± 0.00	16.9 ± 1.68
MF	1.56 ± 0.05	1.56 ± 0.05	0.21 ± 0.00 (A)	0.18 ± 0.01	0.12 ± 0.01	20.3 ± 3.75
C0	1.52 ± 0.02	1.52 ± 0.01	0.20 ± 0.00 (A)	0.17 ± 0.01	0.12 ± 0.01	18.1 ± 2.48
15–30 cm	$P = 0.8108$	$P = 0.0578$	$P = 0.0012$	$P = 0.1635$	$P = 0.7274$	$P = 0.5110$
40-1	1.53 ± 0.07	0.22 ± 0.01	0.20 ± 0.02 (B)	0.19 ± 0.01	0.13 ± 0.01	18.1 ± 2.53
40-2	1.61 ± 0.02	0.21 ± 0.01	0.21 ± 0.02 (A)	0.19 ± 0.01	0.13 ± 0.00	18.9 ± 2.88
40-4	1.59 ± 0.08	0.21 ± 0.02	0.20 ± 0.02 (B)	0.19 ± 0.01	0.13 ± 0.00	18.7 ± 4.33
80-1	1.52 ± 0.06	0.22 ± 0.02	0.21 ± 0.02 (A)	0.20 ± 0.02	0.13 ± 0.00	21.6 ± 2.73
80-2	1.72 ± 0.05	0.18 ± 0.02	0.20 ± 0.02 (B)	0.19 ± 0.01	0.13 ± 0.00	12.5 ± 4.31
80-4	1.62 ± 0.03	0.20 ± 0.02	0.20 ± 0.02 (B)	0.22 ± 0.05	0.12 ± 0.00	16.9 ± 3.41
MF	1.55 ± 0.05	0.21 ± 0.02	0.19 ± 0.02 (C)	0.14 ± 0.06	0.12 ± 0.00	19.5 ± 4.59
C0	1.66 ± 0.03	0.17 ± 0.00	0.19 ± 0.02 (C)	0.19 ± 0.02	0.12 ± 0.00	12.4 ± 1.02

Note. Treatments are defined in the text. Results with different letters are statistically different ($P < 0.05$, Duncan test). Values are the mean ± standard error ($n = 3$).

Table 8.7 Results for Penetration Resistance (PR), Pore Diameter (Ø), Water-Stable Aggregates (WSAs), and Mean Weight Diameter (MWD).

	PR	Pore Ø (%)	WSAs	MWD		
Treatment	(MPa)	<0.2 µm	0.2–0.9 µm	>0.9 µm	(%, 15–30 cm)	(mm, 15–30 cm)
0–15 cm	$P = 0.0460$	$P = 0.7063$	$P = 0.1210$	$P = 0.0084$	$P = 0.0083$	$P = 0.0006$
40-1	457 ± 59 (BC)	37.6 ± 2.02	22.5 ± 1.75	39.9 ± 0.73	78.8 ± 0.75 (C)	0.78 ± 0.04 (AB)
40-2	432 ± 29 (C)	42.5 ± 2.54	20.6 ± 2.38	36.9 ± 0.21	83.3 ± 0.70 (AB)	0.79 ± 0.01 (A)
40-4	512 ± 56 (BC)	47.2 ± 5.64	30.6 ± 2.05	22.2 ± 5.31	86.1 ± 0.41 (A)	0.75 ± 0.02 (AB)
80-1	439 ± 37 (BC)	41.6 ± 5.80	32.4 ± 3.52	25.9 ± 3.78	85.8 ± 0.82 (A)	0.71 ± 0.01 (B)
80-2	460 ± 51 (BC)	50.3 ± 0.75	28.8 ± 0.52	20.9 ± 0.82	81.5 ± 2.81 (B)	0.73 ± 0.02 (B)
80-4	517 ± 80 (CA)	36.1 ± 2.56	24.3 ± 2.02	39.5 ± 1.53	85.9 ± 0.88 (A)	0.79 ± 0.02 (A)
MF	613 ± 70 (BA)	45.0 ± 5.42	32.4 ± 4.52	22.6 ± 1.21	86.0 ± 1.11 (A)	0.77 ± 0.01 (AB)
C0	675 ± 5.3 (A)	40.8 ± 3.14	31.2 ± 2.07	28.0 ± 1.20	83.8 ± 1.21 (AB)	0.82 ± 0.02 (A)
15–30 cm	$P = 0.6730$	$P < 0.0001$	$P = 0.6819$	$P < 0.0001$		
40-1	1,812 ± 163	48.3 ± 3.12 (BC)	28.9 ± 4.67	22.7 ± 6.11 (A)		
40-2	1,753 ± 219	51.5 ± 1.10 (B)	32.1 ± 5.04	16.5 ± 4.06 (BC)		
40-4	1,916 ± 56	50.3 ± 5.19 (B)	30.5 ± 5.94	19.2 ± 4.46 (B)		
80-1	1,656 ± 76	46.1 ± 3.32 (C)	33.6 ± 3.72	20.3 ± 4.81 (B)		
80-2	1,806 ± 149	63.2 ± 3.92 (A)	22.9 ± 7.20	13.9 ± 4.09 (C)		
80-4	1,865 ± 186	51.6 ± 2.38 (B)	28.7 ± 5.20	19.7 ± 3.15 (B)		
MF	2,035 ± 192	45.8 ± 4.05 (C)	30.9 ± 6.29	23.3 ± 2.73 (A)		
C0	1,933 ± 143	54.5 ± 2.69 (B)	22.1 ± 2.42	23.4 ± 5.10 (A)		

Note. All treatments are defined in the text. Results with different letters are statistically different ($P < 0.05$, Duncan test). Values are given as the mean ± standard error ($n = 3$).

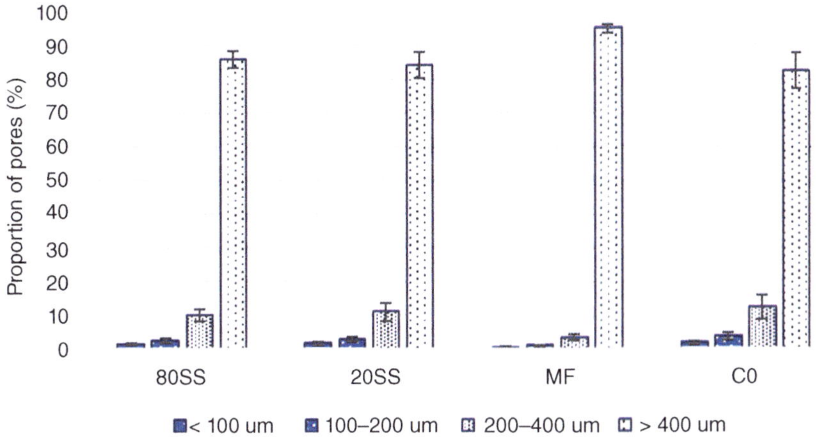

Figure 8.4 Proportion of pores of different equivalent diameter, as calculated from soil thin sections, for the treatments with the highest dose (80SS) and the agronomic dose (20SS) of SS, compared with conventional mineral fertilization (MF) and the unfertilized control (C0). Values are represented as the mean ± standard error ($n = 3$).

differences were observed only at 0–15 cm, with a similar trend of less phosphorus in C0 and MF than in the treatments with SS. A clear correlation was found between SS dose and phosphorus, with 80SS having the highest observed values. For exchangeable potassium, the opposite trend was observed: differences were significant only at 15–30 cm, but not at 0–15 cm. For this parameter and depth, only treatments with 80 Mg/ha spaced by 2 or 4 years (80-2 and 80-4) and MF had lower values than C0 and treatments with 40 Mg/ha (40-1, 40-2, 40-3) or 80 Mg/ha every year (80-1).

With regard to EC and pH, differences were apparent at both depths, with a trend towards higher EC and lower pH with increasing dose of SS compared with C0 and/or MF. As such, the treatment with the highest dose (80-1) displayed the highest EC value and lowest pH, while the results for MF and C0 were the opposite. However, no differences were observed for the total content of carbonates at 0–30 cm, which averaged 161 ± 1.2 g/kg (Table 8.9). For trace metals, differences were observed for manganese, zinc, copper, and nickel (Table 8.9). For the latter three, C0 and MF had the lowest values, and the treatment with the highest dose (80-1) had the highest values. The other SS treatments displayed intermediate values. For manganese, the opposite trend was observed: C0 and 80-2 had the highest values, whereas, MF and other SS rates were associated with intermediate or lower values.

For indicators related to organic carbon and the soil biological indicators (Table 8.10 and Figure 8.5), with the highest SS dose (80-1), TOC was increased by

Table 8.8 Results for Total Nitrogen (TN), Available Phosphorus (Olsen P), Exchangeable Potassium and Sodium, Electrical Conductivity (EC), and pH.

Treatment	TN (mg/kg)	Olsen P (ppm)	Exch. K (g/100 g)	Exch. Na (g/100 g)	EC (µS/cm)	pH
0–15 cm	$P = 0.0010$	$P < 0.0001$	$P = 0.1017$	$P < 0.0001$	$P < 0.0001$	$P < 0.0001$
40-1	1.99 ± 0.03 (B)	83.4 ± 8.31 (BC)	0.35 ± 0.06	0.04 ± 0.01 (A)	183 ± 36.8 (BC)	8.54 ± 0.05 (B)
40-2	1.96 ± 0.07 (B)	73.8 ± 8.35 (B)	0.33 ± 0.02	0.03 ± 0.00 (B)	191 ± 27.2 (ABC)	8.71 ± 0.09 (B)
40-4	2.03 ± 0.06 (B)	96.2 ± 4.78 (CD)	0.34 ± 0.04	0.05 ± 0.01 (A)	203 ± 19.9 (BC)	8.67 ± 0.11 (BC)
80-1	2.22 ± 0.06 (A)	175 ± 16.8 (E)	0.26 ± 0.02	0.02 ± 0.00 (C)	351 ± 26.6 (D)	8.46 ± 0.06 (D)
80-2	1.93 ± 0.08 (B)	114 ± 13.8 (D)	0.32 ± 0.04	0.03 ± 0.01 (B)	214 ± 5.29 (C)	8.61 ± 0.09 (BC)
80-4	1.92 ± 0.06 (B)	78.5 ± 9.33 (B)	0.27 ± 0.00	0.02 ± 0.00 (C)	192 ± 5.29 (BC)	8.65 ± 0.05 (B)
MF	1.83 ± 0.07 (BC)	52.2 ± 6.27 (A)	0.30 ± 0.05	0.02 ± 0.00 (C)	167 ± 16.5 (AB)	8.82 ± 0.02 (A)
C0	1.73 ± 0.05 (C)	37.0 ± 2.18 (A)	0.28 ± 0.04	0.02 ± 0.00 (C)	150 ± 7.89 (A)	8.79 ± 0.03 (AB)
15–30 cm	$P < 0.0001$	$P = 0.1489$	$P = 0.004$	$P = 0.0863$	$P = 0.0287$	$P < 0.0001$
40-1	1.93 ± 0.05 (D)	77.4 ± 4.36	0.30 ± 0.05 (A)	0.04 ± 0.00	224 ± 0.88 (B)	8.29 ± 0.09 (B)
40-2	2.01 ± 0.02 (CD)	70.5 ± 5.59	0.28 ± 0.01 (AB)	0.02 ± 0.00	204 ± 3.11 (BC)	8.45 ± 0.02 (C)
40-4	2.16 ± 0.05 (AB)	93.0 ± 3.91	0.25 ± 0.01 (AB)	0.04 ± 0.01	229 ± 1.76 (B)	8.38 ± 0.02 (BC)
80-1	2.27 ± 0.03 (A)	169 ± 15.2	0.25 ± 0.01 (AB)	0.02 ± 0.00	305 ± 13.8 (A)	8.16 ± 0.06 (A)
80-2	2.10 ± 0.06 (BC)	109 ± 11.1	0.20 ± 0.02 (B)	0.02 ± 0.00	243 ± 1.33 (B)	8.34 ± 0.01 (B)
80-4	1.76 ± 0.07 (B)	65.6 ± 6.47	0.20 ± 0.00 (B)	0.01 ± 0.00	217 ± 11.3 (B)	8.46 ± 0.02 (C)
MF	1.66 ± 0.05 (E)	43.3 ± 6.37	0.19 ± 0.01 (B)	0.02 ± 0.00	195 ± 3.16 (C)	8.55 ± 0.02 (D)
C0	1.99 ± 0.12 (E)	29.5 ± 0.93	0.28 ± 0.04 (AB)	0.02 ± 0.00	180 ± 1.29 (C)	8.57 ± 0.01 (D)

Note. All treatments are defined in the text. Results with different letters are statistically different ($P < 0.05$, Duncan test). Values are given as the mean ± standard error (n = 3).

Table 8.9 Results for Total Carbonates and Available Trace Metals (Mn, Zn, Cu, Ni, Cd, and Pb) at 0–30 cm.

	Carbonates	Available trace metals (mg/kg)					
Treatments	(g/kg)	Mn	Zn	Cu	Ni	Cd	Pb
P-value	0.418	0.002	0.020	0.042	0.028	0.104	0.262
40-1	156 ± 12	8.55 ± 0.1 (AB)	4.02 ± 0.2 (B)	2.74 ± 0.1 (BC)	0.27 ± 0.0 (BC)	0.070 ± 0.0	1.96 ± 0.0
40-2	169 ± 12	8.07 ± 0.0 (B)	3.54 ± 0.2 (BC)	2.65 ± 0.2 (BC)	0.25 ± 0.0 (CD)	0.069 ± 0.0	2.01 ± 0.0
40-4	174 ± 16	7.99 ± 0.5 (B)	4.43 ± 0.2 (B)	2.99 ± 0.1 (B)	0.28 ± 0.0 (BC)	0.070 ± 0.0	2.06 ± 0.1
80-1	160 ± 13	9.00 ± 0.3 (AB)	6.43 ± 0.6 (A)	3.71 ± 0.3 (A)	0.35 ± 0.0 (A)	0.072 ± 0.0	2.08 ± 0.1
80-2	155 ± 11	9.83 ± 1.1 (A)	4.41 ± 0.5 (B)	2.91 ± 0.3 (B)	0.29 ± 0.0 (B)	0.069 ± 0.0	2.02 ± 0.1
80-4	160 ± 9.4	9.43 ± 0.2 (AB)	3.02 ± 0.2 (C)	2.31 ± 0.1 (C)	0.25 ± 0.0 (CD)	0.065 ± 0.0	1.88 ± 0.1
MF	154 ± 10	9.44 ± 0.2 (AB)	2.04 ± 0.1 (D)	1.79 ± 0.1 (D)	0.22 ± 0.0 (DE)	0.062 ± 0.0	1.71 ± 0.0
C0	160 ± 6.4	9.53 ± 0.1 (A)	1.67 ± 0.1 (D)	1.66 ± 0.0 (D)	0.21 ± 0.0 (E)	0.060 ± 0.0	1.76 ± 0.1

Note. Treatments with different letters are statistically different ($P < 0.05$, Duncan test). Results are the mean ± standard error ($n = 3$).

Table 8.10 Results for Total Organic Carbon (TOC), Organic Carbon in Particulate Organic Matter (POM-C), Microbial Biomass Carbon (MBC), and Indicators of Functional Biodiversity as Obtained from Biolog® EcoPlates®.

Treatment	TOC (mg/g)	POM-C (mg/g)	MBC (mg/kg)	AWCD (g/100 g)	NSU (µS/cm)
0–15 cm					
P-value	<0.0001	0.0013	0.001	0.9091	0.3354
40-1	16.55 ± 0.55 (C)	3.21 ± 0.07 (BC)	645 ± 34.6 (B)	0.83 ± 0.05	24.0 ± 1.53
40-2	15.52 ± 0.18 (CD)	3.02 ± 0.11 (BC)	687 ± 38.3 (A)	0.78 ± 0.06	22.7 ± 0.88
40-4	16.29 ± 0.34 (C)	3.29 ± 0.22 (AB)	655 ± 50.6 (B)	0.72 ± 0.05	22.0 ± 0.58
80-1	18.22 ± 0.24 (A)	3.24 ± 0.37 (A)	550 ± 10.8 (D)	0.59 ± 0.10	17.3 ± 2.85
80-2	17.01 ± 0.27 (B)	3.40 ± 0.16 (AB)	614 ± 18.1 (BC)	0.59 ± 0.12	18.7 ± 3.33
80-4	15.92 ± 0.17 (C)	2.92 ± 0.05 (BC)	685 ± 56.2 (A)	0.73 ± 0.01	22.0 ± 0.00
MF	14.49 ± 0.47 (F)	2.60 ± 0.22 (C)	599 ± 1.09 (C)	0.73 ± 0.16	20.0 ± 3.21
C0	13.95 ± 0.40 (F)	2.46 ± 0.09 (C)	600 ± 39.6 (C)	0.65 ± 0.04	22.3 ± 0.88
15–30 cm					
P-value	<0.0001	<0.0001	0.0001	0.9091	0.3354
40-1	14.08 ± 0.08 (B)	2.11 ± 0.14 (BC)	565 ± 13.0 (A)	0.49 ± 0.16	15.7 ± 2.91
40-2	13.50 ± 0.14 (C)	1.84 ± 0.06 (C)	562 ± 11.4 (A)	0.43 ± 0.06	13.8 ± 1.86
40-4	15.34 ± 0.37 (B)	2.68 ± 0.21(B)	599 ± 18.4 (A)	0.40 ± 0.07	12.0 ± 1.53
80-1	16.14 ± 0.35 (A)	3.45 ± 0.58 (A)	469 ± 6.19 (B)	0.55 ± 0.05	13.3 ± 0.88
80-2	15.60 ± 0.84 (A)	2.42 ± 0.50 (B)	570 ± 5.17 (A)	0.71 ± 0.18	20.0 ± 3.79
80-4	13.79 ± 0.39 (C)	2.01 ± 0.10 (BC)	563 ± 33.3 (A)	0.60 ± 0.06	18.7 ± 1.86
MF	13.01 ± 0.44 (D)	2.03 ± 0.01 (BC)	493 ± 43.1 (B)	0.55 ± 0.09	17.7 ± 2.73
C0	13.21 ± 0.19 (D)	2.16 ± 0.08 (B)	461 ± 27.3 (B)	0.53 ± 0.05	15.7 ± 2.91

Note. All treatments are defined in the text. Results with different letters are statistically different ($P < 0.05$, Duncan test). Values are given as the mean ± standard error ($n = 3$). AWCD, available well color development; NSU, number of substrates used.

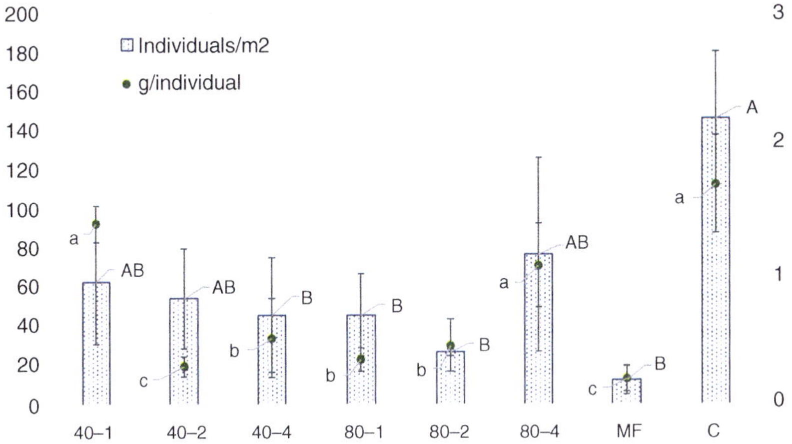

Figure 8.5 Earthworms abundance (bars, left axis, capital letters) and average size (points, right axis, lowercase letters) for the treatments with sewage sludge at different doses, conventional mineral fertilization (MF), and the unfertilized control (C0). Treatments with different letters are statistically different ($P < 0.05$, Duncan test). Values are the mean ± standard error ($n = 3$).

130% at 0–15 cm and by 122% at 15–30 cm compared with C0. MF had no significant differences in TOC concentration compared with C0. The other treatments receiving SS displayed intermediate values. These results were mirrored by those of POM-C, but not completely by MBC. MBC was lower in MF and C0 than in all SS-amended treatments except for 80-1 and 80-2, which showed the lowest values. The highest values were observed in 80-4 and 40-2 (Table 8.10), followed by 40-1 and 80-4. However, no differences were observed in the two indexes of microbial functional diversity considered.

The response of earthworm populations was significant both for density (indi viduals per square meter) and for average size (grams per individual; Figure 8.5). The highest density and average size were observed in C0, while MF showed the lowest values for both parameters. The density in SS-amended treatments was lower (and equivalent to MF) in 80-1, 80-2, and 40-4 treatments. The other treatments (40-1, 40-2, and 80-4) had intermediate values. For the average size of earthworms, only 40-1 and 80-4 had values similar to C0. The other treatments had intermediate values.

Wheat yield, the most relevant indicator of soil provisioning ecosystem services in rainfed extensive agroecosystems such as the one tested here, was also different between treatments in year 25 (Figure 8.6). The highest yields corresponded to MF, followed by 40-1, 40-2, and 40-4. Yield in C0 was less than half that of MF and 40-1, while the highest rate of SS (80-1) showed the second lowest yield.

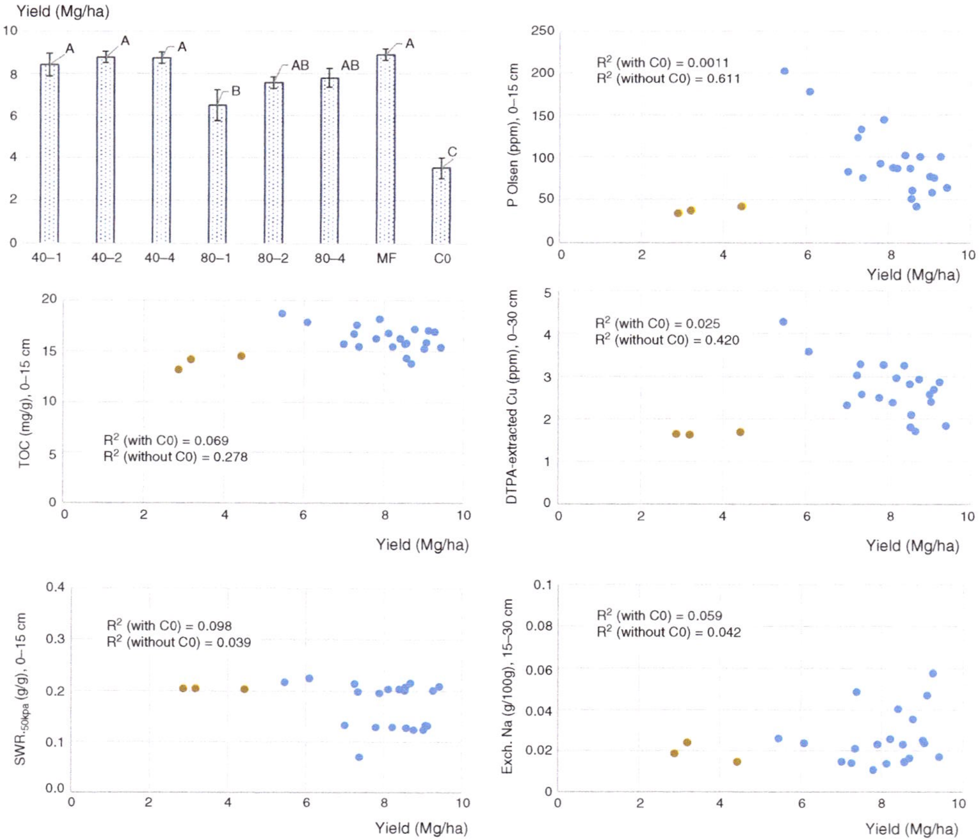

Figure 8.6 Wheat yield for the treatments with sewage sludge (SS) at different doses, conventional mineral fertilization (MF), and the unfertilized control (C0) in year 25 (top left) and the correlation of the most sensitive indicators at 0–15 cm (available phosphorus [P Olsen], total organic carbon [TOC], and soil water retention at −50 kPa [SWR_{-50kPa}]), 0–30 cm (available copper as representative of trace metals), and 15–30 cm (exchangeable sodium). For yield, treatments with different letters had statistically different results ($P<0.05$, Duncan test). Values are the mean ± standard error (n = 3). In the correlation charts, values from the unfertilized treatment (C0) are marked in brown. Other treatments are represented together as blue dots. Pearson's correlation coefficients (R^2) are provided for the correlation with and without C0 data.

Finally, the indicators identified through PCA and ANOVA as most sensitive to treatments were TOC, available phosphorus, trace metals, and water retention at low potentials at 0–15 cm, and TOC, available phosphorus, EC, and exchangeable potassium and sodium at 15–30 cm (Simoes-Mota et al., 2021). The correlations between yield and these selected indicators are shown in Figure 8.6. Greater yields were correlated with lower phosphorus and copper availability and higher exchangeable sodium. In contrast, organic carbon and SWR at low potentials (−50 kPa) had no clear correlation to yield.

Discussion

Effect of SS on Soil Indicators

The first clear finding in this study was that indicators of all groups were sensitive to SS addition and fertilization treatments tested in this study after >25 years of trial. The response was also depth sensitive in many cases, either because differences were observed only at one depth or because the response observed at 0–15 cm was mitigated or became more apparent at 15–30 cm. On one hand, this outcome indicates that the use of SS can have consequences on soil health in the long term in the agronomic and edaphoclimatic conditions of the study. On the other hand, this outcome highlights the need for and adequacy of a multi-indicator approach for assessing the complex interactions of this practice with soil conditions and the performance of the agroecosystem.

However, it is also relevant that some of the preselected indicators were nonresponsive to the dose or frequency of SS or MF, suggesting that use of SS did not affect the soil function or ecosystem service they represented at the time-scale of the study. Therefore, understanding the soil response to changes in practices such as SS fertilization also requires being able to identify the most responsive indicators, as not all the soil properties perceived as sensitive may in fact be responsive to changes in the long term. Identifying the sensitivity and responsivity of soil indicators to changes in management is indeed one of the basic principles of an adequate soil health assessment, and it has been reported as such by the different approaches proposed in the literature. There is consensus that these indicators must meet a series of characteristics, among which the representativeness of the processes and functions to be assessed and the need to be specific to the conditions in which they are used stand out (Karlen et al., 2013). In the current study, chemical and organic matter–related indicators seemed to be the most sensitive ones. This result was expected given that the most relevant impact of SS addition is the incorporation of an organic matter–rich component that also contains other chemicals. In addition, comparable results have been previously reported; for example, Obriot et al. (2016) came to similar conclusions in a similar trial with different soil and climate conditions in France.

Physical indicators were also associated with interesting responses in some cases. The most relevant observations in relation to the physical indicators considered was in relation to soil compaction and structural stability. The addition of SS resulted in a lower PR in the upper layer compared with the use of MF and especially the control (C0; Table 8.7). However, the lower PR did not translate to better stability of the structure or to changes towards a fluffier structure or a larger porosity. Indeed, in contrast to the expected outcome from high doses of organic matter being added over the years (Table 8.1), MWD decreased with the addition of SS, with C0 displaying a higher size of WSAs than the treatments receiving higher doses SS.

The analysis of the results for the chemical indicators sheds some light on the findings on soil physical indicators. Indeed, some parameters indicated that the addition of SS resulted in the accumulation of components that may affect structure directly (soluble salts) and indirectly (by affecting the soil biological activity). As shown in Table 8.8, although EC was within reasonable limits (below 400 μS/cm in a 1 : 2.5 soil: water mix), it was twice as high in 80-1 than in C0, with treatments with intermediate doses displaying intermediate values. Data on exchangeable Na also indicated higher values in some treatments with SS (40-1, 40-4, and 80-2), but not in the 80-1, suggesting that the soluble salts associated higher EC may also include other salts. The accumulation of soluble salts in soils amended with SS with a relatively high level of soluble components (Table 8.3) has been reported in other areas (Hechmi et al., 2021; Zoghlami et al., 2020).

Soil porosity was affected to a lesser extent. The changes observed involved a greater percentage of pores related to water retention at −50 kPa (which was similar in the treatments with high doses, C0 and MF, Table 8.6), but a reduction in pores >0.9 μm in equivalent diameter with a concomitant increase of the smallest quantified pores (<0.2 μm, Table 8.7). Overall, this result indicates that the addition of SS, which may induce some advantages for water retention with low suction owing to the nature of SS itself (Filipovic et al., 2020; Głąb et al., 2018), can also induce deleterious effects on soil structure, implying a more disperse structure (within the generally rather stable structure of this calcareous clayey soil) that may reduce functional porosity at a depth of 15–30 cm.

Changes in soil pH, with a slight acidification with higher doses of SS compared with C0 and MF, can also be related to the addition of acidifying agents and/or to the increase of nitrification with SS addition (Tamir et al., 2013). Such effects have been observed in similar trials on calcareous soils (Mohamed et al., 2018). The amount of NFUs added in some cases (Table 8.3) was in fact much higher than the usual crop needs. Although the elevated carbonate content in this soil (close to 16%, Table 8.5) can explain the lack of response of this indicator (Table 8.9), a change could be expected in the future if continuous addition persists and/or if the same SS is used in soils with a lower buffering capacity (de Soto et al., 2022).

In any case, the potential presence of hotspots of acidification induced by the application of SS can also result in improved availability of some elements, including trace metals (Kuzyakov & Blagodatskaya, 2015).

In relation to nutrients, the two major plant nutrients (phosphorus and nitrogen) studied showed differences with SS addition, resulting in greater contents (nitrogen) and availability (phosphorus) than C0 and MF. In the case of available phosphorus, differences were observed only at 0–15 cm, which can be explained by the high capacity of phosphorus retrogradation of the studied calcareous soil. In these soils, phosphorus retention and immobilization occurred owing to precipitation of insoluble calcium phosphates and adsorption to reactive clay minerals. The availability of phosphorus is thus largely controlled by rates of immobilization and mineralization as biologically mediated processes (Azadi & Baghernejad, 2019; Bamagoos et al., 2021; Strawn et al., 2015). This outcome is expected given the elevated concentrations of nitrogen and phosphorus in SS (Table 8.3), and similar results have been frequently reported for other phosphorus-rich organic sources in Mediterranean soils (Perramon et al., 2016; Yagüe et al., 2011). The situation can be a problem for adjusting SS doses when nitrogen is used as the dosing criterion and phosphorus appears in excess.

In this sense, it is noteworthy that the values under MF were similar to those of C0 for the two nutrients, suggesting that SS addition resulted in their content exceeding the crop needs. These needs were covered by MF in a more precise manner, as soil samples were collected after the completion of the growing season. Understanding the consequences of a potential excess of nutrients in the soil solution was beyond the scope of our study, but it has to be accounted for when assessing the use of SS in agriculture. It is important to note in this sense that, when used in the usual agronomic dose (20 Mg/ha, equivalent to 40-2 or 80-4), no differences were observed in TN compared with MF, but Olsen P was observed to be about 50% higher than in MF or C0 (Table 8.8).

The absence of an effect on available potassium at 0–15 cm and only a moderate effect for some treatments with 80 Mg/ha spaced by 2 or 4 years and MF, which had lower values than C0, seem less relevant owing to the high natural potassium content of this soil. The results also imply that MF does not include this element.

The availability of trace metals with SS addition, which is regularly reported as a risk when SS is used in excess, was also noted in this study. Some elements considered to also be micronutrients (zinc, copper, and nickel) were observed to increase with SS addition, while manganese was less available with low doses (including 40 Mg/ha at different doses) than in MF and C0, with similar amounts only when using 80 Mg/ha (Table 8.9). Zaragüeta et al. (2021) explained this behavior as the result of the affinity that manganese has with carbonates in SS and in soil, coupled with the low manganese contribution normally associated with SS. As such, when incorporated into calcareous soils, manganese undergoes minimal redistribution and mostly remains bound to carbonates. This characteristic could explain the lower efficiency of SS compared with mineral amendments in incorporating manganese into the soil. Other studies have observed similar behavior for this element (Nogueirol et al., 2013).

Available lead and cadmium, which were not observed to be affected by SS addition in this case, were observed to be higher with SS than without SS, however, when all treatments receiving SS were compared together with those without SS (C0 and MF) in a previous study (Zaragüeta et al., 2021). These results indicate again the risk of contamination with some elements when excessive doses of SS are used, although all the observed values remained under the current legal limits in Spain (Ministerio de la Presidencia, 2011). The calcareous nature of the study soil is undoubtedly related to this low availability of the elements analyzed.

In addition, the observed accumulation of some metals and soluble salts can be of use to understand the results observed in the soil biological parameters studied. Among those, the most sensitive ones were earthworm density and size (Figure 8.5) and MBC content (Table 8.10). Despite the high organic carbon concentration and enrichment in the most labile fraction (POM-C) observed with SS addition (Table 8.10) compared with C0 and MF at both depths, earthworms were mostly present in C0. In contrast, MF had the lowest density and average size, and the presence and size of earthworms in SS-amended treatments decreased with increasing dose. This result implies a clear decoupling between TOC and earthworm activity in this soil. The addition of SS can be seen here as linked to the presence of some elements and/or physical condition being more limiting for earthworms than the absence of fertilization. Since C0 was observed to be the most compacted treatment (Table 8.7), the most likely reason for this behavior is the concentration of soluble components, heavy metals, acidifying agents, and/or other elements present in SS not considered in this study, as observed in other studies on the effects of SS on earthworms (Filipi et al., 2023; Valdez et al., 2020). Simoes-Mota et al. (2024) recently reported the possible presence of deleterious effects for biological soil properties of excessive SS application in this soil, which is in line with previously observed effects (Urra, Alkorta, Mijangos, et al., 2019b). These authors also found that most of the accumulated organic carbon added as SS (Table 8.3) was indeed mineralized and not accumulated in the soil after more than 25 years.

However, no changes were observed in the functional diversity of these populations, as expressed by the average well color development and number of substrates used indexes (Table 8.10), suggesting a relatively high resilience and/or redundancy functional groups in this soil (Urra, Alkorta, Mijangos, et al., 2019b).

In relation to MBC, other studies in similar agro-climatic conditions have already shown that excessive doses of SS can cause an increase of the TOC content but also a significant decrease in enzyme activities and MBC content. This was the case of the use of 80 Mg/ha of thermally dried SS in central Spain (Fernández et al., 2009).

In summary, these results show that the use of SS versus MF can be of interest in these soils, as it can induce positive changes in some parameters related mostly to fertility, but that a dose-dependent negative effect is also possible. In addition to the deleterious effects observed here (trace metals accumulation, increased EC, slight acidification), there is increasing evidence that SS can be a source of deleterious elements in the soil (Urra, Alkorta, & Garbisu 2019a; Urra, Alkorta, Mijangos et al., 2019b; Surendran et al. 2023).

Dosage of SS therefore seems of paramount importance in this sense, as observed in other experiments in Mediterranean semi-arid and sub-humid conditions (Fernández et al., 2009; Fuentes et al., 2010). In this framework, while the use of SS seems interesting from the point of view of recovering degraded soils and improving areas with low SOC content (Gomes et al., 2019), the evidence from long-term evaluation of their adverse side effects suggest that additional and efficient SS treatments (Carnevale Miino et al., 2024; Collivignarelli et al., 2019) and/or their combination with other remediation strategies (Maggioli et al., 2022) are needed to reduce the risk associated to their application in agricultural land (Seleiman et al., 2020).

General Soil Health Assessment

As in any agroecosystem, the most relevant soil function assessed in this trial was primary productivity, measured as average crop yield at year 25 of the different treatments. Figure 8.6 shows how only treatments with moderate-to-low doses of SS were able to equal the standard management in the region (MF), whereas no fertilization (C0) and high doses of SS (80-1) resulted in the lowest yields. Two aspects merit discussion in this respect.

On one hand, this result means that, as far as yields are concerned, it is possible to at least partially substitute MF with SS, using doses that can be manageable at the agronomic scale, without significantly compromising other soil functions. It is noteworthy that in the area of study, the usual agronomic recommendation is close to 20 Mg/ha, applied only when needed, most commonly according to nitrogen needs. This value falls within the range of doses used elsewhere; for example, 1–5 Mg/ha on a dry mass basis in Huygens et al. (2022) for Europe, 5–14 t/ha in experiments in Brazil (Melo et al., 2018), or 2–4 Mg/ha dry mass in France (Obriot et al., 2016).

On the other hand, these results provide evidence that the use of SS can carry some risks for soil health. Not only can yields be negatively affected if excess doses are used, but indicators of other soil functions (such as biodiversity sheltering) can also be compromised. Considering the cumulative effect of many of the indicators considered (especially chemical parameters such as the availability of trace metals or the accumulation of phosphorus), long-term monitoring of soil health seems needed in this context.

Our study also identified the most sensitive indicators in rainfed agroecosystem soils (Simoes-Mota et al., 2021), which represent a common combination in sub-humid and semi-arid Spain and many other areas of the Mediterranean (Gómez-Miguel & Badía-Villas, 2016). These indicators were related to nutrient availability, organic carbon, trace metal accumulation, SWR, and salinity. Figure 8.6 shows the relationships of the most representative ones and their relation to yield. Interestingly, the ones showing the most significant correlation with yield when C0 was excluded from the study (available phosphorus and copper) behaved contrary to general expectation, with lower yields as the values increased. For phosphorus, this result can be explained by the observation that Olsen P values in treatments receiving SS exceeded the usual cereal needs and were even beyond the recommend environmental thresholds (Wang et al., 2023). As explained above, an excess of phosphorus in soils can occur when SS doses are calculated on the basis of nitrogen content, which is a common situation. As excessive phosphorus was associated with SS doses resulting in deleterious effects on crop yields and/or soil indicators, this finding indicates that yield cannot be considered the only indicator of soil health when exogenous materials such as SS are added to soil. A holistic approach is needed, with indicators selected based on consideration of the soil functions and/or associated services to be assessed, as well as the local conditions imposed by the soil-climatic and management characteristics at the specific site (Huygens et al., 2022; Seleiman et al., 2020).

For instance, previous studies in the region have shown that changing the conditions from rainfed to irrigation can result in a change in the most sensitive indicators for assessing soil health (Imaz et al., 2010; Apesteguia et al., 2017) in agrosystems. In particular, the response of POC and MWD to tillage system was reduced only 3 years after a rainfed experimental site was changed to irrigated management. Similarly, it can be concluded that there is a risk of overestimating soil health when indicators commonly understood as related to adequate soil health (e.g., TOC, SWR; Bai et al., 2018; Bünemann et al., 2018) are used alone, in the cases in which management conditions compromise their relationship to other indicators used to assess soil functioning. For instance, we observed that the highest yields or the best conditions for earthworm growth in our experiment did not correspond to the treatment showing the highest SOC content.

In summary, soil health can be compromised when SS is used in excess under the conditions tested in this study (rainfed cereal-based annual rotation on a calcareous soil in a Mediterranean sub-humid area). A correct assessment of SS addition to soil requires a holistic approach that at minimum comprises indicators of soil functions related to the soil condition for biological activity and/or the soil physical condition, in addition to yield or other indicators of productivity. In addition, by using this multi-level approach, it is possible to identify the doses and fertilization strategies that can result in an optimal balance between achieving a good yield and protecting soil health.

Conclusions

This work summarizes the study of several soil indicators selected to provide information about soil functioning for the assessment of the use of SS as a fertilizer in a calcareous rainfed Mediterranean agrosystem with a cereal-based rotation. The trial was based on an LTE started in 1992, and the goal of the study was to assess soil health after >25 years of applying SS under different doses and frequencies, compared with conventional MF and an unfertilized control. The hypothesis tested was that the dose of SS can be the leading factor in the response of soil health, which could be assessed through the different indicators tested.

After assessing a collection of 35 indicators comprising physical, chemical, organic matter–related, and biological indicators, the response was found to be indeed related to the use of fertilization as a major factor. Within fertilized treatments, the amount of SS applied determined the evolution of soil indicators. The response was uneven among indicators. In general, while organic matter–related parameters tended to increase with the dose, physical indicators and those related to nutrients and earthworms showed a response based on optimum values. In comparison with MF, the highest dose of SS used (which exceeded by far the usual practice among farmers) seemed to induce negative effects on soil structural stability, salinity, availability of some trace metals, porosity, earthworm density and size, and even crop yield. Moderate doses, closer to those used in agricultural fields, seemed to reduce the risk of soil health loss, while granting yield equal to that supported by conventional fertilization.

It can be concluded that the effect of SS on soil health in the conditions of this study was dependent on the dose and frequency of its application, and it was expressed in different directions, depending on the soil functions assessed and the indicators considered. The assessment of the overall performance of SS showed that the use of lower doses (equivalent to 10–20 Mg/ha) seems to grant basic soil functions, expressed as sustainable yield production, and organic carbon storage and reduced risks of degradation of the physical and chemical conditions of soil.

Acknowledgments

This work was developed with funding from National Institute for Agricultural and Food Research and Technology (INIA) through the grant # RTA2017-00088-C03-01, a predoctoral fellowship and Erasmus+ mobility fundings. A. Simoes-Mota received funding from the European Union's H2020 Research and Innovation programme under Marie Sklodowska-Curie grant agreement No 801586.

Dr. A.D. Bosch-Serra and M. Antúnez, from Universitat de Lleida, are thanked for their assistance in trace metal analysis. Dr. F.J. Arricibita, and Dr. C. Gonzalez, from Universidad Pública de Navarra, are thanked for thin section sampling and lab assistance, respectively.

The Pamplona City Commonwealth public body (Mancomunidad de la Comarca de Pamplona) settled the experimental field in 1992 and keeps it with the assistance of INTIA.

References

Adhikari, K., & Hartemink, A. E. (2016). Linking soils to ecosystem services—A global review. *Geoderma*, *262*, 101–111. http://dx.doi.org/10.1016/j.geoderma.2015.08.009

Alhameid, A., Singh, J., Sekaran, U., Ozlu, E., Kumar, S., & Singh, S. (2019). Crop rotational diversity impacts soil physical and hydrological properties under long-term no- and conventional-till soils. *Soil Research*, *58*(1), 84–94. http://dx.doi.org/10.1071/SR18192

Altieri, M. A., Nicholls, C. I., Henao, A., & Lana, M. A. (2015). Agroecology and the design of climate change-resilient farming systems. *Agronomy for Sustainable Development*, *35*(3), 869–890. http://dx.doi.org/10.1007/s13593-015-0285-2

Andrews, S. S., Karlen, D. L., & Cambardella, C. A. (2004). The soil management assessment framework: A quantitative soil quality evaluation method. *Soil Science Society of America Journal*, *68*(6), 1945–1962. http://dx.doi.org/10.2136/sssaj2004.1945

Apesteguía, M., Virto, I., Orcaray, L., Bescansa, P., Enrique, A., Imaz, M., & Karlen, D. (2017). Tillage effects on soil quality after three years of irrigation in northern Spain. *Sustainability*, *9*(8), 1476. http://dx.doi.org/10.3390/su9081476

Azadi, A., & Baghernejad, M. (2019). Application of kinetic models in describing soil phosphorus release and relation with soil phosphorus fractions across three soil toposequences of calcareous soils. *Eurasian Soil Science*, *52*(7), 778–792. http://dx.doi.org/10.1134/S1064229319070019

Bai, Z., Caspari, T., Gonzalez, M. R., Batjes, N. H., Mäder, P., Bünemann, E. K., de Goede, R., Brussaard, L., Xu, M., Ferreira, C. S. S., Reintam, E., Fan, H., Mihelič,

R., Glavan, M., & Tóth, Z. (2018). Effects of agricultural management practices on soil quality: A review of long-term experiments for Europe and China. *Agriculture, Ecosystems and Environment*, *265*, 1–7. http://dx.doi.org/10.1016/j.agee.2018.05.028

Baker, G. H., & Lee, K. E. (1993). Earthworms. In M. R. Carter (Ed.), *Soil sampling and methods of analysis* (pp. 359–371). Canadian Society of Soil Science and Lewis Publishers.

Bamagoos, A. A., Alharby, H. F., Belal, E. E., Khalaf, A. E. A., Abdelfattah, M. A., Rady, M. M., Ali, E. F., & Mersal, G. A. M. (2021). Phosphate-solubilizing bacteria as a panacea to alleviate stress effects of high soil $CaCo_3$ content in *Phaseolus vulgaris* with special reference to P-releasing enzymes. *Sustainability (Switzerland)*, *13*(13), Article 7063. http://dx.doi.org/10.3390/su13137063

Benyarku, C. A., & Stoops, G. (2005). *Guidelines for preparation of rock and soil thin sections and polished sections*. Universitat de Lleida.

Bescansa, P., Imaz, M. J., Virto, I., Enrique, A., & Hoogmoed, W. B. (2006). Soil water retention as affected by tillage and residue management in semiarid Spain. *Soil and Tillage Research*, *87*(1), 19–27. http://dx.doi.org/10.1016/j.still.2005.02.028

Bolinder, M. A., Janzen, H. H., Gregorich, E. G., Angers, D. A., & VandenBygaart, A. J. (2007). An approach for estimating net primary productivity and annual carbon inputs to soil for common agricultural crops in Canada. *Agriculture, Ecosystems and Environment*, *118*(1–4), 29–42. http://dx.doi.org/10.1016/j.agee.2006.05.013

Bongiorno, G., Bünemann, E. K., Oguejiofor, C. U., Meier, J., Gort, G., Comans, R., Mäder, P., Brussaard, L., & de Goede, R. (2019). Sensitivity of labile carbon fractions to tillage and organic matter management and their potential as comprehensive soil quality indicators across pedoclimatic conditions in Europe. *Ecological Indicators*, *99*, 38–50. http://dx.doi.org/10.1016/j.ecolind.2018.12.008

Bosch-Serra, D., Yagüe, M. R., Poch, R. M., Molner, M., Junyent, B., & Boixadera, J. (2017). Aggregate strength in calcareous soil fertilized with pig slurries. *European Journal of Soil Science*, *68*(4), 449–461. http://dx.doi.org/10.1111/ejss.12438

Bruni, E., Chenu, C., Abramoff, R. Z., Baldoni, G., Barkusky, D., Clivot, H., Huang, Y., Kätterer, T., Pikula, D., Spiegel, H., Virto, I., & Guenet, B. (2022). Multi-modelling predictions show high uncertainty of required carbon input changes to reach a 4 ‰ target. European Journal of Soil Science, November, 1–26. https://doi.org/10.1111/ejss.13330

Bruni, E., Guenet, B., Huang, Y., Clivot, H., Virto, I., Farina, R., Kätterer, T., Ciais, P., Martin, M., & Chenu, C. (2021). Additional carbon inputs to reach a 4 per 1000 objective in Europe: Feasibility and projected impacts of climate change based on Century simulations of long-term arable experiments. *Biogeosciences*, *18*(13), 3981–4004. http://dx.doi.org/10.5194/bg-18-3981-2021

Bünemann, E. K., Bongiorno, G., Bai, Z., Creamer, R. E., De Deyn, G., de Goede, R., Fleskens, L., Geissen, V., Kuyper, T. W., Mäder, P., Pulleman, M., Sukkel, W., van

Groenigen, J. W., & Brussaard, L. (2018). Soil quality—A critical review. *Soil Biology and Biochemistry*, *120*, 105–125. http://dx.doi.org/10.1016/j.soilbio.2018.01.030

Cambardella, C. A., & Elliot, E. T. (1992). Particulate soil organic changes across a grassland sequence. *Soil Science Society of America Journal*, *56*, 777–783.

Carnevale Miino, M., Galafassi, S., Zullo, R., Torretta, V., & Rada, E. C. (2024). Microplastics removal in wastewater treatment plants: A review of the different approaches to limit their release in the environment. *Science of the Total Environment*, *930*, Article 172675. http://dx.doi.org/10.1016/j.scitotenv.2024.172675

Carof, M., Godinot, O., & Le Cadre, E. (2022). Biodiversity-based cropping systems: A long-term perspective is necessary. *Science of the Total Environment*, *838*, Article 156022. http://dx.doi.org/10.1016/j.scitotenv.2022.156022

Clivot, H., Mouny, J. C., Duparque, A., Dinh, J. L., Denoroy, P., Houot, S., Vertès, F., Trochard, R., Bouthier, A., Sagot, S., & Mary, B. (2019). Modeling soil organic carbon evolution in long-term arable experiments with AMG model. *Environmental Modelling and Software*, *118*, 99–113. http://dx.doi.org/10.1016/j.envsoft.2019.04.004

Coll, P., Le Cadre, E., Blanchart, E., Hinsinger, P., & Villenave, C. (2011). Organic viticulture and soil quality: A long-term study in Southern France. *Applied Soil Ecology*, *50*(1), 37–44. http://dx.doi.org/10.1016/j.apsoil.2011.07.013

Collivignarelli, M. C., Abbà, A., Frattarola, A., Miino, M. C., Padovani, S., Katsoyiannis, I., & Torretta, V. (2019). Legislation for the reuse of biosolids on agricultural land in Europe: Overview. *Sustainability*, *11*(21), Article 6015. http://dx.doi.org/10.3390/su11216015

Gobierno de Navarra, 2021. Meteorología y climatología de Navarra [WWW Document]. URL http://meteo.navarra.es/ (accessed 5.30.24).

de Soto, I. S., Zamanian, K., Urmeneta, H., Enrique, A., & Virto, I. (2022). 25 years of continuous sewage sludge application vs. mineral fertilizers on a calcareous soil affected pH but not soil carbonates. *Frontiers in Soil Science*, *2*, 1–10. http://dx.doi.org/10.3389/fsoil.2022.960426

Dominati, E., Mackay, A., Green, S., & Patterson, M. (2014). A soil change-based methodology for the quantification and valuation of ecosystem services from agro-ecosystems: A case study of pastoral agriculture in New Zealand. *Ecological Economics*, *100*, 119–129. http://dx.doi.org/10.1016/j.ecolecon.2014.02.008

Dominati, E., Patterson, M., & Mackay, A. (2010). A framework for classifying and quantifying the natural capital and ecosystem services of soils. *Ecological Economics*, *69*(9), 1858–1868. http://dx.doi.org/10.1016/j.ecolecon.2010.05.002

Doran, J., & Parkin, T. (1996). Quantitative indicators of soil quality: A minimum data set. In *Methods for assessing soil quality* (SSSA Special Publications, pp. 25–37).

Drobnik, T., Greiner, L., Keller, A., & Grêt-Regamey, A. (2018). Soil quality indicators—From soil functions to ecosystem services. *Ecological Indicators*, *94*, 151–169. http://dx.doi.org/10.1016/j.ecolind.2018.06.052

European Comission. (1986). Council Directive of 12 June 1986 on the protection of the environment, and in particular of the soil, when sewage sludge is used in agriculture. *Official Journal of European Communities*, *181*, 6–12.

Fernández, J. M., Plaza, C., García-Gil, J. C., & Polo, A. (2009). Biochemical properties and barley yield in a semiarid Mediterranean soil amended with two kinds of sewage sludge. *Applied Soil Ecology*, *42*(1), 18–24. http://dx.doi.org/10.1016/j.apsoil.2009.01.006

Fernández-Ugalde, O., Virto, I., Bescansa, P., Imaz, M. J., Enrique, A., & Karlen, D. L. (2009). No-tillage improvement of soil physical quality in calcareous, degradation-prone, semiarid soils. *Soil and Tillage Research*, *106*(1), 29–35. http://dx.doi.org/10.1016/j.still.2009.09.012

Filipi, Á. C. K., dos Santos Gonçalves Nascimento, G. C., Bressani, P. A., Oliveira, A. K. G., Santo, D. E., Duarte, C. C. S., Gomes, E. M. V., Ferreira, P. M. P., Pokrywiecki, J. C., da Silva Gonzalez, R., de Souza, F. B., Dusman, E., de Souza, D. C., & Peron, A. P. (2023). Biological effects of sewage sludge—Does its incorporation into agricultural soils in the state of Paraná, Brazil, represent an environmental risk? *Water, Air, and Soil Pollution*, *234*(5), 1–19. http://dx.doi.org/10.1007/s11270-023-06306-8

Filipovic, V., Cerne, M., Simunek, J., Filipovic, L., Romic, M., Ondrašek, G., Bogunovic, I., Mustac, I., Krevh, V., Ferencevic, A., Robinson, D., Palcic, I., Paskovic, I., Ban, S. G., Uila, Z., & Ban, D. (2020). Modeling water flow and phosphorus sorption in a soil amended with sewage sludge and olive pomace as compost or biochar. *Agronomy*, *10*(8), Article 1163. https://doi.org/10.3390/agronomy10081163

Franzluebbers, A. J. (2002). Water infiltration and soil structure related to organic matter and its stratification with depth. *Soil and Tillage Research*, *66*(2), 197–205. http://dx.doi.org/10.1016/S0167-1987(02)00027-2

Fuentes, D., Valdecantos, A., Llovet, J., Cortina, J., & Vallejo, V. R. (2010). Fine-tuning of sewage sludge application to promote the establishment of Pinus halepensis seedlings. *Ecological Engineering*, *36*(10), 1213–1221. http://dx.doi.org/10.1016/j.ecoleng.2010.04.012

Głąb, T., Żabiński, A., Sadowska, U., Gondek, K., Kopeć, M., Mierzwa-Hersztek, M., & Tabor, S. (2018). Effects of co-composted maize, sewage sludge, and biochar mixtures on hydrological and physical qualities of sandy soil. *Geoderma*, *315*, 27–35. http://dx.doi.org/10.1016/j.geoderma.2017.11.034

Gomes, L. A., Gabriel, N., Gando-Ferreira, L. M., Góis, J. C., & Quina, M. J. (2019). Analysis of potentially toxic metal constraints to apply sewage sludge in Portuguese agricultural soils. *Environmental Science and*

Pollution Research, *26*(25), 26000–26014. http://dx.doi.org/10.1007/s11356-019-05796-6

Gómez-Miguel, V. D., & Badía-Villas, D. (2016). Soil distribution and classification. In J. Gallardo (Ed.), *The soils of Spain (World Soil)*. Springer. https://doi.org/10.1007/978-3-319-20541-0_2

Guo, Z., Zhang, J., Fan, J., Yang, X., Yi, Y., Han, X., Wang, D., Zhu, P., & Peng, X. (2019). Does animal manure application improve soil aggregation? Insights from nine long-term fertilization experiments. *Science of the Total Environment*, *660*, 1029–1037. http://dx.doi.org/10.1016/j.scitotenv.2019.01.051

Hechmi, S., Hamdi, H., Mokni-Tlili, S., Zoghlami, I. R., Khelil, M. N., Benzarti, S., Hassen, A., & Jedidi, N. (2020). Carbon mineralization, biological indicators, and phytotoxicity to assess the impact of urban sewage sludge on two light-textured soils in a microcosm. *Journal of Environmental Quality*, *49*(2), 460–471. http://dx.doi.org/10.1002/jeq2.20011

Hechmi, S., Hamdi, H., Mokni-Tlili, S., Zoghlami, R. I., Khelil, M. N., Jellali, S., Benzarti, S., & Jedidi, N. (2021). Variation of soil properties with sampling depth in two different light-textured soils after repeated applications of urban sewage sludge. *Journal of Environmental Management*, *297*, Article 113355. http://dx.doi.org/10.1016/j.jenvman.2021.113355

Huygens, D., Garcia-Gutierrez, P., Orveillon, G., Schillaci, C., Delre, A., Orgiazzi, A., Wojda, P., Tonini, D., Egle, L., Jones, A., Pistoccho, A., & Lugato, E. (2022). *Screening risk assessment of organic pollutants and environmental impacts from sewage sludge management (EUR 31238 EN)*. Publications Office of the European Union. http://dx.doi.org/10.2760/541579

Imaz, M. J., Virto, I., Bescansa, P., Enrique, A., Fernandez-Ugalde, O., & Karlen, D. L. (2010). Soil quality indicator response to tillage and residue management on semi-arid Mediterranean cropland. *Soil and Tillage Research*, *107*(1), 17–25. http://dx.doi.org/10.1016/j.still.2010.02.003

IUSS Working Group WRB. (2022). *World reference base for soil resources. International soil classification system for naming soils and creating legends for soil maps* (4th ed.). International Union of Soil Sciences (IUSS).

Karlen, D. L., Cambardella, C. A., Kovar, J. L., & Colvin, T. S. (2013). Soil quality response to long-term tillage and crop rotation practices. *Soil and Tillage Research*, *133*, 54–64. http://dx.doi.org/10.1016/j.still.2013.05.013

Knudsen, D., Peterson, G. A., & Pratt, P. F. (1982). Lithium, sodium, potassium. In A. L. Page, R. H. Miller, & D. R. Keeney (Eds.), *Methods of soil analysis* (pp. 225–246). Soil Science Society of America.

Kuzyakov, Y., & Blagodatskaya, E. (2015). Microbial hotspots and hot moments in soil: Concept & review. *Soil Biology and Biochemistry*, *83*, 184–199. http://dx.doi.org/10.1016/j.soilbio.2015.01.025

Lal, R. (2016). Potential and challenges of conservation agriculture in sequestration of atmospheric CO_2 for enhancing climate-resilience and improving productivity of soil of small landholder farms. *CABI Reviews*, *11*, 1–16. http://dx.doi.org/10.1079/PAVSNNR201611009

Land, M., Haddaway, N. R., Hedlund, K., Jørgensen, H. B., Kätterer, T., & Isberg, P.-E. (2017). How do selected crop rotations affect soil organic carbon in boreo-temperate systems? A systematic review protocol. *Environmental Evidence*, *6*(1), Article 9. http://dx.doi.org/10.1186/s13750-017-0086-y

Lorenz, K., Lal, R., & Ehlers, K. (2019). Soil organic carbon stock as an indicator for monitoring land and soil degradation in relation to United Nations' Sustainable Development Goals. *Land Degradation and Development*, *30*(7), 824–838. http://dx.doi.org/10.1002/ldr.3270

Maggioli, L., Chamizo, S., Román, R., Asensio-Grima, C., & Cantón, Y. (2022). Coupling sewage sludge amendment with cyanobacterial inoculation to enhance stability and carbon gain in dryland degraded soils. *Agriculture*, *12*(12), Article 1993. https://doi.org/10.3390/agriculture12121993

Martínez-Mena, M., Boix-Fayos, C., Carrillo-López, E., Díaz-Pereira, E., Zornoza, R., Sánchez-Navarro, V., Acosta, J. A., Martínez-Martínez, S., & Almagro, M. (2021). Short-term impact of crop diversification on soil carbon fluxes and balance in rainfed and irrigated woody cropping systems under semiarid Mediterranean conditions. *Plant and Soil*, *467*(1–2), 499–514. http://dx.doi.org/10.1007/s11104-021-05101-w

Melo, W., Delarica, D., Guedes, A., Lavezzo, L., Donha, R., de Araújo, A., de Melo, G., & Macedo, F. (2018). Ten years of application of sewage sludge on tropical soil. A balance sheet on agricultural crops and environmental quality. *Science of the Total Environment*, *643*, 1493–1501. http://dx.doi.org/10.1016/j.scitotenv.2018.06.254

Ministerio de la Presidencia, R., & con las, Democrática, C., y. M. (2011). Ley 22/2011, de 28 de Julio, de Residuos y Suelos Contaminados. *Agencia Estatal Oficial Del Estado*, *181*, 85650–85705.

Mohamed, B., Mounia, K., Aziz, A., Ahmed, H., Rachid, B., & Lotfi, A. (2018). Sewage sludge used as organic manure in Moroccan sunflower culture: Effects on certain soil properties, growth and yield components. *Science of the Total Environment*, *627*, 681–688. http://dx.doi.org/10.1016/j.scitotenv.2018.01.258

Nelson, D. W., & Sommers, L. E. (1982). Total carbon, organic carbon and organic matter. In A. L. Page, R. H. Miller, & D. R. Keeney (Eds.), *Methods of soils analysis* (pp. 539–579). Soil Science Society of America and Agronomy Society of America.

Nogueirol, R. C., De Melo, W. J., & Alleoni, L. R. F. (2013). Testing extractants for Cu, Fe, Mn, and Zn in tropical soils treated with sewage sludge for 13 consecutive years. *Water, Air, and Soil Pollution*, *224*(5), Article 1557. https://doi.org/10.1007/s11270-013-1557-6

Obriot, F., Stauffer, M., Goubard, Y., Cheviron, N., Peres, G., Eden, M., Revallier, A., Vieublé-Gonod, L., & Houot, S. (2016). Multi-criteria indices to evaluate the effects of repeated organic amendment applications on soil and crop quality. *Agriculture, Ecosystems and Environment*, *232*, 165–178. http://dx.doi.org/10.1016/j.agee.2016.08.004

Olsen, S. R., & Sommers, L. E. (1982). Phosphorus. In *Methods of soil analysis. Part 2* (pp. 403–430). American Society of Agronomy.

Papadakis, J. (1961). Climatic tables for the world.

Perramon, B., Bosch-Serra, A. D., Domingo, F., & Boixadera, J. (2016). Organic and mineral fertilization management improvements to a double-annual cropping system under humid Mediterranean conditions. *European Journal of Agronomy*, *76*, 28–40. http://dx.doi.org/10.1016/j.eja.2016.01.014

Rasband, W. (2008). *ImageJ 1.40*.

Rodrigues, L. A. T., Dieckow, J., Giacomini, S., Ottonelli, A. S., Zorzo, G. P. P., & Bayer, C. (2022). Carbon sequestration capacity in no-till soil decreases in the long-term due to saturation of fine silt plus clay-size fraction. *Geoderma*, *412*, Article 115711. http://dx.doi.org/10.1016/j.geoderma.2022.115711

Romanyà, J., & Rovira, P. (2011). An appraisal of soil organic C content in Mediterranean agricultural soils. *Soil Use and Management*, 27(3), 321–332. http://dx.doi.org/10.1111/j.1475-2743.2011.00346. x

Rose, C. W. (1966). *Agricultural physics*. Pergamon.

Sánchez-Navarro, V., Shahrokh, V., Martínez-Martínez, S., Acosta, J. A., Almagro, M., Martínez-Mena, M., Boix-Fayos, C., Díaz-Pereira, E., & Zornoza, R. (2022). Perennial alley cropping contributes to decrease soil CO_2 and N_2O emissions and increase soil carbon sequestration in a Mediterranean almond orchard. *Science of the Total Environment*, *845*, Article 157225. http://dx.doi.org/10.1016/j.scitotenv.2022.157225

Seleiman, M. F., Santanen, A., & Mäkelä, P. S. A. (2020). Recycling sludge on cropland as fertilizer—Advantages and risks. *Resources, Conservation and Recycling*, *155*, Article 104647. http://dx.doi.org/10.1016/j.resconrec.2019.104647

Sherrod, L. A., Dunn, G., Peterson, G. A., & Kolberg, R. L. (2002). Inorganic carbon analysis by modified pressure-calcimeter method. *Soil Science Society of America Journal*, *66*(1), 299–305. http://dx.doi.org/10.2136/sssaj2002.2990

Shukla, M. K., Lal, R., & Ebinger, M. (2006). Determining soil quality indicators by factor analysis. *Soil and Tillage Research*, *87*, 194–204. http://dx.doi.org/10.1016/j.still.2005.03.011

Simoes-Mota, A., Barré, P., Baudin, F., Poch, R. M., Bruni, E., Anton, R., Enrique, A., & Virto, I. (2024). Effects of long-term sewage sludge addition to a calcareous soil on soil organic C fractions and soil functions. *Geoderma*, *445*, Article 116868. http://dx.doi.org/10.1016/j.geoderma.2024.116868

Simoes-Mota, A., Poch, R. M., Enrique, A., Orcaray, L., & Virto, I. (2021). Soil quality assessment after 25 years of sewage sludge vs. mineral fertilization in a calcareous soil. *Land*, *10*(7), Article 727. http://dx.doi.org/10.3390/land10070727

Simoes-Mota, A., Virto, I., & Poch, R. M. (2022). Effects of long-term sewage sludge application to a calcareous soil structure. *Soil Use and Management*, *38*(4), 1693–1704. http://dx.doi.org/10.1111/sum.12838

SITNA: Sistema de Información Territorial de Navarra. (2023). *Ortofoto color 2003* and *2023*. Gobierno de Navarra Available via SITNA at: https://sitna.navarra.es/navegar/

Sousa, M. G., Araujo, J. K. S., Fracetto, G. G. M., Ferreira, T. O., Fracetto, F. J. C., de Araújo Filho, J. C., Otero, X. L., dos Santos, J. C. B., da Silva, A. H. N., & de Souza Junior, V. S. (2022). Changes in organic carbon and microbiology community structure due to long-term irrigated agriculture on Luvisols in the Brazilian semi-arid region. *Catena*, *212*, Article 106058. http://dx.doi.org/10.1016/j.catena.2022.106058

Strawn, D. G., Bohn, H. L., & O'Connor, G. A. (2015). *Soil chemistry*. Wiley Blackwell.

Surendran, U., Jayakumar, M., Raja, P., Gopinath, G., & Chellam, P. V. (2023). Microplastics in terrestrial ecosystem: Sources and migration in soil environment. *Chemosphere*, *318*, Article 137946. http://dx.doi.org/10.1016/j.chemosphere.2023.137946

Tamir, G., Shenker, M., Heller, H., Bloom, P. R., Fine, P., & Bar-Tal, A. (2013). Organic N mineralization and transformations in soils treated with animal waste in relation to carbonate dissolution and precipitation. *Geoderma*, *209–210*, 50–56. https://doi.org/10.1016/j.geoderma.2013.05.028

Torri, S. I., Corrêa, R. S., & Renella, G. (2014). Soil carbon sequestration resulting from biosolids application. *Applied and Environmental Soil Science*, 2014, Article 21768. https://doi.org/10.1155/2014/821768

Urra, J., Alkorta, I., & Garbisu, C. (2019a). Potential benefits and risks for soil health derived from the use of organic amendments in agriculture. *Agronomy*, *9*(9), Article 542. http://dx.doi.org/10.3390/agronomy9090542

Urra, J., Alkorta, I., Mijangos, I., Epelde, L., & Garbisu, C. (2019b). Application of sewage sludge to agricultural soil increases the abundance of antibiotic resistance genes without altering the composition of prokaryotic communities. *Science of The Total Environment*, *647*, 1410–1420. http://dx.doi.org/10.1016/j.scitotenv.2018.08.092

Valdez, A. S., Bosch-Serra, À. D., Yagüe, M. R., Poch, R. M., & Puigpinós, E. (2020). Earthworm community and soil microstructure changes with long-term organic fertilization. *Archives of Agronomy and Soil Science*, *66*(7), 957–970. http://dx.doi.org/10.1080/03650340.2019.1648792

Vance, E. D., Brookes, P. C., & Jenkinson, D. S. (1987). An extraction method for measuring soil microbial biomass C. *Soil Biology and Biochemistry*, *19*(6), 703–707. http://dx.doi.org/10.1016/0038-0717(87)90052-6

Veerman, C., Pinto Correia, T., Bastioli, C., Biro, B., Bouma, J., Cienciala, E., Emmett, B., Frison, E. A., Grand, A., Hristov, L., Kriauciuniene, Z., Soussana, J.-F., Vela, C., & Wittkowski, R. (2020). Caring for soil is caring for life—Ensure 75% of soils are healthy by 2030 for healthy food, people, nature and climate. In *European Commission* (Issue KI-02-20-463-EN-C). Publications Office of the European Union. https://doi.org/doi/10.2777/821504

Virto, I., Fernández-Ugalde, O., Barré, P., Imaz, M. J., Enrique, A., Bescansa, P., & Poch, R. M. (2013). Micromorphological analysis on the influence of the soil mineral composition on short-term aggregation in semi-arid Mediterranean soils. *Spanish Journal of Soil Science*, *3*(2), 116–129. http://dx.doi.org/10.3232/SJSS.2013.V3.N2.07

Virto, I., Imaz, M., Fernández-Ugalde, O., Gartzia-Bengoetxea, N., Enrique, A., & Bescansa, P. (2015). Soil degradation and soil quality in Western Europe: Current situation and future perspectives. *Sustainability*, *7*(1), 313–365. http://dx.doi.org/10.3390/su7010313

Vogel, H. J. (2008). *Quantim4 C/C++ Library for Scientific Image Processing*.

Wang, Y., Cui, Y., Wang, K., He, X., Dong, Y., Li, S., Wang, Y., Yang, H., Chen, X., & Zhang, W. (2023). The agronomic and environmental assessment of soil phosphorus levels for crop production: a meta-analysis. *Agronomy for Sustainable Development*, *43*(2), Article 35. http://dx.doi.org/10.1007/s13593-023-00887-8

Yagüe, M. R., Andraski, T. W., & Laboski, C. A. M. (2011). Manure composition and incorporation effects on phosphorus in runoff following corn biomass removal. *Journal of Environmental Quality*, *40*(6), 1963–1971. http://dx.doi.org/10.2134/jeq2010.0505

Yagüe, M. R., Domingo-Olivé, F., Bosch-Serra, À. D., Poch, R. M., & Boixadera, J. (2016). Dairy cattle manure effects on soil quality: porosity, earthworms, aggregates and soil organic carbon fractions. *Land Degradation and Development*, *27*(7), 1753–1762. http://dx.doi.org/10.1002/ldr.2477

Yost, J. L., & Hartemink, A. E. (2019). Effects of carbon on moisture storage in soils of the Wisconsin Central Sands, USA. *European Journal of Soil Science*, *70*(3), 565–577. http://dx.doi.org/10.1111/ejss.12776

Zak, J. C., Willig, M. R., & Moorhead, D. L. (1994). Functional diversity of microbial communities: A quantitative approach. *Soil Biology and Biochemistry*, *26*, 1101–1108. http://dx.doi.org/10.1016/0038-0717(94)90131-7

Zaragüeta, A., Enrique, A., Virto, I., Antón, R., Urmeneta, H., & Orcaray, L. (2021). Effect of the long-term application of sewage sludge to a calcareous soil on its total and bioavailable content in trace elements, and their transfer to the crop. *Minerals*, *11*(4), Article 356. http://dx.doi.org/10.3390/min11040356

Zoghlami, R. I., Hamdi, H., Mokni-Tlili, S., Hechmi, S., Khelil, M. N., Ben Aissa, N., Moussa, M., Bousnina, H., Benzarti, S., & Jedidi, N. (2020). Monitoring the variation of soil quality with sewage sludge application rates in absence of rhizosphere effect. *International Soil and Water Conservation Research*, *8*, 245–252. http://dx.doi.org/10.1016/j.iswcr.2020.07.007

9

Soil Health in Forest and Agroforestry Systems in Portugal

Manuel Madeira, Fernando Raimundo, Ana Raquel Rodrigues, Afonso Martins, António Fabião, and Paula Soares

Chapter Overview

Soil conditions in forests and agroforestry systems may differ from those in agricultural areas owing to vegetation type and management methods. Further, changes in these methods may affect productive, ecological, and economic functions of such systems. In Portugal, forest plantations (mostly *Eucalyptus* spp.) used to produce wood for industry occur in regions with a strong ocean influence. The eucalypt plantations are intensively managed as coppiced stands in a short rotation forestry (10- to 12-year rotations), generally for two or three rotations, often with the use of weed control techniques and tree fertilization. Afterwards, the conventional method of site preparation consists of the partial removal or incorporation of logging residues and stump removal or crushing, typically followed by ripping. While agroforestry systems of chestnut for intensive fruit production are prevalent in the northern and central inland areas, evergreen oak woodlands (mostly cork oak and holm oak) are predominant in the southern area, where livestock, exploitation of local products (e.g., cork, honey, aromatics, mushrooms), hunting, and tourism are main activities. Traditionally, chestnut orchards were intensively cultivated (two to four tillage operations every year), but presently minimum tillage or no-tillage, and soil vegetation cover becomes dominant practices, allowing fruit production, grazing, and mushroom production. Evergreen oak woodlands, associated with natural/improved pastures or natural shrub vegetation, are mostly managed for production of cork and grazing (swine, sheep, and cattle) or for facilitating natural tree regeneration through shrub

Soil Health Series: Volume 5 Soil Health and Sustainability in Spain and Portugal, First Edition.
Edited by Iñigo Virto and Rodrigo Antón.

Published 2025 by Soil Science Society of America, Inc. and John Wiley & Sons, Inc.

encroachment or prevention of grazing. Accordingly, this chapter addresses issues regarding soil health perspectives in the above-mentioned forests and agroforestry systems, considering the effect of different forest or agroforest management strategies on soil functional processes, experimentally assessed at plot scale, for a rotation period or for several years.

Introduction

Plantations of eucalypt (mostly *Eucalyptus globulus* Labill.) presently cover about 845,000 ha in Portugal (ICNF, 2019), prevailing in areas with a strong oceanic influence (Figure 9.1a) and within a wide range of mean annual rainfall (600–2,000 mm). Most of the large-scale establishment of eucalypt plantations took place prior to the end of the 1980s and mainly involved the use of existing forestry land, degraded cropland, and grasslands (Costa et al., 2014). They occur on flat to strongly undulating landscapes, where soils are mostly Regosols, Cambisols, and Arenosols (IUSS Working Group, 2015) developed from several parent materials (e.g., schists, granites, sandstones). The massive emergence of eucalypt plantations was associated with intensive establishment practices, such as deep plowing, ripping, and bench terracing. This generalized site preparation over large areas raised concerns among researchers, policy-makers, and environmentalists about the potential negative impacts on soil degradation, such as acidification, compaction, organic carbon and soil fertility losses, and erosion (Madeira, 1989; Madeira & Araújo, 2015). To address such concerns, legislation was enacted to establish general guidelines to manage the plantation of such species and preserve natural vegetation. Accordingly, afforestation with fast-growing species, namely eucalypts, needs previous authorization by the Forest Services (Decree-Law no. 175/88), and legislation has established specific norms and constraints regarding plantations of those species, for example, the obligation to preserve oak woodlands, natural vegetation, and the functions of main watercourses (Decree-Law no. 528/89). Similarly, additional legislation forbids some mechanical practices, such as the use of excavator bucket or backhoe and movements in the direction of the steepest slope (Decree-Law no.15-A/2018).

Harvest residues (HRs) in eucalypt plantations are of great importance owing to their roles in soil fertility (Achat et al., 2015), organic carbon sequestration (Mayer et al., 2020), soil protection in terms of reducing erosion risks (Fernández et al., 2004), temperature in the topsoil (Andrade et al., 2010), biological activity (Hartmann et al., 2012), and productive sustainability (Jones et al., 1999; Madeira & Araújo, 2015). Therefore, their management draws great concern. However, in the Mediterranean region, HRs also pose a potential risk for forest fires (Petriccione et al., 2006) and eventually a source for fuel. Consequently, Portuguese legislation

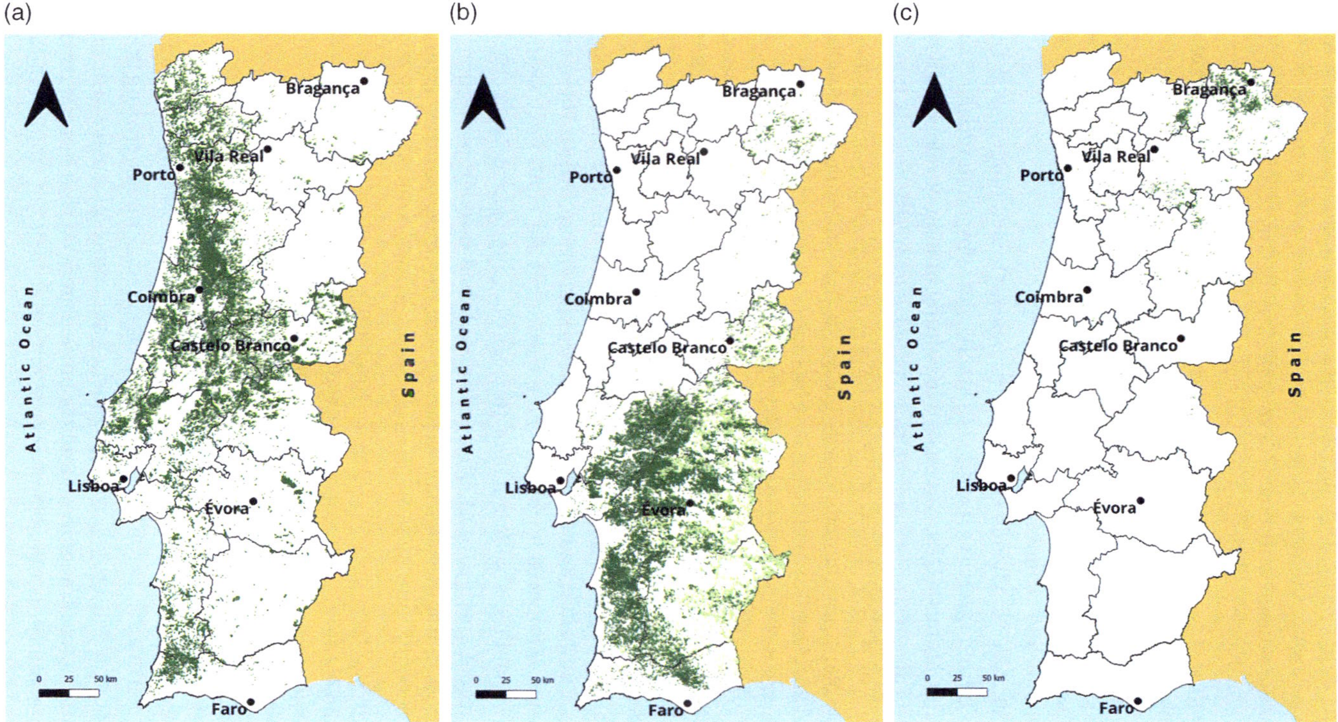

Figure 9.1 Distribution of eucalypt plantations (a), evergreen oak woodlands (b), and chestnut groves (c) in Inland Portugal, in 2015 (ICNF, 2019).

requires the removal of HRs or at least their breakdown and incorporation into the soil (Decree-Law no. 82/2021).

Inadequate management practices may negatively affect soil characteristics in intensive eucalypt plantations, especially at establishment and during the inter-rotation period (i.e., the time between harvesting and the establishment of a new plantation), owing to the use of heavy machinery and the loss of tree cover and HRs. It is then essential to analyze whether intensive practices for eucalypt reforestation/afforestation and HR management threaten soil functions and to identify management systems that ensure provision of sustainable productive, environmental, and economic services.

Evergreen oak woodlands (*montados* or *dehesas*), including mostly *Quercus suber* L. (QS) and *Quercus rotundifolia* Lam. [*Q. ilex* subsp. *rotundifolia* (Lam.) O. Schw. ex Tab. Morais] (QR), correspond to cultural human-developed landscapes (Manning et al., 2006). They are multifunctional agroforestry systems that consist of a mosaic of widely spaced scattered oaks combined with crops, pastures, or shrubs. Despite strong human activity pressures, they still have high biodiversity values and are classified as at-risk native biodiversity hotspots (Myers et al., 2000). Evergreen oak woodlands occupy large areas (1,063,000 ha, about 58% being QS; ICNF, 2019) mostly in southern Portugal. While QS is dominant in areas with some oceanic influence, QR occurs in areas with a wide range of temperatures and stronger drought periods (Figure 9.1b).

Montados occur in flat to strongly undulating landscapes made of sedimentary, magmatic, and metamorphic rocks, where developed soils are mostly Regosols, Arenosols, Cambisols, and Luvisols (IUSS Working Group, 2015). These agroforestry systems, influenced by former and present land use, show great spatial and temporal variability, a wide range of tree density (approximately 10–180 trees/ha), and recent emergence of QS stands established through plantation. Their understory was traditionally composed of annual crops, pastures (natural or seeded), or fallow, usually in rotation schemes, and used for extensive livestock rearing (sheep, pigs, cattle, and goats), with harrowing being used to avoid shrub encroachment. Recent changes in management of oak woodlands have mainly been related to stressing environmental conditions (e.g., soil type and hydrological regime; Costa et al., 2010), disturbance regimes (e.g., wildfire occurrence; Moreira et al., 2001), and the vulnerable agricultural economy associated with changes in the incentive structure of successive EU's Common Agricultural Policy reforms. These changes usually place pressure on soil conditions and therefore soil quality. Through land intensification (e.g., higher grazing pressure, modern permanent-crop farming systems) and more extensive use of land (e.g., shrub encroachment in set aside or burned areas), a shift has occurred from a traditional multifunctional landscape to different single-functional landscapes. According to recent changes in management of evergreen oak woodlands, the evaluation of improved pasture establishment, grazing, and shrub encroachment on soil conditions is a pivotal issue.

Stands of chestnut (*Castanea sativa* Mill.), a perennial and traditional staple food crop, are mostly present in northeastern Portugal and to a lesser extent in the central region (Figure 9.1c), occupying an area of approximately 48,300 ha (ICNF, 2019). They occur in flat to gently undulating landscapes of highlands, where oceanic influence is low and soils are mostly Cambisols and Regosols (IUSS Working Group, 2015) developed on schists (Raimundo, 2003). This agroforestry system is an important feature of the landscape, and one of the most profitable productive systems of those regions, constituting an essential source of income for rural communities. A great concern emerged during the 1990s about the effect of intensive practices commonly used in their conventional management (e.g., intensive soil tillage) on soil status and system sustainability (Abreu et al., 1993; Portela et al., 1999; Martins et al., 2005). Soil tillage carried out three or four times a year with a tine cultivator is aimed at incorporating organic residues into the soil, avoiding soil surface crust, controlling weeds, and saving soil water (Portela et al., 1999). Such management systems raise concerns about the negative impacts on soil organic matter (OM) status and biodiversity; the risks of soil compaction and erosion, soil nutrient losses, tree root system damage, and the occurrence of ink disease; and ultimately, the decline of chestnut stands (Portela et al., 1999; Marcelino et al., 2000). The predicted increases in the length, severity, and frequency of summer droughts (Miranda et al., 2002) pose another threat to chestnut stands, and water availability is becoming a main concern for producers who are aiming to sustain or increase productivity (Martins et al., 2010). In these circumstances, it is crucial to evaluate whether the replacement of conventional soil management practices by other practices, such as no tillage (Raimundo, 2003) with the inclusion of irrigation, can affect the physical, chemical, and biological soil condition and the system's productivity. It is also crucial to develop guidelines for productivity improvement and long-term sustainability of chestnut stands.

Study Sites and Experiments

Eucalypt Plantations

For evaluation of the effects on soil status and tree growth from intensive and deep soil disturbance related to site preparation and HR management (removal, incorporation into the soil, or maintenance on the soil surface), experimental trials were established and monitored over a rotation period in representative areas of the eucalypt distribution range (Table 9.1). These trials were established

Table 9.1 Location, Climate Characteristics, Landscape, and Soil of the Different Study Sites Used to Evaluate Soil Conditions in Eucalypt Plantations, Evergreen Oak Woodlands, and Chestnut Plantations

Sites	Coordinates	MAR (mm)	MAT (°C)	ALT (m)	Landscape	Parent material	Soils classification (IUSS Working Group, 2015)	Texture
Eucalypt plantations (mainly *Eucalyptus globulus*)								
QF	39°21′01″N, 09°14′00″W	600	15.2	76	Gently undulating	Sandstones	Regosols/Cambisols	Sandy clay loam
VP	39°21′10″N, 08°19′05″W	820	16.0	178	Flat	Sandstones/clays	Luvisols/Planosols	Clay
PN	39°20′0″N, 08°51′38″W	890	15.1	83	Undulating	Sands/sandstones	Regosols/Arenosols	Loam sandy
Evergreen oak woodlands (*Quercus suber* and *Quercus rotundifolia*)								
HM[ab]	38°31′29″N, 08°01′15″W	665	15.4	241	Undulating	Gneisses	Regosols/Leptosols	Sandy loam
HE[ab]	3°07′02″N, 07°29′50″W	625	15.0	362	Undulating	Gneisses	Regosols/Leptosols	Sandy loam
HO[ab]	38°51′53″N, 07°32′24″W	670	16.2	390	Undulating	Schists	Cambisols/Regosols	Sandy loam/Loam
TV[ab]	38°48′30″N, 07°24′22″W	685	16.2	310	Gently undulating	Schists	Regosols/Leptosols	Silt loam
MG[a]	39°08′24″N, 08°20′08″W	685	15.2	163	Gently undulating	Sands	Arenosols/Regosols	Loamy sand
Chestnut plantations (*Castanea sativa*)								
LP	41°35′47″N, 06°56′38″W	818	9.9	749	Gently undulating	Schists	Regosols/Cambisols	Loam
TR	41°52′53″N, 06°50′12″W	758	12.3	906	Gently undulating	Schists	Cambisols	Silt loam

Note. MAR, mean annual rainfall; MAT, mean annual temperature; ALT, altitude; ab, woodland with holm oak (dominant) and cork oak; a, woodland with cork oak. Sites: QF, Quinta do Furadouro; VP, Vale Pequeno de Cima; PN, Penegral; HM, Herdade da Mitra; HE, Herdade dos Esquerdos; HO, Herdade do Olival; TV, Tapada Real de Vila Viçosa; MG, Machoqueira do Grou; LP, Lamas de Podence; TR, Terroso.

in areas with flat to undulating landscapes based on sedimentary materials to avoid the effects of erosion and the restriction of soil depth owing to underlying compact siliceous rocks. These trials enabled disentangling the effects associated with initial soil conditions and those of soil perturbations linked to site preparation for establishing new plantations.

Deep soil disturbance was evaluated at a site (Penegral [PN]) that was previously occupied by a maritime pine stand. Two soil-treatment types were compared (Madeira et al., 1989): surface disc harrowing and deep plowing. The surface disc harrowing soil management system included the clear-cutting and removal of the shrub layer followed by disc harrowing to 25-cm depth, so that the soil surface became rough. The deep plowing system involved clear-cutting the shrub layer, followed by surface disc harrowing and plowing to 80-cm depth and then additional surface disc harrowing to level out the soil surface. Through this treatment, deep soil layers that were poor in OM were brought to the surface.

At two other sites (Quinta do Furadouro [QF] and Vale Pequeno de Cima [VP]; Table 9.1), after coppice of a previous first rotation eucalypt plantation, an evaluation was conducted to test whether HR management options (removal, incorporation into the soil, or maintenance on the soil surface) affected soil organic carbon and nutrient status, soil physical conditions, and tree growth (Jones et al., 1999). Sites differed in terms of climate conditions and soil type: the oceanic influence was stronger at QF (Csb, Köppen-Geiger classification) than at VP (Csa), and the dominant soils were, respectively, Regosols and Planosols (IUSS Working Group, 2015). At QF, the experimental design consisted of four treatments: (R) removal of HR; (S) broadcast of HR on soil surface; (I) incorporation of HR into the soil by harrowing (~20-cm depth); and (Is) I followed by ripping (~50-cm depth). At the VP site, owing to soil type, treatments R, S, and I included ripping up to 80- to 85-cm depth. In addition, an I treatment without ripping was included (Ih) to assess the effect of deep ripping. At both sites, the experimental design was replicated in five different blocks. Each replicate was a plot of 30 m × 30 m, with a core of 36 trees (18 m × 18 m) surrounded by two buffer tree rows on each of the four sides. Sprouting of stumps of former plantations was controlled by a systemic herbicide (glyphosate). Eucalypt seedlings were from a single clone provenance, and each was supplied at out planting with ~100 g of a fertilizer containing 14% nitrogen, 16% phosphorus, and 17% potassium (Jones et al., 1999). The tree density was at approximately 1,111/ha (3 m × 3 m spacing). Installation practices occurred under optimal soil moisture for machinery operations.

Oak Woodlands

Assessment of soil status in oak woodlands had several objectives and thus involved several systems and sites (Table 9.1). It took place in QS and QR mixed stands or in QS pure stands, all under Csa climate (Köppen-Geiger classification).

The first aim was to analyze the effect of scattered oak trees on local-scale soil functions (Manning et al., 2006). Next, the effect of the management changes on soil physical, chemical, and biological parameters were closely evaluated. The studies encompassed the replacement of natural pastures or shrub understory by improved pastures, grazing systems (sheep versus cattle), and shrub encroachment by grazing prevention, favoring natural vegetation.

The effect of scattered trees on soil health was evaluated at the Tapada Real de Vila Viçosa (TV) site (Rodrigues, 2019; Table 9.1), an estate traditionally used for hunting of small game species. After 1994, the oak stand was exclusively oriented for browsing by herbivore ungulate production, specifically, red deer and fallow deer at approximately 0.35 and 0.1 animals/ha, respectively.

The assessment of improved pastures relative to natural pastures was done at the Herdade dos Esquerdos (HE) site (Table 9.1), which was under a rotation of cereal crops, grazing (natural pastures), and fallow until the 1970s. After abandonment of cereal cropping in 1973, the estate was used for extensive sheep grazing, and in 1973 and 1979, some areas were sown with improved permanent pastures, which were intensively grazed (5–8 sheep·ha^{-1}·yr^{-1}; Crespo, 2006; Gómez-Rey et al., 2012).

Replacement of natural shrub understory by improved pastures was evaluated at the Machoqueira do Grou (MG) site (Rodrigues et al., 2019; Table 9.1), a pure QS stand established in 1965 with approximately 170 trees/ha. In 1992, the site held two estates with different management regimes. One was converted into a permanent natural pasture associated with shrubs for extensive cattle grazing, while the other retained its natural understory, comprising herbaceous and shrub layers, without grazing. In the former, a pasture mixture (*Trifolium* spp., *Ornithopus* spp., and *Biserrula pelecinus* L.) was sown in 2009 after tillage with a disking harrow, and it has since been grazed by about 1.4 cows/ha for 1 month every year (0.1 livestock units [LU]·ha^{-1}·yr^{-1}). In the latter, management is exclusively oriented for cork production, and the shrub understory control is done with a rotary mower every 4–6 years, without fertilizer application.

The effect of grazing system was evaluated at both HE and Herdade do Olival (HO) sites (Rodrigues, 2019, Table 9.1), which have natural and improved pastures. The HE site had two management systems: an improved pasture (IP) sown in 1979 and grazed by 5–8 sheep·ha^{-1}·yr^{-1} (0.9–1.2 LU·ha^{-1}·yr^{-1}) and a natural pasture (NP) grazed by less than 1 sheep·ha^{-1}·yr^{-1} (~0.15 LU·ha^{-1}·yr^{-1}). Sheep grazed in both pastures intermittently throughout the year. At the HO site, an IP, sown in 1998, was grazed by 0.7 cows·ha^{-1}·yr^{-1} (0.7 LU·ha^{-1}·yr^{-1}), and an NP was grazed by the same cattle herd at a similar stocking rate. Both areas were grazed alternately throughout the year.

Finally, the effect of shrub encroachment on soil health was evaluated at the Herdade da Mitra (HM) site (Gómez-Rey et al., 2013; Table 9.1), a mature open woodland (10–45 trees/ha) of QS and QR with a shrub understory, dominated by *Cistus salviifolius* L. (CS) and *Cistus ladanifer* L (CL). The study area, used for goat grazing until 1991, was fenced to prevent grazing and protect vegetation and kept undisturbed (Simões et al., 2009). Soil samples were taken in selected cells without shrubs and with tall shrubs of CL (higher than 1 m) and CS (larger than 1 m of crown diameter) at least to 5 m from the crown limit of mature QS and QR trees.

Chestnut Groves

The impacts of soil management practices on soil health in chestnut stands were assessed in two experimental trials established at the Lamas de Podence (LP) site (Table 9.1): one monitored from 1995 to 2003 (Raimundo, 2003) and another from 2001 to 2008 (Martins et al., 2010; Martins et al., 2011). The former was established in a 50-year-old chestnut stand, where trees were spaced about 12 m × 12 m. Treatments were conventional tillage (CT) with a chisel plow up to approximately 15-cm depth and harrowing up to approximately 10-cm depth (HR), no-tillage with spontaneous vegetation (NV), and no-tillage with seeded pasture (NP). Treatments, replicated three times, were applied in 30 m × 40 m plots. The latter included the treatments CT, NV, NP, and NP with irrigation. Evaluation of soil health monitoring activity at the farmer level was extended to chestnut stands located at the Terroso (TR) site (Table 9.1), managed for some decades using tillage and no-tillage practices (Borges et al., 2018).

Concepts and Indicators

Soils have been traditionally been qualified using terms such as “good,” “poor,” “fertile,” or “infertile” (Brady & Weil, 2008). Soil health is a relatively new concept arising from long-term attempts to define “good” soils; it is associated with other terms such as soil fertility, soil quality, and soil security (Wood & Blankinship, 2022). Soil health and soil quality are commonly considered to be synonymous and are used interchangeably (Bünemann et al., 2018). Soil quality has been defined as “the capacity of a specific kind of soil to function, within natural or managed ecosystems boundaries, to sustain plant and animal productivity, maintain or enhance water and air quality, and support human health and habitation” (Karlen et al., 1997; pag. 6). It has also been defined as “a measure of the condition of soil relative to the requirements of one or more biological species and/or to any human purpose” (Johnson et al., 1997). The term “soil health” is sometime preferred because it portrays soil as a living dynamic system whose functions are mediated by a diversity of living organisms that require management and conservation (Doran & Zeiss, 2000). Therefore, soil health is used in the sense of "dynamic soil quality," referring to soil properties that may change or be affected according to soil use and management over the human lifetime scale. That is, natural composition and properties of soils, generally not influenced by human management, are not considered (Moebius-Clune et al., 2016). Within this chapter, we consider soil health as "the continued capacity of the soil to function as a vital living ecosystem that sustains plants, animals and humans" (USDA-NRCS, 2001). We focused on soil health assessment by measuring dynamic soil properties both at the surface soil and at 0- to 20-cm depth (oak woodlands and chestnut groves) or by considering the soil profile as a control section (e.g., eucalypt plantations).

Maintenance of soil quality or soil health is a critical component of sustainable forest or agroforest management because soil plays a major role in regulating most ecological processes (forest growth, nutrient cycling, water cycling, biodiversity). Management practices that preserve the sustainability of the soil are essential to the maintenance of ecosystem productive capacity (Raison & Rab, 2001). Soil quality is a complex concept related to the fulfilment of several economic and ecological functions, and it requires the study of a wide range of physical, chemical, and biological properties through temporal and spatial comparisons.

Soil quality evaluation is a tool to assess management-induced changes in the soil and to link existing resource concerns to environmentally sound land management practices (USDA-NRCS, 2001). Soil quality assessments are thus used to evaluate the effects of management on the health of the soil. In the present chapter, they include (1) comparing the effects of different management systems or practices on soil quality in field trials or sites with similar soil types, and (2) monitoring trends over time to determine the impact of management on indicators of soil quality and to identify soil resource status and effects on productive outcomes. For that, soil quality indicators (physical, chemical, and biological) were chosen according to their relationship to specific soil properties and soil functions (Table 9.2). Soil OM (and labile fractions) was considered separately because it commonly provides information on a wide range of properties such as soil fertility, soil structure, soil stability, nutrient retention, and biological activity.

Because it is impractical to monitor most soil processes and to measure every soil property, emphasis was placed on identifying a set of key soil properties (indicators) that can reflect soil quality or soil health (Raison & Rab, 2001; (USDA-NRCS, 2001) in study systems. The set of indicators presented in Table 9.2 includes key soil properties or indicators that were considered potentially relevant for monitoring and understanding soil quality, mostly at plot and treatment level, to demonstrate how soil quality can change following soil and management practices

Table 9.2 Indicators Used Among Experiments and Study Areas for the Assessment of Soil Quality and Soil Health

Indicators	Sites	Relationship to soil health
Soil organic matter		
Organic-C	All sites	Nutrient availability, structure stability, soil biota and erosion protection, water and nutrient retention
C_{POM}	HE, MG, HM, LP	
C_{HW}	HE, MG, HM, LP	
C_{AC}	TR	
Physical		
Bulk density	All sites	Porosity, aeration, rooting, water movement
Infiltration rate	QF, HE, HO, LP	Water movement
Compaction	PN, QF, VP, LP	Plant rooting, aeration, water movement
Available water	QF, VP, LP	Plant available water
Aggregate stability	QF, VP, HE, HO	Water movement, soil erosion, microbe habitat
Chemical		
pH	All sites	Acidification, plant nutrient supply
Extractable P and K	All sites	Plant nutrient supply
Extractable Ca	All sites	Plant nutrient supply
Biological		
Microbial biomass	HE, HM, MG, LP	Habitat biodiversity, microbial catalytic potential
N mineralization	HE, HM, MG, LP	Soil productivity and N supply
Soil respiration	HE, HM, MG, LP	Microbial activity
Enzymatic activity	HE	Element and organic matter cycling
Arthropods	QF	Organic matter dynamics and soil structure
Sporocarps	LP	Soil suppressiveness and nutrient availability

Note. Adapted from USDA-NRCS (2001), Brady and Weil (2008), and Bünemann et al. (2018). C_{POM}, carbon of the particulate organic matter fraction; C_{HW}, hot-water-soluble carbon; C_{AC}, active carbon or organic carbon oxidized by 0.02 M $KMnO_4$. Site abbreviations are defined in Table 9.1.

(Karlen et al., 1997), that is, according to their sensitivity to changes and according to soil types and land uses of the study areas. Obtaining monitoring data constitutes a first step in compiling or designing a minimum data set of relevant indicators adequate for the study systems and applicable to other areas with different soil types and ecological conditions.

Soil Quality in Forestry and Agroforestry Systems

Eucalypt Plantations

Intensive Site Preparation: Deep Plowing Versus Superficial Harrowing

Disturbance associated with site preparation for establishing new eucalypt plantations may lead to substantial changes in soil condition that depend on the intensity of soil disturbance, as revealed by the results from the field trial at the PN site (Table 9.3; Figure 9.2). Indeed, disc harrowing up to 20-cm soil depth significantly decreased bulk density and compaction and increased aeration porosity in the 0- to 10-cm topsoil layer, while the deep plowing had no effect on these indicators and significantly decreased water infiltration rate. In the short term (18 and 32 months), soil disturbance led to decreases in the organic carbon concentration, the sum of Ca-Mg-K-Na, and the extractable phosphorus. These decreases were much more pronounced in deep-plowed soil, probably owing to the more intensive soil disturbance (inversion of soil layers) and the higher degree of OM mineralization. This trend was also observed in areas of eucalypt plantations established using the bench terracing technique (Martinho et al., 2023). As the organic carbon concentration was still

Table 9.3 Bulk density (BD), Aeration Porosity (AP), Water Infiltration Rate (IR), Compaction (SC), Organic Carbon (OC), Carbon in Particulate Organic Matter (C_{POM}), Extractable Phosphorus, and the Sum of Ca-Mg-K-Na (SB) in the 0- to 10-cm Soil Layer at 18, 32, and 132 Months After a Eucalypt Plantation Installation

	BD	AP	IR	SC	OC	C_{POM}	Ext P_2O_5	SB
Treatment	g/cm^3	% (v/v)	cm/h	MPa	g/kg		mg/kg	$cmol_c/kg$
18 months								
US	1.54[a]	11.7[a]	16.4[a]	1.77[a]	18.2[a]	—	8.6[a]	3.13[a]
DH	1.37[b]	23.4[b]	18.0[a]	0.56[b]	12.5[b]	—	7.8[a]	2.36[b]
DP	1.59[a]	10.7[a]	6.7[b]	1.63[a]	5.2[c]	—	4.8[b]	1.62[c]
32 months								
DH	1.36[a]	—	—	—	12.6[a]	—	7.7[a]	2.43[a]
DP	1.63[b]	—	—	—	4.6[b]	—	1.8[b]	1.58[b]
132 months								
DH	1.30[a]	—	—	—	14.4[a]	5.7[a]	4.6[a]	2.20
DP	1.51[b]	—	—	—	8.7[b]	2.5[b]	1.8[b]	2.28

Note. Adapted from Madeira et al. (1989, 2002). US, undisturbed soil; DH, disc harrowing; DP, deep plowing. For each period, different letters in the same column indicate significant differences ($P < .05$) between soil practices by the Tukey test.

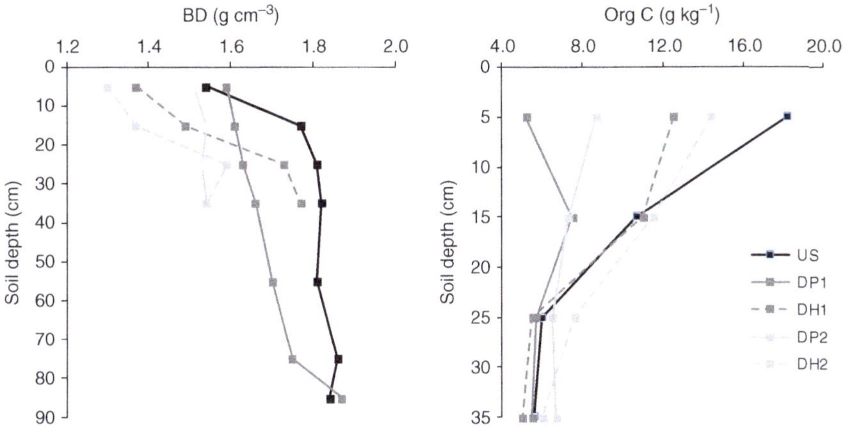

Figure 9.2 Bulk density (BD) and organic carbon (Org C) concentration with soil depth along the first rotation of an eucalypt plantation at the Penegral site. US, undisturbed soil; DH, disc harrowing; DP, deep plowing; 1, 32 months; 2, 132 months. Adapted from Madeira et al. (1989, 2002).

lower at the end of the rotation (11 years) than at the start, longer periods than the common rotation appear to be necessary to recover initial soil carbon status. Results clearly show that deep plowing is more deleterious than disc harrowing regarding the soil physical condition, carbon accumulation, and fertility in the 0- to 10-cm layer.

Deep plowing also affected the properties of soil deeper layers. Despite higher values of bulk density in the topsoil, conventional deep plowing alleviated bulk density (and compaction) in subsoil layers along the rotation period (Figure 9.2), creating an improved environment for root growth. However, organic carbon losses in these layers were higher in deep plowing than in disc harrowing, and the increment of organic carbon was negligible over the rotation period. As tree productivity in disc harrowing (252.4 m^3/ha) was higher than in deep plowing (202.4 m^3/ha; Madeira et al., 2002), subsoil compaction was presumably not harmful to the extent of hampering tree growth. Thus, in such sites, less intensive practices, such as disc harrowing or ripping, are preferable with regard to soil quality status and economic results. Meanwhile, monitoring periods longer than one rotation are necessary for assessing changes in soil parameters and tree productivity, considering the practices associated with HR management.

Harvest Residue Management Effects

As HRs contain large amounts of nutrients and organic carbon (Achat et al., 2015), their maintenance on site, either incorporated or left on the soil surface, is expected to protect the soil surface and improve soil carbon status and fertility. However, at the end of rotation period at the QF site (MADRP, 2005), HR removal (38.8 t/ha) did not significantly affect the soil organic carbon concentration in the topsoil layer (Table 9.4). This trend suggests that most of the existing carbon in HRs was returned to the atmosphere, as observed in a lysimetric experiment (Gómez-Rey et al., 2008), in which removal or addition of HRs did not lead to significant variations in the soil organic carbon concentration. Removal of HRs led to lower exchangeable calcium than in the I treatment, and it is notable that exchangeable calcium and extractable potassium were lower in the S treatment, as compared with the I treatment, suggesting an effect of residue placement on the decomposition and nutrient dynamics (Magalhães, 2000).

At the term of the rotation, the aggregation coefficient, aggregate mean weight diameter, and available water were similar in tested HR treatments (Table 9.4), but incorporation of HRs by harrowing led to a significantly lower bulk density than in the R treatment. Despite similar physical conditions in R and S treatments (Table 9.4), low water infiltration rates (<21 mm/h) were more frequent (55%) in areas in which organic residues were removed (Magalhães, 2000), suggesting that

Table 9.4 Bulk Density (BD), Aggregation Coefficient (AC), Mean Weight Diameter (MWD), Available Water (AW), pH (H_2O), and Concentrations of Organic Carbon (Org C), Total Nitrogen (N), Exchangeable Ca^{2+}, and Extractable Phosphorus and Potassium in the 0- to 10-cm and 10- to 20-cm Soil Layers at the End of Rotation Period in the Experiment at the Quinta do Furadouro Site

	BD	AC	MWD	AW	pH	N	Org C	Ca^{2+}	Ext P	Ext K
Treatment	g/cm^3	%	mm	% (v/v)		g/kg		$cmol_c/kg$	mg/kg	
0–10 cm										
R	1.37[a]	37.3	0.86	10.1	5.33[a]	0.84	14.8	3.53[a]	5.1	128[ab]
S	1.32[ab]	31.5	0.60	13.5	5.18[ab]	0.79	13.9	3.10[a]	5.3	97[b]
I	1.20[b]	37.0	1.00	13.0	5.10[b]	0.82	16.0	5.22[b]	3.5	145[a]
Is	1.30[ab]	34.8	0.95	10.6	5.07[b]	0.84	15.4	3.77[ab]	5.7	139[ab]
10–20 cm										
R	1.45	—	—	14.1	5.12[b]	0.65	8.97	3.68[ad]	4.8	88[b]
S	1.54	—	—	15.6	5.25[a]	0.63	9.38	2.44[d]	4.8	71[b]
I	1.50	—	—	17.2	5.10[b]	0.62	9.62	4.22[a]	4.4	117[a]
Is	1.49	—	—	11.3	5.00[b]	0.58	8.52	2.58[d]	5.7	94[ab]

Note. Values are means ($n = 24$). R, harvest residue removal; S, broadcast of harvest residues; I, harvest residue incorporation; Is, harvest residue incorporation with ripping. For each soil layer, different letters in the same column indicate significant differences ($P < .05$) between treatments by the Tukey test.

the absence of HRs negatively affected the soil surface characteristics to some extent. Our results indicated that HR removal is possible if the site is sufficiently resistant and resilient to carbon and nutrient losses, but HR incorporation seems to be a better option. Although ripping applied after disc harrowing (treatment Is) did not induce significant changes in soil down to 20-cm depth relative to other treatments (Table 9.4), it significantly reduced compactness up to about 40- to 50-cm depth (Figure 9.3). Long-term monitoring periods over several rotations are necessary for a consistent evaluation of HR management effects on soil status and productivity in eucalypt plantations.

Dominant soils in the experiment carried out at the VP site were close to Planosols (sensu IUSS Working Group, 2015), showing an abrupt textural discontinuity at 20- to 30-cm depth and harmful subsoil compaction (Figure 9.3) more intense than at the PN site (Figure 9.2). In these circumstances, the application of HR management (29.4 t/ha) was associated with ripping down to 80- to 85-cm depth, as it is a useful practice to alleviate compaction and improve drainage and aeration, tree rooting and stability, soil biological habitat, and therefore, productivity of forest plantations; thus, for necessary comparisons, treatment I was applied with and without ripping (MADRP, 2005). The results highlight that the incorporation of organic residues with ripping effectively led to a significant decrease in subsoil compaction (Figure 9.3) and an approximately 63% increase in tree growth in relation to the same treatment without ripping (Magalhães, 2000). Ripping, when applied in the context of adequate soil moisture conditions, is a proper technique to remove an inherent soil

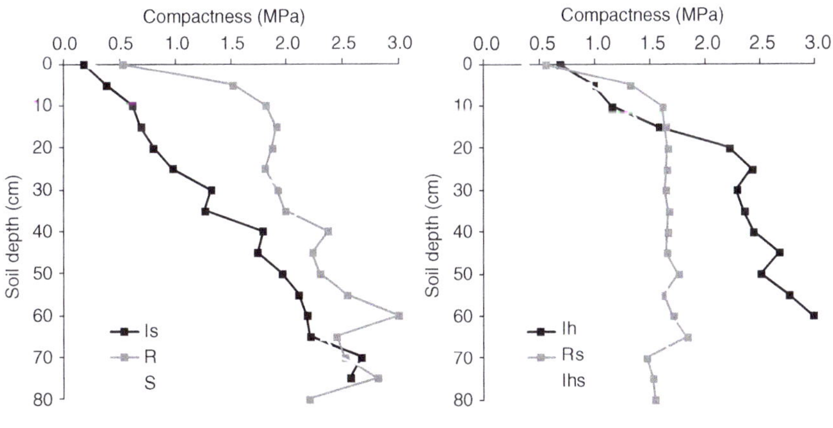

Figure 9.3 Soil resistance to penetration as affected by soil ripping and harvest residue management in the Quinta do Furadouro (left) and Vale Pequeno de Cima (right) sites (unpublished data). R, harvest residue removal; I, harvest residue incorporation; S, broadcast of harvest residues; h, harrowing; s, ripping.

limitation for forest productivity, avoiding inversion of soil layers and potential deleterious changes in soil physical conditions.

In the experiment carried out at the VP site, topsoil showed a lower nutrient status than at the QF site (Magalhães, 2000). Different methods of managing HRs at the end of the rotation period did not significantly affect topsoil physical conditions, such as bulk density, the aggregation coefficient, aggregate mean weight diameter, and available water (Table 9.5). Although treatments with HR incorporation showed higher organic carbon and total nitrogen concentrations, no significant differences occurred between tested treatments, as observed at the QF site. Incorporation of HRs (without ripping) into the soil induced significantly higher concentrations of extractable calcium down to 20-cm depth and extractable phosphorus down to 10-cm depth, as compared with HR removal (Table 9.5). At this site, it was evident that HR removal was deleterious for soil nutrient status, suggesting that the effects of HR management are site specific and are stronger in soils with a lower nutrient status. Therefore, deep information on soil conditions and nutrient status is essential to support decisions on soil and HR management in order to avoid degradation threats.

Table 9.5 Bulk Density (BD), Aggregation Coefficient (AC), Mean Weight Diameter (MWD), Available Water (AW), pH in Water (pH_W), Concentrations of Organic Carbon (Org C), Total Nitrogen (N), Exchangeable Ca^{2+}, and Extractable Phosphorus and Potassium in the 0- to 10-cm and 10- to 20-cm Soil Layers in the Experiment at the Vale Pequeno de Cima Site at the End of the Rotation Period

	BD	AC	MWD	AW		N	Org C	Ca^{2+}	Ext P	Ext K
Treat	g/cm³	%	mm	% (v/v)	pH_W	g/kg		cmol/kg	mg/kg	
0–10 cm										
Rs	1.44	35.8	0.283	9.6	5.16	0.55	11.66	0.13[a]	0.64[a]	29
Ss	1.46	28.1	0.245	8.6	5.11	0.55	13.90	0.49[ab]	1.44[ab]	25
Ihs	1.44	29.4	0.230	8.2	5.15	0.60	15.76	0.65[ab]	1.20[ab]	28
Ih	1.45	22.4	0.152	8.8	4.90	0.61	16.87	0.91[b]	1.97[b]	34
10–20 cm										
Rs	1.46	—	—	10.1	5.29	0.40	8.11	0.05[a]	0.87	24
Ss	1.48	—	—	9.6	5.16	0.34	7.80	0.16[a]	1.28	17
Ihs	1.44	—	—	9.2	5.21	0.37	8.79	0.20[ab]	1.18	21
Ih	1.51	—	—	10.7	4.98	0.43	10.14	0.51[b]	1.93	19

Note. Values are means ($n = 25$). R, Harvest residue removal; S, broadcast of harvest residues I, harvest residue incorporation; h, harrowing; s, ripping. For each soil layer, different letters in the same column indicate significant differences ($P < .05$) between treatments by the Tukey test.

Harvest Residue Management and Biological Indicators

In a review regarding the *Collembola* communities in eucalypt plantations in Portugal, Sousa et al. (2000) concluded that such communities are dependent on both the plantations themselves and their respective management systems. Thus, the effects of HR management (removal, R; maintenance on the soil surface, S; and incorporation, I) on indexes of diversity, equitability, and richness for great groups of Arthropods and *Collembolla* were evaluated in the trial conducted at the QF site (Table 9.6). Results indicate that such indexes mostly showed no significant differences related to residue treatments but showed strong variations between sampling dates. Differences occurred only in some samplings, as observed for treatment S in May 2000, regarding *Collembolla* communities (Table 9.6). In addition, in the same field trial, HR treatments did not show significant differences in the percentage of active ectomycorrhiza or in the number of ectomycorrhizal morphotypes (Marchante et al., 2001).

Nitrogen mineralization at the QF site was determined by using aerobic and anaerobic laboratory incubations. These results also showed no significant differences associated with HR management (data not shown), and net nitrogen mineralization in aerobic incubation was largely associated with N-NH_4^+, following data reported in a study that included the same site (Gómez-Rey et al., 2010).

Table 9.6 Indexes of Diversity of Shannon (DIV), Equitability of Pielou (EQI), and Richness of Margalef (RIC) for the Great Groups of Arthropods and Collembolla According to Harvest Residue Management for Samplings on Different Dates

		Arthropods			Collembolla		
Date	**Treatment**	**DIV**	**EQI**	**RIC**	**DIV**	**EQI**	**RIC**
April 1998	R	1.00	0.26	1.98	3.34	0.79	3.92
	S	0.94	0.28	1.31	2.11	0.50	3.39
	I	1.06	0.31	1.44	1.92	0.48	2.93
May 2000	R	1.27	0.45	1.00	1.05[b]	0.66	0.57[b]
	S	1.46	0.42	1.39	2.18[a]	0.51	3.21[a]
	I	1.17	0.45	0.93	1.92[a]	0.96	1.67[ab]
May 2003	R	0.91	0.26	1.31	2.79[a]	0.67	3.46
	S	1.07	0.31	1.31	2.94[a]	0.67	3.40
	I	0.88	0.24	1.59	1.44[b]	0.55	3.88

Note. Data from Seabra-Ferreira et al. (2001) and unpublished data. Treatments as in Table 9.4. For each date, different letters in the same column indicate significant ($P < .05$) differences between treatments by the Tukey test.

These results may be site specific, as suggested by a study encompassing five eucalypt plantations within a precipitation gradient (500–1,800 mm). In that study, it was found that net nitrogen mineralization as well as carbon and nitrogen mineralization rates mostly depended on the level of precipitation (see Gómez-Rey et al., 2010).

Soil Quality in Evergreen Oak Woodlands

Effect of Scattered Oak Trees

Scattered single trees in *montado* systems accumulate high amounts of organic carbon in their biomass (Ruiz-Peinado et al., 2012), influence the microclimate (Nunes et al., 2011), and strongly influence soil functions in a gradient that changes with the distance from the tree trunk relative to the neighboring open areas (Manning et al., 2006). Indeed, at the TV site, a strong increase of soil organic carbon concentration was observed beneath scattered QR canopies at least down to 20 cm in the areas closer to the tree trunk (Figure 9.4). In the 0- to 5-cm soil layer, the concentration doubled from a distance twice the tree crown radius to the

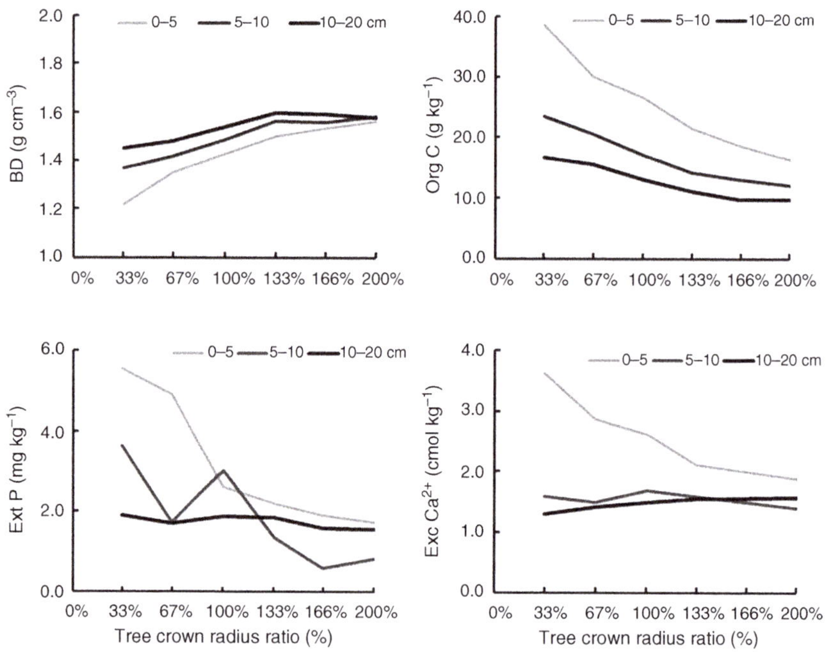

Figure 9.4 Variation of soil parameters with the distance to the tree trunk (% of tree crown ratio) and soil depth, in a *Quercus rotundifolia* woodland located at the Tapada Real de Vila Viçosa site (Table 9.1). Adapted from Rodrigues (2019).

neighborhood of the tree trunk. This pattern followed that reported for QR woodlands, grazed or non-grazed, at the HM (Nunes et al., 2001) and HE (Gómez-Rey et al., 2011) sites and in California oak woodlands (Dahlgren et al., 1997, 2003), *dehesas* (Simón et al., 2013), and an alley cropping agroforestry system (Cardinael et al., 2015). This increment of organic carbon in the soil beneath trees might be related to root litter (David, 2000) and organic residues deposited on the soil surface by litter fall (Escudero et al., 1985; Sá et al., 2001; Nunes et al., 2001).

In turn, soil bulk density beneath trees was lower than in the open areas (Figure 9.4), agreeing with trends reported for other oak woodlands (Dahlgren et al., 1997, 2003; Gómez-Rey et al., 2012) and agroforestry systems (Cardinael et al., 2015). The lowering of soil bulk density beneath tree canopies is most likely associated with the higher soil OM content, which, coupled with increased mixing soil fauna/flora activity, may enhance soil structure and result in increased total porosity (Dahlgren et al., 2003). Although the accumulation of OM itself contributes to a decrease in soil bulk density, such a decrease underneath the trees may be mostly due to the increasing of soil porosity. Indeed, soil porosity in the 0- to 5-cm layer increased from 41% (v/v) in the open areas to 54% (v/v) in areas closer to the tree trunk. This increase may enhance soil aeration and drainage and improve plant rooting and activity of soil organisms (Fisher & Binkley, 2000).

Soil fertility indices (e.g., extractable P, exchangeable Ca^{2+}) indicated that soils underneath the trees are more fertile than in the open areas (Figure 9.4), following trends reported for oak woodlands (Dahlgren et al., 2003; Gallardo, 2003). This difference is primarily associated with nutrient recycling via decomposition of tree litter fall (Dahlgren et al., 1997; Andivia et al., 2010; Sá et al., 2001) and decreased nutrient loss from the soil profile caused by increased transpiration and rainfall interception by tree canopies (David et al., 2006). Thus, the increment of soil fertility beneath tree crowns is strongly associated with the carbon cycle and driven by OM accumulation, which also leads to a higher concentration of nitrogen (Gallardo et al., 2000) and higher capacity to retain cations (Amorim et al., 2023). These trends highlight the role of tree islands at the landscape level, which may depend on tree density, soil type, land and soil management, and the history of site management.

Replacement of Natural Pastures by Improved Pastures

At the HE site, long-term IP grazed by sheep improved the soil bulk density and soil carbon status with varying intensities according to soil depth and tree presence (Figure 9.5). In fact, values of soil bulk density for each soil depth were lower in IP than in NP both beneath trees and in the open areas, but differences were only significant beneath tree crowns in the 0- to 10-cm soil layer. Conversely, concentrations of organic carbon were higher in IP than in NP, and significant differences occurred only in the 0- to 10-cm soil layer and underneath trees.

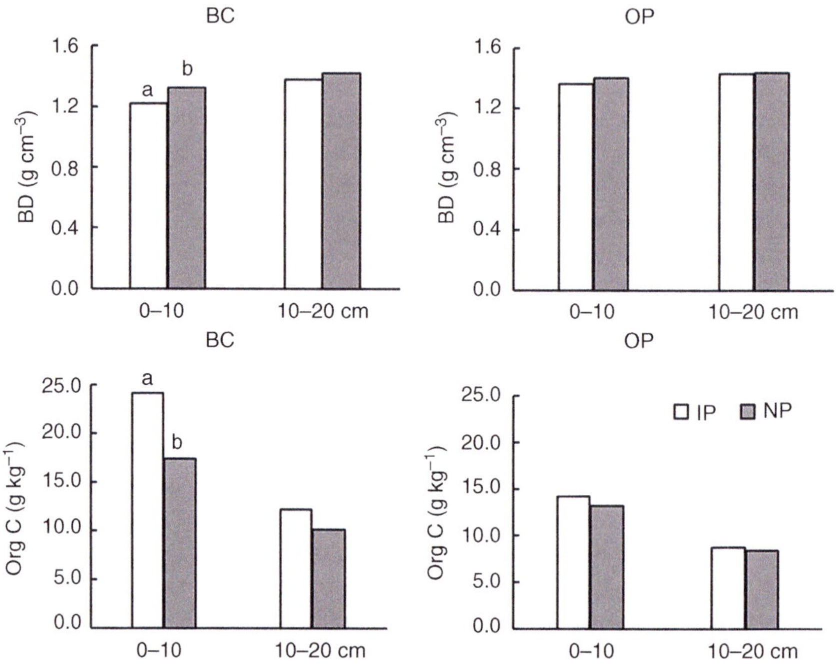

Figure 9.5 Bulk density (BD), and organic C (Org C) concentrations in improved (IP) and natural (NP) pastures, beneath trees (BC) and in the open areas (OP). Adapted from Gómez-Rey et al. (2012). Different letters in the same depth indicate significant ($P<.05$) differences.

Improved pastures enhanced the labile OM pool (Table 9.7), as the increase in carbon in the particulate OM (C_{POM}) and hot-water-soluble carbon (C_{HW}) in the open areas was stronger than for total organic carbon, leading to wider differences between studied pastures. Indeed, C_{POM} in the 0- to 10-cm soil layer in IP beneath the trees was about twice that measured in NP, and the concentration of the carbon organic fraction extracted by hot water was four times higher. This trend may be associated with tree inputs and interaction between trees and pastures, and measured labile fractions could be early indicators of management effects on soil. The carbon in microbial biomass (C_{mic}) showed no significant differences between study pastures, but its proportion relative to the soil organic carbon (C_{mic}/C) was significantly higher in NP, both beneath trees and in the open areas. Conversely, the metabolic coefficient (qCO_2) was significantly higher in IP than in NP, both beneath trees and in the open, but the potentially mineralizable carbon (*Co*) and the carbon mineralization rate (CMR) did not show significant differences between pasture types. Despite the lower soil C_{mic}/C ratio under IP, the CMR was similar in IP and NP, which may have been associated with the greater qCO_2 measured for the former (Table 9.7).

Table 9.7 Carbon in Particulate Organic Matter (C_{POM}) and Extractable by Hot Water (C_{HW}) Fractions, Microbial C (C_{mic}), Metabolic Coefficient (qCO_2), Potentially Mineralizable Carbon (*Co*), and Carbon Mineralization Rate (CMR) in the 0- to 10-cm Soil Layer of an QR Woodland Under Improved (IP) and Natural (NP) Pastures at the Herdade dos Esquerdos Site

Treat	C_{POM} (g/kg)	C_{HW} (g/kg)	C_{HW}/C (mg/kg)	C_{mic} (mg/kg)	C_{mic}/C	qCO_2 (mg C-CO_2·g^{-1} C_{mic}·h^{-1})	*Co* (mg C-CO_2/ kg soil)	CMR (mg C-CO_2/g C)
Beneath tree canopy								
IP	9.8[a]	1.12[a]	4.64[a]	491	1.83[b]	2.68[a]	3782	141.9
NP	5.2[b]	0.28[b]	1.61[b]	426	2.62[a]	1.93[b]	2647	162.2
In the open								
IP	3.9	0.29	2.04[a]	363	2.46[b]	2.14[a]	1876	122.0
NP	2.8	0.15	1.14[b]	388	3.10[a]	1.33[b]	1643	130.9

Note. Adapted from Gómez-Rey et al. (2012). Different letters in the same column indicate significant differences ($P < .05$) between pastures by least square differences (LSD) when significant ANOVA results occurred.

Table 9.8 Extractable Phosphorus (Ext P), Total Nitrogen (TN), Microbial Nitrogen (N_{mic}), Net Nitrification (NN), Net Nitrogen Mineralization (NNM), Nitrogen Mineralization Rate (NMR) in the 0- to 10-cm Soil Layer of Improved (IP) and Natural (NP) Pastures Beneath the Tree Canopy and in the Open at the Herdade dos Esquerdos Site

Treat	Ext P (mg/kg)	TN (g/kg)	C/N	N_{mic} (mg/kg)	N_{mic}/ TN	NN (mg N-NO_3^-/ kg soil)	NNM (mg N/ kg soil)	NMR (mg N g/ N total)
Beneath the canopy								
IP	25.7[a]	1.91[a]	12.6	42.6	1.94	266.7[a]	261.8[a]	108.4[a]
NP	3.0[b]	1.34[b]	13.0	41.7	2.52	64.7[b]	57.9[b]	42.9[b]
In the open								
IP	14.2[a]	1.22[a]	11.6	34.9	2.71	107.9[a]	105.9[a]	80.8[a]
NP	1.2[b]	1.04[b]	12.6	37.9	2.98	28.2[b]	25.5[b]	25.5[b]

Note. Adapted from Gómez-Rey et al. (2012). For each position, different letters in the same column indicate significant differences ($P < .05$) between pastures by least square differences (LSD) when significant ANOVA results occurred.

Total nitrogen concentration significantly increased under IP as compared with NP, both under tree crowns and in the open areas (Table 9.8). Extractable phosphorus exhibited a similar trend, which was in accordance with long-term fertilizer practices followed for the establishment and maintenance of IP. Microbial nitrogen biomass (N_{mic}) was also similar among studied pastures, and the net nitrification and the net nitrogen mineralization were significantly higher in IP than in NP, both underneath trees and in the open. Naturally, the nitrogen mineralization rate (NMR) was much higher in both positions (respectively, about 2.5 and 3.0 times) for IP. That is, the stronger soil nitrogen availability in IP was associated with the higher total nitrogen content and higher NMR. This trend was confirmed by in situ incubation for 2 years, where the annual average NMR (mg N mineralized/g N total) in the open areas was about 5.2 for IP and 3.6 for NP (unpublished data).

Soil enzymatic activities (β-glucosidase, urease, protease, and acid phosphomonoesterase) in the studied pastures were evaluated for 1 year by using soil samples seasonally collected in the 5-cm topsoil layer of IP and NP at the HE site (Rodrigues et al., 2015). Enzyme activities associated with carbon, nitrogen, and phosphorus cycles did not respond consistently to pasture management, but were rather influenced by tree islands. In fact, β-glucosidase activity and nitrogen cycle-related soil enzymatic activity were positively enhanced under the tree canopy throughout the study period, whereas enzymatic acid phosphomonoesterase activity was negatively affected. In conclusion, further evaluations should consider different environmental factors, Mediterranean climate seasonality, and interannual variability.

Replacement of Shrub Understory by Improved Pastures

At the MG site, soils of a cork oak stand with periodically controlled natural shrubs (Figure 9.6, left) showed a notable high organic carbon concentration (Table 9.9), considering their coarse texture (Arenosols), and a high C : N ratio and very low concentrations of extractable phosphorus and exchangeable Ca^{2+} (Table 9.9).

After 5 years, the replacement of natural shrubs by an IP (Figure 9.6, right) did not change soil bulk density values and led to a nonsignificant decrease of soil

Figure 9.6 Cork oak stand with natural shrub understory (NU) and with a 5-year-old improved pasture (IP) at the Machoqueira do Grou site. Photos by Ana Raquel Rodrigues.

Table 9.9 Bulk Density (BD), pH Values, and Concentrations of Organic Carbon (Org C), Nitrogen, Particulate Organic Matter (C_{POM}), Carbon Extracted By Hot Water (C_{HW}), Extractable Phosphorus, and Exchangeable Calcium at the MG Site

	BD	Org C	N	C/N	C_{POM}	C_{HW}	pH	Ext P	Ca^{2+}
Treat	g/cm^3	g/kg			g/kg			mg/kg	cmol$_c$/kg
0–10 cm									
IP	1.18	17.5	0.91[a]	19.5[b]	8.2	0.63[a]	5.68[a]	6.8[a]	1.87[a]
NU	1.18	21.4	0.65[b]	32.6[a]	9.6	0.48[b]	5.23[b]	3.3[b]	0.69[b]
10–20 cm									
IP	1.27	7.5	0.41[a]	18.9[b]	2.2	0.34	5.51	6.0[a]	0.45
NU	1.26	9.0	0.30[b]	30.0[a]	2.2	0.24	5.33	2.1[b]	0.30

Note. Adapted from Rodrigues et al. (2019). NU, natural shrub understory; IP, improved pasture. For each soil layer, different letters in the same column indicate significant differences ($P < .05$) between treatments by the Student *t*-test.

organic carbon concentration and C_{POM} fraction (Table 9.9). This trend did not follow that observed for the long-term replacement of NP by IP, and it may have been associated with the soil disturbance associated with establishment of the pasture and the decrease of above- and belowground residue inputs from shrub vegetation (Rodrigues et al., 2019). The short period of evaluation may be another reason for such differences.

In contrast, IP significantly increased total nitrogen (in parallel with a strong decrease of the C : N ratio), extractable phosphorus and calcium concentrations, and soil pH values, mostly in the 0- to 10-cm soil layer (Table 9.9) and potentially in association with the role of pasture leguminous and fertilizer inputs. Despite the lower concentrations of organic carbon and particulate OM, IP led to an increase of about 30% of the concentration of organic carbon extracted by hot water, indicating the usefulness of this indicator to detect changes in surface soil layers (Ghani et al., 2003; Chen et al., 2009) driven by the establishment of such pastures.

Improved pastures did not significantly increase microbial carbon biomass and CMR relative to natural shrub understory, but significantly increased the initial metabolic coefficient (Table 9.10), following trends observed in the evaluation of IP compared with NP. Improved pastures showed no significant effect on microbial nitrogen biomass, but led to significantly higher cumulative nitrogen mineralization (and to a large predominance of nitrification) and NMR (Table 9.10). That is, short-term IP had enhanced nitrogen cycling and soil fertility, following the trend observed for older improved pastures.

Table 9.10 Initial Content of Microbial Carbon and Nitrogen (C_{mic}, N_{mic}), Microbial C:N Ratio, Initial Metabolic Coefficient (qCO_2), Carbon Mineralization Rate (CMR), Cumulative Net Nitrification and Ammonification, and Nitrogen Mineralization Rate (NMR) After 112 Days of Aerobic Incubation in Soil from the 5-Year-Old Improved Pasture (IP) and Natural Understory (NU)

Treat	C_{mic} (mg/kg)	N_{mic} (mg/kg)	C_{mic} : N_{mic}	qCO_2 (mg C-CO_2· g^{-1} C_{mic}·h^{-1})	CMR (mg C-CO_2/g C)	Net mineralized N (mg N/kg soil) NO_3^--N	NH_4^+-N	NMR (mg N/g N total)
IP	172.9	31.9	5.4^a	3.0^a	26.0	93.7^a	−5.4^b	69.6^a
NU	177.5	26.1	6.9^b	2.1^b	30.1	17.5^b	45.2^a	45.5^b

Note. Adapted from Rodrigues et al. (2019). Different letters in each column indicate significant differences ($P < .05$) between treatments by the Student t-test.

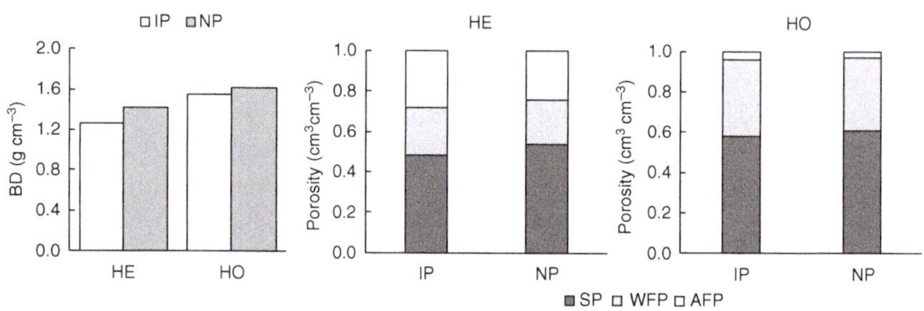

Figure 9.7 Bulk density (BD), and volume of air-filled porosity (AFP) and water-filled porosity (WFP), at −10 kPa, and solid phase (SP), for the 0- to 10-cm soil layer under improved (IP) and natural (NP) pastures, at the Herdade dos Esquerdos (HE) and Herdade do Olival (HO) sites. Adapted from Rodrigues (2019).

Grazing Systems

The soil bulk density values observed down to 10-cm depth at the HO site in the IP and NP grazed by cattle (1.55 and 1.63 g/cm^3, respectively) were much higher than those measured in similar pastures grazed by sheep (1.27 and 1.42 g/cm^3, respectively) at the HE site (Figure 9.7). Accordingly, values of total soil porosity were lower in the former (0.42 and 0.39 cm^3/cm^3, respectively) than in the latter (0.52 and 0.46 cm^3/cm^3, respectively). Such a difference suggests the presence of soil compaction, a documented process associated with the increase of grazing intensity in pastures (Greenwood & McKenzie, 2001), silvopastures (Sharrow, 2007), and holm oak woodlands (Pulido et al., 2017, 2018). This trend led to strong changes in soil aeration porosity. Indeed, in contrast to the HE site, where the estimated air-filled porosity at field capacity (at −10 kPa soil water potential) in the open areas was 0.28 and 0.24 cm^3/cm^3 in the IP and NP, respectively (Figure 9.7), at the HO site, values were much lower or negligible, varying between 0.04 and 0.03 cm^3/cm^3, respectively. Air-filled porosity at −33 kPa was higher but corresponded only to 0.08–0.11 cm^3/cm^3 at the HO site (data not shown).

Pastures (IP and NP) and position relative to trees did not significantly influence mean values of soil saturated hydraulic conductivity (K_S) at either the HE or HO site (data not shown; Rodrigues, 2019). At the HE site, grazed by sheep, soils under IP had generally higher and less variable K_S values than under NP (Figure 9.8). At the HO site, grazed by cattle, soils showed a wider range of K_S (0–11,142 mm/h) than at the HE site (10–5,019 mm/h; Rodrigues, 2019). Soils of IP at the HO site more frequently had K_S values between 1 and 10 mm/h, while values in NP were mostly below 1 mm/h (Figure 9.8). In contrast with the trend observed for the HE site, soils in the HO site showed a high variability of K_S values, with a greater proportion of those below 10 mm/h in both IP and NP. These data agree with the above-mentioned trends for bulk density and aeration porosity and emphasize that intensively managed grazing can actually lead to the degradation of soil (Billota et al., 2007), hindering positive effects associated with the replacement of NP by IP.

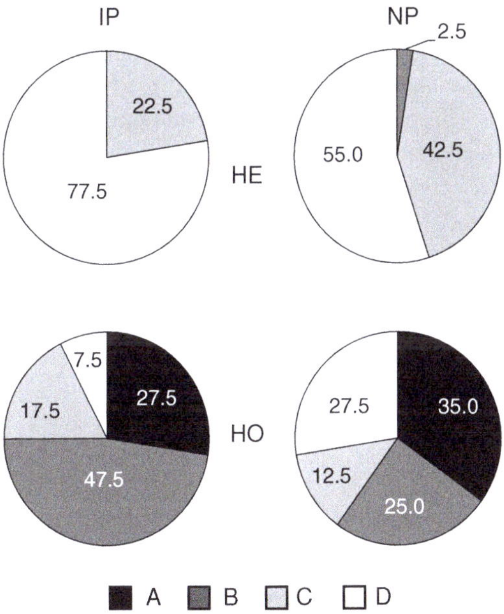

Figure 9.8 Proportion of classes of soil saturated hydraulic conductivity, measured in improved (IP) and natural (NP) pastures at the Herdade dos Esquerdos (HE) and Herdade do Olival (HO) sites. A, <1 mm/h; B, 1–10 mm/h; C, 11–100 mm/h; D, >100 mm/h. Adapted from Rodrigues (2019).

Shrub Encroachment in Degraded Soil

At the HM site, the effect of shrub encroachment (CS and CL) on soil condition was assessed in the 0- to 10-cm soil layer through organic carbon concentration (and labile fractions) and biological indicators (Gómez-Rey et al., 2013). Although higher beneath *Cistus* shrubs, concentrations of organic carbon, total nitrogen, and carbon of the particulate OM were not significantly different from those observed in the bare soil (NE in Table 9.11), following the trend reported in a previous study (Simões et al., 2009). Conversely, C_{HW} was 2.6 times higher beneath shrubs, suggesting that this small fraction, representing a readily decomposable fraction and a main substrate for microorganisms (Ghani et al., 2003), is an early indicator of changes related to soil vegetation cover. The significantly higher potentially mineralizable carbon beneath shrubs (20 weeks of incubation; Table 9.11) followed the trend reported by Moro et al. (1997) for other shrubs and by Gómez-Rey et al. (2012) for Mediterranean oak woodlands. The increment in cumulative carbon mineralized may be explained by the higher proportion of C_{HW}, highlighting the role of this labile carbon fraction on soil carbon dynamics. Thus, shrub encroachment enhances soil carbon cycling, in which the higher carbon mineralization may also limit carbon accumulation in soil.

The enhanced soil carbon pool beneath shrubs sustained a larger soil microbial carbon concentration (although not significant for CS), while the soil microbial nitrogen was not significantly different (Table 9.11). The variation in the soil C_{mic} : OC ratios suggests that a higher microbial biomass was supported per unit of soil organic carbon under shrubs. This possibility agrees with the similar microbial metabolic quotient (qCO_2), which implies that the substrate use efficiency by the soil microbial community was not altered (Wardle & Ghani, 1995). These trends suggest that shrub encroachment does not affect the substrate-carbon use efficiency, and the increment in microbial activity beneath shrub canopies seems to be the consequence of a higher growth rate of soil microorganisms rather than a more efficient microbial community. As shrub encroachment did not affect the soil nitrogen content, the increase in nitrogen mineralization may be in relation to faster turnover of soil nitrogen (Table 9.11).

Soil Quality in Chestnut Groves

Monitoring activity in intensively managed chestnut groves at the LP site assessed conventional soil tillage with disc harrowing and other less intensive practices, such as no-tillage with spontaneous vegetation (NV) and seeded improved pastures (NS; Figure 9.9). Conventional tillage and disc harrowing treatments were practiced three times per year (beginning and end of spring, beginning of autumn), including the incorporation of litter fall (and other organic residues) into the soil, whereas in the no-tillage treatments, residues were kept on the soil surface (Raimundo, 2003).

Table 9.11 Concentrations of Soil Organic Carbon (OC) and Total Nitrogen (TN), Carbon in Particulate Organic Matter (C_{POM}) and Hot Water Extracts (C_{HW}), and Microbial Carbon and Nitrogen (C_{mic} and N_{mic}), Initial Metabolic Coefficient (qCO_2), Cumulative Carbon Mineralization (C_{min}), Carbon Mineralization Rate (CMR), Cumulative Net Nitrogen Mineralization (N_{min}), and Nitrogen Mineralization Rate (NMR) as Influenced by Shrub Encroachment

	OC	TN	C_{POM}	C_{HW}	C_{min}	C_{mic}	qCO_2	CMR	N_{mic}	N_{min}	NMR
Treatment	g/kg				mg $C\text{-}CO_2$/kg soil	mg/kg	mg $C\text{-}CO_2 \cdot g^{-1}$ $C_{mic} \cdot h^{-1}$	mg-CO_2/ g C	mg/kg	mg N/kg soil	mg N/g N
NE	7.7	0.69	2.21	0.15^{a}	703^{a}	328^{a}	0.58	90^{a}	33.2	43.9^{a}	69.5^{a}
CS	10.3	0.75	3.57	0.39^{b}	1.790^{b}	505^{ab}	0.46	186^{b}	41.4	84.1^{b}	116.8^{b}
CL	10.7	0.80	3.71	0.37^{b}	1.796^{b}	637^{b}	0.38	163^{b}	53.7	83.0^{b}	101.7^{b}

Note. Adapted from Gómez-Rey et al. (2013). Different letters in the same column indicate significant differences ($P < .05$) according to soil cover by the Tukey test.

Figure 9.9 Chestnut groves subjected to conventional tillage (left) and no tillage with seeded pastures (right) at the Lamas de Podence site. Photos by Fernando Raimundo.

Table 9.12 Bulk Density (BD), Aeration Porosity (AP), Resistance to Penetration (RP), Infiltration Rate (IR), Available Water (AW) According to Soil Depth in Conventional Tillage (CT), Harrowing (HR), No-Tillage (NV), and No-Tillage with Seeded Pasture (NS) Treatments; Months After Treatment Application: 44 (A), 48 (B), 92 (C), and 100 (D)

	BD (g/cm³)		AP (cm³/100 cm³)	RP (MPa)		IR (cm/h)		AW (cm³/cm³)
Treatment	A	C	A	B	D	A	D	A
0–10 cm								
CT	1.09[a]	1.12	23.1[a]	1.14[a]	0.95[a]	36.9[a]	79.9[a]	0.17
HR	1.14[a]	1.20	17.9[b]	1.65[ab]	1.63[ab]	—	13.7[c]	0.16
NV	1.29[b]	1.31	13.5[c]	2.12[b]	1.98[b]	7.2[b]	66.8[ab]	0.15
NS	—	1.32	—	2.07[b]	1.90[b]	—	34.5[bc]	—
10–20 cm								
CT	1.28	1.34	18.8[a]	3.07	2.18	—	—	0.17
HR	1.35	1.36	12.0[b]	3.37	2.44	—	—	0.13
NV	1.32	1.33	9.1[b]	3.16	2.41	—	—	0.14
NS	—	1.42	—	3.40	2.38	—	—	—

Note. Different letters in the same column, for each soil depth, indicate significant differences ($P < .05$) between soil management practices by the Tukey test.

Significant differences in soil physical indicators among treatments during a 9-year period mostly occurred in the 0- to 10-cm soil layer (Table 9.12). Surprisingly, values of bulk density, aeration porosity, resistance to penetration, available water, and infiltration rate indicated better soil quality and health under conventional tillage than in no-tillage with spontaneous vegetation or improved pastures. This trend may have been associated with the annual and seasonal soil disturbance in the conventional treatment, and it suggests that longer periods are necessary to consistently evaluate effects of no-tillage systems. Although conventional tillage did not appear to exert a negative influence on physical soil indicators, it might have had deleterious effects on the tree root system and soil biological conditions (Raimundo, 2003).

Soil under different treatments showed high organic carbon concentrations up to 10-cm depth. About 8 years after the application of treatments, it was noted that the soil organic carbon concentration did not significantly differ between them (Figure 9.10). A slight increment was found in the 0- to 5-cm soil depth in no-tillage treatments, but longer evaluation periods are necessary to obtain a clearer trend in this regard. Although conventional tillage enhances OM mineralization, the incorporation of organic residues into the soil may counteract such an effect. That is, in the short term, the placement (incorporated and maintenance on the soil surface) of organic residues may influence the trend of soil organic carbon concentration in study treatments.

A similar tendency occurred on farms that followed the conventional or the no-tillage management system at a site (TR site) with ecological conditions similar to the LP site (Table 9.13), where no significant differences in soil organic carbon concentrations were detected down to 20-cm depth (Borges et al., 2018). Nevertheless, the labile organic carbon fractions, in the 0- to 10-cm soil layer, showed significant differences between management systems, with higher concentrations in the no-tillage treatment (Table 9.13). Indeed, the particulate organic carbon fraction, a precursor of other labile soil organic carbon fractions (Haynes, 2005) and activity of microorganisms (Six et al., 2002), increased about 42% in the no-tillage system relative to the conventional tillage system. The active carbon (i.e., the fraction of organic carbon oxidized by 0.02 M $KMnO_4$) and the C_{HW} showed an increment of 36% and 32%, respectively.

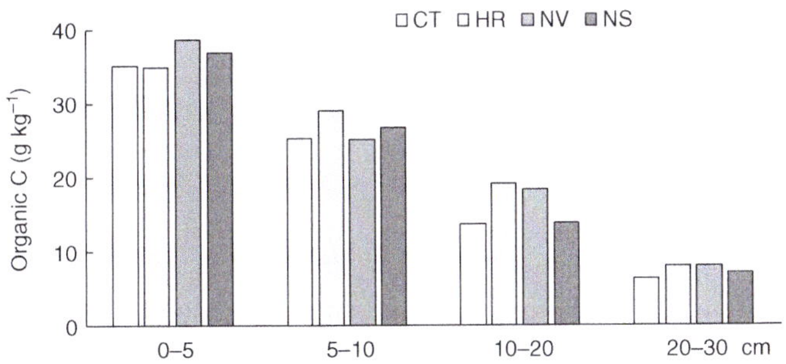

Figure 9.10 Concentration of organic carbon, according to soil depth (cm), in conventional tillage (CT), harrowing (HR), no-tillage (NV), and no-tillage with seeded pasture (NS), measured 94 months after treatment application.

Table 9.13 Concentration of Soil Organic Carbon (Org C) and the Particulate Organic Carbon (POM), Active Carbon (AC), and Hot-Water Extractable Carbon (C_{HW}) Fractions in Soils of Chestnut Groves Under No-Tillage (NT) and Conventional Tillage (CT) Management at the Terroso Site

Soil depth	Treatments	Org C	C_{POM}	AC	C_{HW}
cm		g/kg		mg/kg	
0–10	NT	15.31 ± 3.82	4.57 ± 1.48[a]	340 ± 90[a]	660 ± 210[a]
	CT	14.80 ± 1.85	3.22 ± 0.90[b]	250 ± 40[b]	500 ± 50[b]
10–20	NT	9.77 ± 1.34	1.58 ± 0.31	160 ± 40	330 ± 50
	CT	10.89 ± 1.83	1.44 ± 0.24	150 ± 60	310 ± 60

Note. Adapted from Borges et al. (2018). Values are mean ± standard deviation ($n = 16$). For each soil depth, different letters in the same column indicate significant differences ($P < .05$) between soil management practices by the Tukey test.

Table 9.14 Concentrations of Organic Carbon (Org C), Total Nitrogen (TN), Carbon Microbial Biomass (CBM), Nitrogen Microbial Biomass (NBM), and Soil Respiration (SR) in the 0- to 15-cm Soil Depth Measured 45 Months After Treatment Installation

Treat	Org C	CBM	SR	TN	NBM
	mg/g soil	µg/g soil	µg C-CO_2·g^{-1}·day^{-1}	mg/g soil	µg/g soil
CT	21.5	545.0	49.9[a]	1.32	59.5
HR	22.7	464.1	46.4[a]	1.36	54.6
NV	22.4	533.9	63.8[b]	1.37	57.3
NS	23.6	561.1	51.3[a]	1.55	70.2

Note. Adapted from Raimundo et al. (2004). CT, conventional tillage; HR, harrowing; NV, no-tillage; NS, no tillage and improved pasture. Different letters in the same column indicate significant differences ($P < .05$) between soil management practices by the test of Fisher PLSD.

Following the trend observed for the concentration of soil organic carbon, concentrations of nitrogen and carbon and nitrogen microbial biomass obtained from soil incubation carried out 45 months after treatment application (at the LP site) also showed no significant differences between tillage and no-tillage treatments (Table 9.14). However, soil respiration was significantly higher in the no-tillage (with spontaneous vegetation) treatment than in the others.

The assessment of treatments on soil nitrogen mineralization dynamics (and availability) took place in the field at the LP site by using the in situ sequential coring method (Raison et al., 1987). Incubation was done from March to November in both 1999 and 2004, and tubes were incubated in the 0- to 15-cm soil layer.

Table 9.15 Amount of Nitrogen, Cumulative Mineral Nitrogen (NH_4^+-N + NO_3-N) from February to November, and Nitrogen Mineralization Rate up to 15-cm Soil Depth, and Foliar Nitrogen (FN) Concentration in Chestnut Groves According to Soil Management Treatments

Year	Treatment	Total N	Mineral N	N Mineralization rate	FN
		kg/ha	kg/ha	g N/100 g N	mg/g
1999	CT	2175.8	46.4[c]	2.12	—
	NV	2406.7	58.3[b]	2.42	—
	NS	2585.3	71.7[a]	2.77	—
2004	CT	2175.7	51.8[b]	2.38	15.1[b]
	NV	2423.5	65.9[ab]	2.72	17.3[a]
	NS	2536.5	85.1[a]	3.36	17.6[a]

Note. Adapted from Raimundo et al. (2015). Treatments as in Table 9.14. For each year, different letters in the same column indicate significant differences ($P < .05$) between soil management practices by the Tukey test.

Results showed that total nitrogen was higher in the no-tillage than in the conventional treatment, but it was not significantly different (Table 9.15), and annual cumulative mineralized nitrogen was 27–28% and 55–64% higher, respectively, in NV and NS relative to the conventional management system. Such a difference was partially associated with a higher NMR in the no-tillage treatments. The significantly higher nitrogen concentration in the foliage of chestnut trees subjected to the no-tillage treatments followed the differences observed in mineral nitrogen availability (Table 9.15).

Although no-tillage treatments (with natural and improved seeded pastures) in chestnut groves did not show clear trends in physical and some biological indicators of soil quality, and even apparently indicated negative effects, in another similar trial they significantly increased nut productivity, forage, and sporocarp biomass and diversity relative to conventional practices, especially when associated with irrigation (Table 9.16). However, variations in productivity owing to annual drought conditions were much more strongly important than those associated with treatments (Martins et al., 2010). Conservative soil management practices also significantly increased production of edible mushrooms species (with a high amount of *Boletus edulis*), contributing to sustainable management of wild edible mushrooms that are increasingly collected and commercialized. No-tillage treatments increased species richness, the occurrence of sporocarps, and the diversity of fungal sporocarps based on the Shannon diversity index (Table 9.16). These increases may have been associated with more crop residues (mainly from roots) providing a greater and heterogeneous substrate to support fungal

Table 9.16 Production of Nuts, Forage, Total Sporocarps (TS), and Commercial Edible mushrooms (CEM); Species Richness (SR); Sporocarp Occurrence (SO); and Shannon Index H (SIH) in a Chestnut Grove Subjected to Several Managements Systems

Treatment	Nuts	Forage	TS	CEM	SR	SO	SIH
	kg/tree	g/m^2	kg/ha^2				
CT	18.8	—	22.5[a]	6.1[a]	11.0[a]	20.3[a]	2.9[a]
NV	27.2	228.3	120.0[c]	43.0[b]	17.9[b]	45.0[b]	3.4[b]
NS	19.6	320.1	85.0[b]	43.0[b]	23.4[c]	58.7[c]	3.5[b]
NSI	24.4	381.0	151.4[d]	82.0[c]	27.4[c]	76.3[d]	3.5[b]

Note. Adapted from Martins et al. (2010). CT, conventional tillage; NV, no tillage and natural vegetation; NS, no tillage and improved pasture; NSI, same as NS but irrigated. Different letters in the same column indicate significant differences ($P < .05$) between soil management practices by the Tukey test.

metabolic activity and development of sporocarps. Although sporocarp data represent an incomplete view of the total soil fungal diversity, differences in sporocarp abundance between treatments probably reflect different mycelial abundance in the soil, as a result of the reduced soil disturbance intensity. The no-tillage treatment with spontaneous herbaceous vegetation seemed to be the most favorable management system to ensure productivity and sustainability of chestnut plantations, as it also led to the highest annual net income (Martins et al., 2010).

General Considerations and Perspectives

Research and monitoring developed at the study sites demonstrated how various management practices can affect indicators of soil quality in forestry and agroforestry systems. The research and monitoring have also advanced knowledge about how these indicators respond to land and soil management, creating perspectives for their sustainable management. Nevertheless, temporal changes of soil quality indicators suggest the need for long-term monitoring (e.g., for several rotations) in stable research sites across a range of soil and forest ecosystem types, in order to consistently assess trends in soil physical, chemical, and biological conditions; organic carbon stocks; and soil nutrient status (Raison & Rab, 2001).

Monitoring activities in eucalypt plantations established by using former deep and intensive site preparation practices, such as deep plowing, disc harrowing, and certainly, bench terracing, revealed that soil physical and chemical conditions are negatively affected beyond the first rotation period. Monitoring also showed that physical and chemical changes in the topsoil of eucalypt plantations as being affected by different options of HR management (removal, incorporation, maintenance on the soil surface), with or without ripping, were nonexistent or occurred at a low level. Although HR management options did not significantly affect topsoil conditions (Magalhães, 2000), tree productivity increased (a soil-based ecosystem service) from removal to incorporation of HRs at the QF site but not at the VP site (Figure 9.11), where tree growth was much lower. These differences may have been driven by variations in ecological conditions and inherent soil characteristics. Monitoring findings also highlight the need to assess the effect of soil management practices at deeper layers, especially when soils exhibit impeding rooting layers. Indeed, in both the QF and the VP sites, the alleviation of soil compactness by ripping (Figure 9.3) led to a significant increase in timber production (Figure 9.11), which might have been facilitated by the use of resources (water and nutrients). That is, the use of ripping is a promising soil technique for adapting eucalypt plantations to climate change and stronger drought. These results also emphasize that in eucalypt plantations it is essential to know the extent to which changes in soil parameters can affect the desirable productive outcomes (and the relationship between them) and the environmental and economical sustainability (Wood & Blankinship, 2022), as well as how they vary with soil conditions. We are aware that soil health evaluations in eucalypt plantations should consider their location in strongly sloping terrains, where soil loss through erosion and even landslides may occur, especially during the inter-rotation period when the soil is unprotected (Madeira & Araújo, 2015).

In the short term, soil monitoring in chestnut groves showed that the replacement of conventional tillage by no-tillage practices can benefit soil carbon and nitrogen cycling, enhancing their multifunctionality by improving soil fertility and increasing forage and edible mushrooms production and soil biological diversity (especially with irrigation). Following the results of research and monitoring

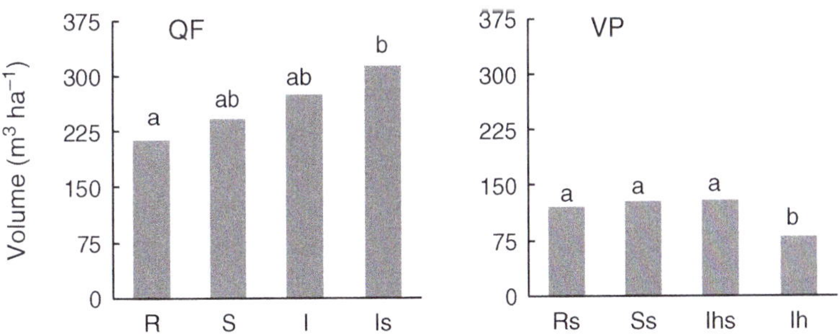

Figure 9.11 Effects of harvesting residue management (and ripping) on timber production in eucalypt plantations, in the first rotation (12 years), at Quinta do Furadouro (QF) and Vale Pequeno de Cima (VP) sites. Different letters in treatments indicate significant differences ($P<.05$) by the Tukey test. Treatments as in Tables 9.4 and 9.5, respectively.

and the use of new mechanical systems for nut collection, farmers and producers are progressively replacing the conventional intensive cultivation system with the no-tillage system associated with natural or seeded improved pastures, which are usually harvested before nut collection. The no-tillage system treatment (with spontaneous herbaceous vegetation) seems to be the most favorable management system to ensure productivity and sustainability of chestnut plantations, as it leads to the highest annual net income and may contribute to reconcile productive, environmental, economic, and social issues (Martins et al., 2010). The multifunctional management of chestnut plantations may create a sustainable land-use system, promoting soil carbon accumulation, fertility increase, and protection against erosion; enhancing biodiversity and environmental quality; and leading to economic diversification (Martins et al., 2011). Such management also presents a strategy to cope with rising temperatures, increasing drought events, and challenges related to the control of emergent pests and diseases (e.g., chestnut blight).

Several physical, chemical, and biological soil quality indicators revealed variations in the topsoil conditions in evergreen oak woodlands, following changes in their management. Undoubtedly, these indicators show that widely spaced trees create "islands of healthier soil" that enhance soil functions and ecosystem services; however, their existence in the landscape closely depends on tree density and soil type. On one hand, such indicators show the positive effects on soil conditions related to the replacement of natural pastures and natural shrubs by improved pastures as well as the potential of natural shrubs to recover soils. On the other hand, they demonstrate that intensively managed grazing by cattle can actually lead to soil degradation of grassland environments (Billota et al., 2007; Eldridge et al., 2017), hampering positive effects of improved pastures. Thus, it is vital to evaluate how variation in soil health indicators affect pasture productivity and diversity, tree nutrition (see Table 9.17) and health, and tree productivity (acorns and cork). Specifically, relationships of cork quality in cork oak woodlands with management changes are of great importance. Moreover, oak woodlands are complex systems with different land use history and different economic and ecological aims, and their ecological stability and sustainability constitute a crucial issue under climate change and economic scenarios (Simões et al., 2016). Therefore, it is indispensable to align trends in soil health indicators with effective conservation for oak woodlands in order to predict long-term effects of different management practices to ensure natural oak regeneration in a wide range of these systems.

A broad set of parameters were used in each study system for monitoring soil changes, exceeding the minimum data set (11) proposed by Bünemann et al. (2018), considering physical, chemical, and biological soil properties. The use of some parameters makes assessing the soil status difficult, as they require complicated and/or expensive laboratory procedures (Milazzo et al., 2023).

Table 9.17 Nutrient Foliar Concentration in *Quercus suber* (QS) and *Quercus rotundifolia* (QR) Growing in Areas with Improved (IP) and Natural (NP) Pastures at the Herdade dos Esquerdos Site in 2010 and 2012

			N	P	K	Ca	Mg	Mn
Trees	Year	Pastures	mg/g					
QS	2010	IP	14.77	0.96	4.23[b]	2.47	1.14	0.19
		NP	15.46	0.97	6.52[a]	2.25	1.09	0.17
	2012	IP	15.69	0.94	4.43	3.57	1.32	0.30
		NP	16.06	0.95	6.47	3.35	1.24	0.25
QR	2010	IP	13.67[a]	0.84	3.08	4.19[a]	1,52[a]	0.13[a]
		NP	12.32[b]	0.82	3.32	2.78[b]	1.16[b]	0.05[b]
	2012	IP	13.44	0.85	3.41	4.28	1.68	0.11
		NP	12.69	0.83	3.62	3.62	1.34	0.05

Note. Unpublished data. Different letters in the same column indicate significant differences ($P < .05$) in nutrient foliar concentration according type of pastures by the Tukey test.

Many parameters are interrelated (e.g., bulk density and hydraulic conductivity), however, and the measurement of one can suffice for determining the trend of another and thus allow a reduction in monitoring parameters. Monitoring results from eucalypt plantations and agroforestry systems suggest that the minimum data set of soil health indicators may include organic carbon (and/or labile organic carbon fractions); bulk density and resistance to penetration; and total nitrogen, extractable phosphorus, potassium, and calcium. Depending on the study system, evaluations of nitrogen mineralization, soil respiration, and microbial biomass are of great usefulness. Soil measurements allowing participation of farmers and foresters are of great importance in the conversion to sustainable forest and agroforestry systems (Hughes et al., 2023). For instance, visual soil assessment, such as the visual evaluation of soil structure (Mueller et al., 2013), is cost effective and practical for providing rapid and reliable information on soil physical quality indicators (Milazzo et al., 2023) that influence soil functions, such as fertility, biological activity, and root development. In addition, the evaluation of the status and morphology of the soil surface (e.g., surface integrity, degree of litter incorporation, erosion signals) is useful in establishing indices of soil health to define soil resistance to disturbances and the infiltration of water (Eldridge et al., 2017). Visual evaluations in combination with laboratory procedures will meet the needs of farmers and foresters to assess soil quality in the field and avoid laborious and costly laboratory procedures (Hughes et al., 2023).

Indicator data obtained in each system are in relation to specific ecological and soil conditions; however, as reference values are not available, the interpretation of the data remains open (Bünemann et al., 2018). Therefore, there is a need to establish references that can be associated, for instance, with a soil with maximum production and/or environmental performance. Although reported data can be an auxiliary for the development of indicator references, they may constitute a guideline for monitoring activity enlargement, aiming at the development of scoring functions for each system, in representative ecological and soil conditions.

Acknowledgment

This study was funded by the Forest Research Centre (CEF), a research unit supported by Fundação para a Ciência e Tecnologia I.P. (FCT), Portugal, through project reference UID/00239: Centro de Estudos Florestais. Data were obtained following several former research projects: PAMAF-4029-95 (1995–1998), PRAXIS XXI-3.2/FLOR/2123/95 (1997–1999), PAMAF-8003-1997 (1997–2000), Project AGRO 19 (2002–2005), AGRO 537 (2004–2006), POCTI/AGG/42985/2001 (2003–2006), POCTI/AGG/48914/2002 (2004–2007), and PTDC/AGR-AAM/102369/2008 (2010–2013). Authors are grateful to Juan Carrabs for the figures regarding species distribution and Paulo Marques for additional figures.

References

Abreu, C. G., Coutinho, J. F., Cardoso, A. O., & Campos, J. A. (1993). *Suppressive soils and chestnut ink disease*. In E. Antognozzi (Ed.), Proceedings of the International Congress on Chestnut (pp. 533–536), Spoleto.

Achat, D. L., Deleuze, C., Landmann, G., Pousse, N., Ranger, J., & Augusto, L. (2015). Quantifying consequences of removing harvesting residues on forest soils and tree growth—A meta-analysis. *Forest Ecology and Management*, *348*, 124–141. https://doi.org/10.1016/j.foreco.2015.03.042

Amorim, H. C. S., Ashworth, A. J., O'Brien, P. L., Thomas, A. L., Runkle, B. R. K., & Philipp, D. (2023). Temperate silvopastures provide greater ecosystem services than conventional pasture systems. *Scientific Reports*, *13*, Article 18658. https://doi.org/10.1038/s41598-023-45960-0

Andivia, E., Fernández, M., Vázquez-Piqué, J., González-Pérez, A., & Tapias, R. (2010). Nutrients return from leaves and litterfall in a Mediterranean cork oak (*Quercus suber* L.) forest in southwestern Spain. *European Journal of Forest Research*, *129*, 5–12. https://doi.org/10.1007/s10342-009-0274-6

Andrade, J. A. V., Abreu, F. M. G., & Madeira, M. (2010). Influence of litter layer removal on the soil thermal regime of a pine forest in a Mediterranean climate. *Revista Brasileira de Ciência do Solo*, *34*(5), 1481–1490. https://doi.org/10.1590/S0100-06832010000500001

Billota, G. S., Brazier, R. E., & Haygarth, P. M. (2007). The impacts of grazing animals on the quality of soils, vegetation, and surface waters in intensively managed grasslands. *Advances in Agronomy*, *94*, 237–280. https://doi.org/10.1016/S0065-2113(06)94006-1

Borges, O., Raimundo, F., Martins, A., & Madeira, M. (2018). Carbon fractions as indicators of organic matter dynamics in chestnut orchards under different soil management practices. *Agroforestry Systems*, *92*, 301–310. https://doi.org/10.1007/s10457-017-0088-3

Brady, N. C., & Weil, R. R. (2008). *The nature and properties of soils* ((14th ed). ed.). Pearson Prentice Hall.

Bünemann, E. K., Bongiorno, G., Bai, Z., Creamer, R. E., De Deyn, G., de Goede, R., Fleskens, L., Geissen, V., Kuyper, T. W., Mäder, P., Pulleman, M., Sukkel, W., van Groenigen, J. W., & Brussaard, L. (2018). Soil quality—A critical review. *Soil Biology and Biochemistry*, *120*, 105–125. https://doi.org/10.1016/j.soilbio.2018.01.030

Cardinael, R., Chevallier, T., Barthès, B. G., Saby, N. P. A., Parent, T., Dupraz, C., Bernoux, M., & Chenu, C. (2015). Impact of alley cropping agroforestry on stocks, forms and spatial distribution of soil organic carbon—A case study in a Mediterranean context. *Geoderma*, *259–260*, 288–299. https://doi.org/10.1016/j.geoderma.2015.06.015

Chen, H., Hou, R., Gong, Y., Li, H., Fan, M., & Kuzyakov, Y. (2009). Effects of 11 years of conservation tillage on soil organic matter fractions in wheat monoculture in Loess Plateau of China. *Soil Tillage Research*, *106*, 85–94. https://doi.org/10.1016/j.still.2009.09.009

Costa, A., Madeira, M., & Lima Santos, J. (2014). Is cork oak (*Quercus suber* L.) woodland loss driven by eucalyptus plantation? A case-study in southwestern Portugal. *iForest –. Biogeosciences and Forestry*, *7*(3), 193–203. https://doi.org/10.3832/ifor1031-007

Costa, A., Pereira, H., & Madeira, M. (2010). Analysis of spatial patterns of oak decline in cork oak woodlands in Mediterranean conditions. *Annals of Forest Sciences*, *67*, Article 204. https://doi.org/10.1051/forest/2009097

Crespo, D. G. (2006). The role of pasture improvement in the rehabilitation of the "montado/dehesa" system and in developing its traditional products. In J. M. C. Ramalho, A. E. M. Horta, C. Mosconi, & A. Rosati (Eds.), *Animal products from the Mediterranean area* (pp. 185–197). Wageningen Academic Publishers.

Dahlgren, R., Horwath, W., Tate, K., & Camping, T. (2003). Blue oak enhances soil quality in California oak woodlands. *California Agriculture*, *57*(2), 42–47. https://doi.org/10.3733/ca.v057n02p42

Dahlgren, R. A., Singer, M. J., & Huang, X. (1997). Oak tree and grazing impacts on soil properties and nutrients in California oak woodland. *Biogeochemistry, 39*, 45–64. https://doi.org/10.1023/A:1005812621312

David, T. M. S. B. S. (2000). *Intercepção da precipitação e transpiração em árvores isoladas de Quercus rotundifolia Lam*. Instituto Superior de Agronomia, Universidade Técnica de Lisboa.

David, T. S., Gash, J. H. C., Valente, F., Pereira, J. S., Ferreira, M. I., & David, J. S. (2006). Rainfall interception by an isolated evergreen oak tree in a Mediterranean savannah. *Hydrological Processes, 20*(13), 2713–2726. https://doi.org/10.1002/hyp.6062

Doran, J. W., & Zeiss, M. R. (2000). Soil health and sustainability: Managing the biotic component of soil quality. *Applied Soil Ecology, 15*, 3–11. https://doi.org/10.1016/S0929-1393(00)00067-6

Eldridge, D. J., Delgado-Baquerizo, M., Travers, S. K., Val, J., & Oliver, I. (2017). Do grazing intensity and herbivore type affect soil health? Insights from a semi-arid productivity gradient. *Journal of Applied Ecology, 54*, 976–985. https://doi.org/10.1111/1365-2664.12834

Escudero, A., Garcia, B., Gomez, J. M., & Luis, E. (1985). The nutrient cycling in *Quercus rotundifolia* and *Quercus pyrenaica* ecosystems ("dehesas') of Spain. *Acta Oecologica, 6*, 73–86.

Fernández, C., Veja, J. A., Gras, J. M., Fonturbel, T., Cuiñas, P., Dambrine, E., & Alonso, M. (2004). Soil erosion after *Eucalyptus globulus* clearcutting: Differences between logging slash disposal treatments. *Forest Ecology and Management, 195*(1–2), 85–95. https://doi.org/10.1016/j.foreco.2004.02.052

Fisher, R. F., & Binkley, D. (2000). *Ecology and management of forest soils* (3rd ed.). John Wiley & Sons, Inc.

Gallardo, A. (2003). Effect of tree canopy on the spatial distribution of soil nutrients in a Mediterranean Dehesa. *Pedobiologia, 47*, 117–125. https://doi.org/10.1078/0031-4056-00175

Gallardo, A., Rodríguez-Saucedo, J. J., Covelo, F., & Fernández-Alés, R. (2000). Soil nitrogen heterogeneity in a Dehesa ecosystem. *Plant and Soil, 222*, 71–82.

Ghani, A., Dexter, M., & Perrott, K. W. (2003). Hot-water extractable carbon in soils: A sensitive measurement for determining impacts of fertilization grazing and cultivation. *Soil Biology and Biochemistry, 35*, 1231–1243. https://doi.org/10.1016/S0038-0717(03)00186-X

Gómez-Rey, M. X., Garcês, A., & Madeira, M. (2011). Organic C distribution and N mineralization in soil of oak woodlands with improved pastures. *Revista de Ciências Agrárias, 34*(1), 80–92.

Gómez-Rey, M. X., Garcês, A., & Madeira, M. (2012). Soil organic C accumulation and N availability determined by improved pastures in oak woodlands in Southern Portugal. *Soil Use and Management, 28*(4), 497–507. https://doi.org/10.1111/j.1475-2743.2012.00428.x

Gómez-Rey, M. X., Madeira, M., Gonzalez-Prieto, S. J., & Coutinho, J. (2010). Soil C and N dynamics within a precipitation gradient in Mediterranean eucalypt plantations. *Plant and Soil*, *336*(1), 157–171. https://doi.org/10.1007/S11104-010-0456-5

Gómez-Rey, M. X., Madeira, M., Gonzalez-Prieto, S. J., & Coutinho, J. (2013). Soil C and N dynamics in a Mediterranean oak woodland with shrub encroachment. *Plant and Soil*, *371*(1–2), 339–354. https://doi.org/10.1007/s11104-013-1695-z

Gómez-Rey, M. X., Madeira, M., & Vasconcelos, E. (2008). Effects of eucalypt residue management on nutrient leaching and soil properties. *European Journal of Forest Research*, *127*(5), 379–386. https://doi.org/10.1007/s10342-008-0217-7

Greenwood, K. L., & McKenzie, B. M. (2001). Grazing effects on soil physical properties and the consequence for pastures—A review. *Animal Production Science*, *41*(8), 1231–1250. https://doi.org/10.1071/EA00102

Hartmann, M., Howes, C. G., VanInsberghe, D., Yu, H., Bachar, D., Christen, R., Nilsson, R. H., Hallam, S. J., & Mohn, W. W. (2012). Significant and persistent impact of timber harvesting on soil microbial communities in Northern coniferous forests. *The International Society for Microbial Ecology Journal*, *6*, 2199–2218. https://doi.org/10.1038/ismej.2012.84

Haynes, R. J. (2005). Labile organic matter fractions as central components of the quality of agricultural soils: An overview. *Advances in Agronomy*, *85*, 221–268. https://doi.org/10.1016/s0065-2113(04)85005-3

Hughes, H. M., Koolen, S., Kuhnert, M., Baggs, E. M., Maund, S., Mullier, G. W., & Hillier, J. (2023). Towards a farmer-feasible soil health assessment that is globally applicable. *Journal of Environmental Management*, *345*, Article 118582. https://doi.org/10.1016/j.jenvman.2023.118582

ICNF. (2019). *6° Inventário Florestal Nacional*. Relatório Final.

IUSS Working Group. (2015). *World Reference Base for Soil Resources 2014, update 2015. International soil classification system for naming soils and creating legends for soil maps* (World Soil Resources Reports No. 106). FAO.

Johnson, D. L., Ambrose, S. H., Bassett, T. J., Bowen, M. L., Crummey, D. E., Isaacson, J. S., Johnson, D. N., Lamb, P., Saul, M., & Winter-Nelson, A. E. (1997). Meanings of environmental terms. *Journal Environmental Quality*, *26*, 581–589. https://doi.org/10.2134/jeq1997.00472425002600030002x

Jones, H. E., Madeira, M., Herraez, L., Dighton, J., Fabião, A., Gonzales-Rio, F., Fernandez-Marcos, M., Gomez, C., Tomé, M., Feith, H., Magalhães, M. C., & Howson, G. (1999). The effect of organic-matter management on the productivity of *Eucalyptus globulus* stands in Spain and Portugal: Tree growth and harvest residue decomposition in relation to site and treatment. *Forest Ecology and Management*, *122*(1–2), 73–86. https://doi.org/10.1016/S0378-1127(99)00033-X

Karlen, D. L., Mausbach, M. J., Doran, J. W., Cline, R. G., & Harris, R. F. (1997). Soil quality: A concept, definition and framework for evaluation. *Soil Science Society America Journal*, *61*, 4–10. https://doi.org/10.2136/sssaj1997.03615995006100010001x

Madeira, M. (1989). Changes in soil properties under *Eucalyptus* plantations in Portugal. In J. S. Pereira & J. J. Landsberg (Eds.), *Biomass production by fast-growing trees* (pp. 81–99). Kluver Academic Publishers.

Madeira, M., & Araújo, C. (2015). Soil degradation risks and prevention measures in planted forests. The case of eucalyptus plantations in Portugal. In A. A. González & N. G. Bengoetxea (Eds.), *Soil degradation risks in planted forests* (pp. 107–117). Administration of the Autonomous Community of the Basque Country, Department of Economic Development and Competitiveness. Servicio Central de Publicaciones del Gobierno Vasco, Donostia-San Sebastian. Legal Record: VI 410-2015

Madeira, M., Azevedo, A., Soares, P., Tomé, M., & Araújo, M. C. (2002). Efeitos da lavoura profunda e da gradagem nas características do solo e na produtividade de plantações de *Eucalyptus globulus*. *Revista de Ciências Agrárias*, *25*(1–2), 158–169.

Madeira, M., Melo, G., Alexandre, C., & Steen, E. (1989). Effects of deep ploughing and superficial disc harrowing on physical and chemical soil properties and biomass in a new plantation of *Eucalyptus globulus*. *Soil & Tillage Research*, *14*(2), 163–175. https://doi.org/10.1016/0167-1987(89)90029-9

MADRP. (2005). Optimização produtiva, ambiental e económica de sistemas florestais intensivos (OFLORINT). In *Project Agro 19-2001.09.003826.4. Relatório final. Ministério da Agricultura, Desenvolvimento Rural e das Pescas.*

Magalhães, M. C. S. (2000). *Efeitos de técnicas de preparação do solo e gestão dos resíduos orgânicos em características físico-químicas do solo de plantações florestais.* Ph. D. Thesis. Instituto Superior de Agronomia, Universidade Técnica de Lisboa.

Manning, A. D., Fischer, J., & Lindenmayer, D. B. (2006). Scattered trees are keystones—Implications for conservation. *Biological Conservation*, *132*(3), 311–321. https://doi.org/10.1016/j.biocon.2006.04.023

Marcelino, V., Torres, N., Portela, E., & Martins, A. (2000). Soil physical properties and the occurrence of chestnut ink disease: A micromorphological study. *Ecologia Mediterranea*, *26*(1–2), 129–135. http://hdl.handle.net/10348/8710

Marchante, E., Madeira, M., & Freitas, H. (2001). Avaliação da colonização ectomicorrízica em plantações experimentais. *Revista de Ciências Agrárias*, *24*(3–4), 205–212.

Martinho, M. A. S., Oliveira, B. R. F., Machado, A. I., & Keizer, J. (2023). Immediate changes in soil chemical and biological fertility and short-term changes in soil biological activity following bench terrace construction for eucalypt reforestation. *Catena*, *233*, Article 107462. https://doi.org/10.1016/j.catena.2023.107462

Martins, A., Linhares, I., Raimundo, F., Borges, O., Coutinho, J. P., Gomes-Laranjo, J., & Sousa, V. (2005). The importance of deep soil layers to supply water to agroforestry systems: A case study of a mature chestnut orchard in Northern Portugal. *Acta Horticulture*, *693*, 663–670. https://doi.org/10.17660/ActaHortic.2005.693.89

Martins, A., Marques, G., Borges, O., Portela, E., Lousada, J., Raimundo, F., & Madeira, M. (2011). Management of chestnut plantations for a multifunctional land use under Mediterranean conditions: effects on productivity and sustainability. *Agroforestry Systems*, *81*(2), 175–189. https://doi.org/10.1007/s10457-010-9355-2

Martins, A., Raimundo, F., Borges, O., Linhares, I., Sousa, V., Coutinho, J. P., Gomes-Laranjo, J., & Madeira, M. (2010). Effects of soil management practices and irrigation on plant water relations and productivity of chestnut stands under Mediterranean conditions. *Plant and Soil*, *327*(1), 57–70. https://doi.org/10.1007/s11104-009-0031-0

Mayer, M., Prescott, C. E., Abaker, W. E. A., Augusto, L., Cécillon, L., Ferreira, G. W. D., James, J., Jandl, R., Katzensteiner, K., Laclau, J.-P., Laganière, J., Nouvellon, Y., Paré, D., Stantur, J. A., Vanguelova, E. I., & Vesterdal, L. (2020). Tamm Review: Influence of forest management activities on soil organic carbon stocks: A knowledge synthesis. *Forest Ecology and Management*, *466*, Article 118127. https://doi.org/10.1016/j.foreco.2020.118127

Milazzo, F., Francksen, R. M., Abdalla, M., Enri, S. R., Zavattaro, L., Pittarello, M., Hejduk, S., Newell-Price, P., Schils, R. L. M., Smith, P., & Vanwalleghem, T. (2023). An overview of permanent grassland grazing management practices and the impacts on principal soil quality indicators. *Agronomy*, *13*(5), Article 1366. https://doi.org/10.3390/agronomy13051366

Miranda, P. M. A., Coelho, F. E. S., Tomé, A. R., Valente, M. A., Carvalho, A., Pires, C. A. L., Pires, H. O., Pires, V. C., & Ramalho, C. (2002). 20th century Portuguese climate and climate scenarios. In F. D. Santos, K. Forbes, & R. Moita (Eds.), *Climate change in Portugal: Scenarios, impacts and adaptation measures (SIAM project)* (pp. 23–83). Gradiva.

Moebius-Clune, B. N., Moebius-Clune, D. J., Gugino, B. K., Idowu, O. J., Schindelbeck, R. R., Ristow, A. J., van Es, H. M., Thies, J. E., Shayler, H. A., McBride, M. B., Kurtz, K. S. M., Wolfe, D. W., & Abawi, G. S. (2016). *Comprehensive assessment of soil health—The Cornell framework manual* ((3rd ed.). ed.). Cornell University.

Moreira, F., Rego, F. C., & Ferreira, P. G. (2001). Temporal (1958-1995) pattern of change in a cultural landscape of Northwestern Portugal: Implications for fire occurrence. *Landscape Ecology*, *16*, 557–567. https://doi.org/10.1023/A:1013130528470

Moro, M. J., Pugnaire, M. I., Haase, P., & Puigdefábregas, J. (1997). Mechanisms of interactions between a leguminous shrub understory in a semi-arid environment. *Ecography*, *20*, 175–184. https://doi.org/10.1111/j.1600-0587.1997.tb00360.x

Mueller, L., Shepherd, G., Schindler, U., Ball, B. C., Munkholm, L. J., Hennings, G., Smolentseva, E., Rukhovic, O., Lukin, S., & Hu, C. (2013). Evaluation of soil structure in the framework of an overall soil quality rating. *Soil Tillage Research*, *127*, 74–84. https://doi.org/10.1016/j.still.2012.03.002

Myers, N., Mittermeier, R. A., Mittermeier, C. G., Fonseca, G. A. B., & Kent, J. (2000). Biodiversity hotspots for conservation priorities. *Nature*, *403*, 853–858.

Nunes, J., Andrade, J., Abreu, F., Gazarini, L., & Madeira, M. (2011). Microclimatic specificity of a Mediterranean oak woodland (montado) in context of global change. *Folia Oecologica*, *38*(1), 96–103. http://hdl.handle.net/10174/7677

Nunes, J., Sá, C., Soares-David, T., Madeira, M., & Gazarini, L. (2001). Influência de árvores isoladas de *Quercus rotundifolia* Lam. nas características do solo. *Revista de Ciências Agrárias*, *24*(3–4), 213–226.

Petriccione, M., Moro, C., & Rutigliano, F. A. (2006). Preliminary studies on litter flammability in Mediterranean region. *Forest Ecology and Management*, *234*, S128. https://doi.org/10.1016/j.foreco.2006.08.169

Portela, E., Aranha, J., Martins, A., & Pires, A. L. (1999). Soil factors, farmer's practices and chestnut ink disease: Some interactions. *Acta Horticulture*, *494*, 433–441. https://doi.org/10.17660/ActaHortic.1999.494.65

Pulido, M., Schnabel, S., Contador, J. F. L., Lozano-Parra, J., & Gómez-Gutiérrez, A. (2017). Selecting indicators for assessing soil quality and degradation in rangelands of Extremadura (SW Spain). *Ecological Indicators*, *74*, 49–61. https://doi.org/10.1016/j.ecolind.2016.11.016

Pulido, M., Schnabel, S., Contador, J. F. L., Lozano-Parra, J., & González, F. (2018). The impact of heavy grazing on soil quality and pasture production in rangelands of SW Spain. *Land Degradation & Development*, *29*, 219–230. https://doi.org/10.1002/ldr.2501

Raimundo, F., Coutinho, J., Martins, A., & Madeira, M. (2015). Soil management system effects on N availability and tree N status in chestnut plantations under Mediterranean conditions. *Revista de Ciências Agrárias*, *38*(4), 547–562. https://doi.org/10.19084/RCA15139

Raimundo, F., Madeira, M., Coutinho, J., & Martins, A. (2004). Efeito de sistemas de mobilização do solo na mineralização do N, na biomassa microbiana e na respiração do solo de soutos do Nordeste de Portugal. *Revista de Ciências Agrárias*, *27*(1), 361–387.

Raimundo, F. P. F. (2003). *Sistemas de mobilização do solo em soutos. Influência na produtividade de castanha e nas características físicas e químicas do solo*. Ph.D. Thesis. Universidade de Trás-os-Montes e Alto Douro.

Raison, R. J., Connell, M. J., & Khanna, P. K. (1987). Methodology for studying fluxes of soil mineral–N in situ. *Soil Biology and Biochemistry*, *19*, 521–530. https://doi.org/10.1016/0038-0717(87)90094-0

Raison, R. J., & Rab, M. A. (2001). Guiding concepts for the application of indicators to interpret change in soil properties and processes in forests. In R. J. Raison, A. G. Brown, & D. W. Flinn (Eds.), *Criteria and indicators for sustainable forest management* (pp. 231–258). CAB International.

Rodrigues, A. R. (2019). *Management practices and soil quality patterns in evergreen woodlands (montado)*. Ph.D. Thesis,. Instituto Superior de Agronomia, Universidade de Lisboa. https://www.repository.utl.pt/bitstream/10400.5/18325/1/ARRodrigues_PhD.pdf

Rodrigues, A. R., Costa e Silva, F., Correia, A. C., Bicho, M. C., Madeira, M., & Coutinho, J. (2019). Do improved pastures enhance soil quality of cork oak woodlands in the Alentejo Region (Portugal)? *Agroforestry Systems*, *94*, 125–136. https://doi.org/10.1007/s10457-019-00376-6

Rodrigues, A. R., Silva, E., Santos, G., Coutinho, J., & Madeira, M. (2015). Do improved pastures affect enzymatic activity and C and N dynamics in soils of the montado system? *Revista de Ciências Agrárias*, *38*(4), 535–546. https://doi.org/10.19084/RCA15138

Ruiz-Peinado, R., Montero, G., & Del Rio, M. (2012). Biomass models to estimate carbon stocks for hardwood tree species. *Forest Systems*, *21*, 42–52. https://doi.org/10.5424/fs/2112211-02193

Sá, C., Madeira, M., & Gazarini, L. (2001). Produção e decomposição da folhada de *Quercus suber* L. *Revista de Ciências Agrárias*, *24*(3,4), 245–256.

Seabra-Ferreira, C., Vasconcelos, T. M., & Madeira, M. (2001). Estudo das comunidades de artrópodes edáficos de um povoamento de *Eucalyptus globulus* (Labill.). *Revista de Ciências Agrárias*, *24*(3–4), 257–270.

Sharrow, S. (2007). Soil compaction by grazing livestock in silvopastures as evidenced by changes in soil physical properties. *Agroforestry Systems*, *71*, 215–223. https://doi.org/10.1007/s10457-007-9083-4

Simões, M. P., Belo, A. F., Fernandes, M., & Madeira, M. (2016). Regeneration patterns of *Quercus suber* according to *Montado* management systems. *Agroforestry Systems*, *90*, 107–115. https://doi.org/10.1007/s10457-015-9818-6

Simões, P., Madeira, M., & Gazarini, L. (2009). Ability of *Cistus* L. shrubs to promote soil rehabilitation in extensive oak woodlands of Mediterranean areas. *Plant and Soil*, *323*(1), 249–265. https://doi.org/10.1007/s11104-009-9934-z

Simón, N., Montes, F., Díaz-Pinés, E., Benavides, R., Roig, S., & Rubio, A. (2013). Spatial distribution of the soil organic carbon pool in a holm oak dehesa in Spain. *Plant and Soil*, *366*, 537–549. https://doi.org/10.1007/s11104-012-1443-9

Six, J., Contant, R. T., Paul, E., & Paustian, K. (2002). Stabilization mechanisms of soil organic matter: Implications for C-saturation of soils. *Plant and Soil*, *241*(2), 155–176. https://doi.org/10.1023/A:1016125726789

Sousa, J. P., Da Gama, M. M., Ferreira, C., & Barrocas, H. (2000). Effects of eucalyptus plantations on Collembola communities in Portugal: A review. *Belgian Journal of Entomology*, *2*, 187–201.

United States Department of Agriculture. (2001). *Guidelines for soil quality assessment in conservation planning*. Natural Resources Conservation Service.

Wardle, D. A., & Ghani, A. (1995). A critique of the microbial metabolic quotient (qCO_2) as a bioindicator of disturbance and ecosystem development. *Soil Biology and Biochemistry*, *27*, 1601–1610. https://doi.org/10.1016/0038-0717(95)00093-T
Wood, S. A., & Blankinship, J. C. (2022). Making soil health science practical: Guiding research for agronomic and environmental benefits. *Soil Biology and Biochemistry*, *172*(14), Article 108776. https://doi.org/10.1016/j.soilbio.2022.108776

Printed and bound by CPI Group (UK) Ltd, Croydon, CR0 4YY

30/06/2025

14696465-0001